T0380073

VARIORUM COLLECTED STUDIES SERIES

Numerals and Arithmetic in the Middle Ages

For Menso Folkerts
in gratitude and respect

Charles Burnett

Numerals and Arithmetic in the Middle Ages

Published in the Variorum Collected Studies Series by

Ashgate Publishing Limited
Wey Court East
Union Road
Farnham, Surrey
GU9 7PT
England

Ashgate Publishing Company
Suite 420
101 Cherry Street
Burlington, VT 05401–4405
USA

www.ashgate.com

ISBN 978–1–4094–0368–5

British Library Cataloguing in Publication Data
Burnett, Charles (Charles S.F.)
   Numerals and arithmetic in the Middle Ages.
   – (Variorum collected studies series)
   1. Numerals – History – To 1500. 2. Numeration – History – To 1500.
   3. Mathematics, Medieval.
   I. Title II. Series
   513.5'0902–dc22

   ISBN 978–1–4094–0368–5

Library of Congress Control Number: 2010936535

VARIORUM COLLECTED STUDIES SERIES CS967

Mixed Sources
Product group from well-managed forests and other controlled sources
www.fsc.org Cert no. SGS-COC-2482
© 1996 Forest Stewardship Council

Printed and bound in Great Britain by
TJ International Ltd, Padstow, Cornwall

CONTENTS

This volume contains x + 370 pages

PREFACE

This volume brings together articles on the different numeral forms used in the Middle Ages, and their use in mathematical and other contexts. Some articles study the introduction of Hindu-Arabic numerals into Western Europe between the late tenth and the early thirteenth centuries, documenting, in more detail than anywhere else, the different forms in which they are found, before they acquired the standard shapes with which we are familiar today (articles I, V, VI, VII, VIII, IX, XI). Others deal with experiments with other forms of numeration within Latin script, that are found in the twelfth century: e.g., using the first nine Roman numerals as symbols with place value (III), abbreviating the Roman numerals (IV), and using the Latin letters as numerals (X). Different types of numerals are used for different purposes: for numbering folios, dating coins, symbolizing learning and mathematical games, as well as for practical calculations and advanced mathematics. The application of numerals to the abacus (I, II), and to calculation with pen and paper (or stylus and parchment) is discussed (VII, IX), and several Latin texts have been critically edited for the first time (III, VII, VIII, IX).

CHARLES BURNETT

London
24 August 2010

PUBLISHER'S NOTE

The articles in this volume, as in all others in the Variorum Collected Studies Series, have not been given a new, continuous pagination. In order to avoid confusion, and to facilitate their use where these same studies have been referred to elsewhere, the original pagination has been maintained wherever possible.

Each article has been given a Roman number in order of appearance, as listed in the Contents. This number is repeated on each page and is quoted in the index entries.

Asterisks in the margins are to alert the reader to additional information supplied at the end of the volume in the Addenda and corrigenda.

ACKNOWLEDGEMENTS

Grateful acknowledgement is made to the following institutions, journals and publishers for their kind permission to reproduce the papers included in this volume: SCIAMVS (for articles I and VIII); Centre National de la Recherche Scientifique, Paris (II); Herzog August Bibliothek, Wolfenbüttel (III); Deutsche Akademie der Naturforscher Leopolina, Nationale Academy der Wissenschaften, Saale (IV); Franz Steiner Verlag, Stuttgart (V); Bibliotheca Herziana, Rome, and Hirmer Verlag, Munich (VI); CSLI publications, Stanford, CA (VII); Asociación Matemática Venezolana, Caracas (IX); Harrassowitz Verlag, Wiesbaden (X); and Fabrizio Serra editore, Pisa and Rome (XI). For additional photographs I am very grateful to the Bibliothèque nationale de Luxembourg, the Bodleian Library, Oxford, and St John's College, Oxford. Kurt Lampe and Ji-Wei Zhao have kindly agreed to reprint article VIII.

I

The Abacus at Echternach in ca. 1000 A.D.

The single sheet of manuscript from the Benedictine monastery of Echternach, catalogued as no. 770 in the Bibliothèque nationale de Luxembourg, is probably the nearest thing we have to the abacus board devised by the famous mathematician and educator of the late tenth century, Gerbert d'Aurillac (d. 1003). This article describes the board in the light of what we know of Gerbert's own instrument and also introduces another smaller example of exactly the same board. Both examples have hitherto been unnoticed by scholars.[1]

*   *   *   *

1 The Gerbertian Abacus

From the late tenth century until at least the mid twelfth century the principal method of studying practical arithmetic in the schools of Western Europe was that of the abacus with marked counters. This was a a board or sheet of parchment ('abacus board') on which parallel lines were drawn to provide columns for powers of ten. Counters marked with each of the numerals were placed in the columns and successively replaced in the course of an arithmetical calculation. Rules were written to describe the sequences of procedures, especially for division and multiplication.

---

[1] I am very grateful to Luc Deitz, the curator of the manuscripts and rare books collection (Réserve précieuse) of the Bibliothèque nationale de Luxembourg for introducing me to the manuscript-sheet, and inviting me to give a talk at the library on its relation to the early history of the introduction of Arabic numerals into Europe, on 19 October 2001, as well as for showing me a preliminary study of his on the logical schemata contained on the sheet and correcting an early draft of this paper. Throughout this article I am indebted to the research of Menso Folkerts, whose article, 'Frühe Darstellungen des Gerbertschen Abakus' (in *Itinera mathematica: Studi in onore di Gino Arrighi per il suo 90a compleanno*, eds R. Franci, P. Pagli and L. Toti Rigatelli, Siena 1996, pp. 23–43) constitutes the starting point for the study of early depictions of the abacus, and who has generously shared his knowledge and research materials with me. I have also benefited from the advice of David Juste, Daibhi O Croinin, Alison Peden and Thomas Falmagne, and from the services of Reiner Nolden of the Stadtarchiv of Trier. The contents of the manuscript sheet as a whole will be the subject of a monograph written jointly by Luc Deitz, Thomas Flamagne and the present author.

The purpose of the exercise was to demonstrate how numbers interacted with each other rather than to facilitate calculation, which could more easily be done on the fingers, and the abacus was regarded as an instrument of geometry rather than of arithmetic. When the counters were marked with Arabic numerals, this kind of abacus mimicked with columns and marked counters the appearance of calculation with the same Arabic numerals but using pen and paper (or quill and parchment) — a process later called the 'algorism'. This mimicry extended to the use of a counter for zero, which is strictly not necessary when columns are drawn for each of the decimal places: one simply leaves a column empty.[2]

The origin of this form of abacus is still unknown.[3] Our earliest testimonies rather associate a *revival* of its use with Gerbert d'Aurillac, especially with his period as a teacher at Reims (972 to 983). It seems likely that Gerbert introduced the practice of marking the counters with Arabic numerals (which he would have come across when he studied in Catalonia, before coming to Reims), and established a form of the abacus board that became an exemplar for most subsequent teachers of the abacus. However, although several texts on calculating with the abacus (starting with one by Gerbert himself) survive,[4] they do not include diagrams of the instrument itself. The 'Gerbertian' form of the table can only be deduced from two descriptions of the table, and a few early manuscript depictions of the table occurring separately from the texts on calculation.[5] Hence the newly-discovered Echternach manuscript sheet

---

[2]For examples of how to calculate on this abacus see K. Vogel, 'Gerbert von Aurillac als Mathematiker', *Acta historica Leopoldina*, 16, 1985, pp. 9–23, G. Ifrah, *The Universal History of Numbers*, English translation by D. Bellos etc., London, 1998, pp. 579–85, and G. Beaujouan, 'Les Chiffres arabes selon Gerbert: L'abaque du Pseudo-Boèce', in *Autour de Gerbert d'Aurillac*, 1996, pp. 322–8; see also G.R. Evans, '*Difficillima et ardua*: theory and practice in treatises on the abacus 950–1150', *Journal of Medieval History*, 3, 1977, pp. 21–38.

[3]This problem is most fully discussed in W. Bergmann, *Innovationen im Quadrivium des 10. und 11. Jahrhunderts*, Wiesbaden, 1985. Bergmann, however, did not take account of the evidence of the early depictions of the abacus tables occurring separately from the texts on calculation.

[4]Several of these are edited in N. Bubnov, *Gerberti postea Silvestri II papae Opera mathematica* (972–1003), Berlin, 1899, pp. 1–22 and 197–245.

[5]These are: Bern, Burgerbibliothek, Cod. 250, fol. 1r, s. x^{ex} (**B**); Paris, Bibliothèque nationale de France, lat. 8663, fol. 49v, s. xi^{in} (**P**); ibid., lat. 7231, fol. 85v, s. xi^{in} (**Q**); Vatican, Biblioteca apostolica Vaticana, lat. 644, fols 77v–78r, s. xi^{in} (**V**); Rouen, Bibliothèque municipale, 489, fols 68v–69r, s. xi^{ex} (**R**); Oxford, St John's College 17, fols 48v–49r, s. xii, (**J**); and Chartres, Bibliothèque municipale 498, fols 165v–167r, ca. 1140 (**C**). **BPV** are illustrated and described in M. Folkerts, 'Frühe Darstellungen' (n. 1 above), **R** in idem, 'The names and forms of the numerals on the abacus in the Gerbert tradition', to be published in the proceedings of a conference on Gerbert d'Aurillac held at Bobbio, and **J** in Evans, '*Difficilis et ardua*' (n. 2 above). **C** survives in microfilm only.

may be a unique example of the artefact itself.

2 The Echternach abacus board

The Echternach abacus board (Plate 1) is a sheet of thick parchment 603 millimetres wide and 420 millimetres long. Since the sheet was used as a paste down in the binding of a large Echternach Bible that was written for the Abbot Regimbertus (1051–81) and has survived intact as the 'Riesenbibel' ('Bible géante'; now Bibliothèque nationale de Luxembourg, no. 264), the absolute terminus *ante quem* of the writing on the sheet is 1081.[6] However, since the material on the sheet must have been regarded as obsolete when it was reused as a paste down, it could have been produced much earlier, as the handwriting also suggests. The sheet was cut down to fit the dimensions of the bible and was once at least one-eighth wider, and somewhat longer.[7] There is no evidence that the sheet was ever a folio of a manuscript. The original parchment appears to have been cut especially in order to produce an abacus table that could be displayed in a teaching situation or used as a board for calculating. The letters are all capitals, written clearly in a strong black ink. The columns of the abacus table were originally drawn vertically across the whole sheet. A little later another scribe, using a lighter coloured ink, reused the bottom half of the sheet in order to draw a representation of the Boethian monochord. Perhaps at the same time, other scribes used the verso of the sheet to represent the whole of logic in the form of schemata. Both these additions, of course, took place before the sheet was bound into the Riesenbibel, and the logical schemata are difficult to read because they are on the side of the sheet that was pasted onto the binding. The *mensura monochordi* does not obscure any writing on the abacus board, but only shortens the columns, giving less room for making calculations. A scholar may simply have wished to adapt the sheet into a compendium of information on logic, geometry (to which the abacus was regarded as belonging) and music. That this is an adaptation of the original plan is clear. For the full-sheet abacus was obviously meant to be a display copy or to be laid out on a table in class; the music and logic, on the other hand, are written in small letters, and the music includes some continuous text that cannot be read at a distance, which suggests that both these additional elements were meant for individual reading. The priority of the abacus

---

[6]The only description of this manuscript sheet up to now is J. Leclercq, 'Un nouveau manuscrit d'Echternach à Luxembourg', *Scriptorium*, 7, 1953, pp. 219–25. Leclercq briefly mentions the paste down which was still attached to the bible when he saw it (p. 225), but he refers only to the *mensura monochordi* and the schemata of logic written on it, and not to the abacus. The paste down was detached and then mislaid some time after Leclercq's visit to the library.

[7]See section 3 ii below. The original dimensions were probably ca. 680 x 440 mm, as is also indicated by the vestige of a fold 340 mm from the left hand side (i.e. it is probable that the sheet was once folded in half). I owe this observation to Luc Deitz.

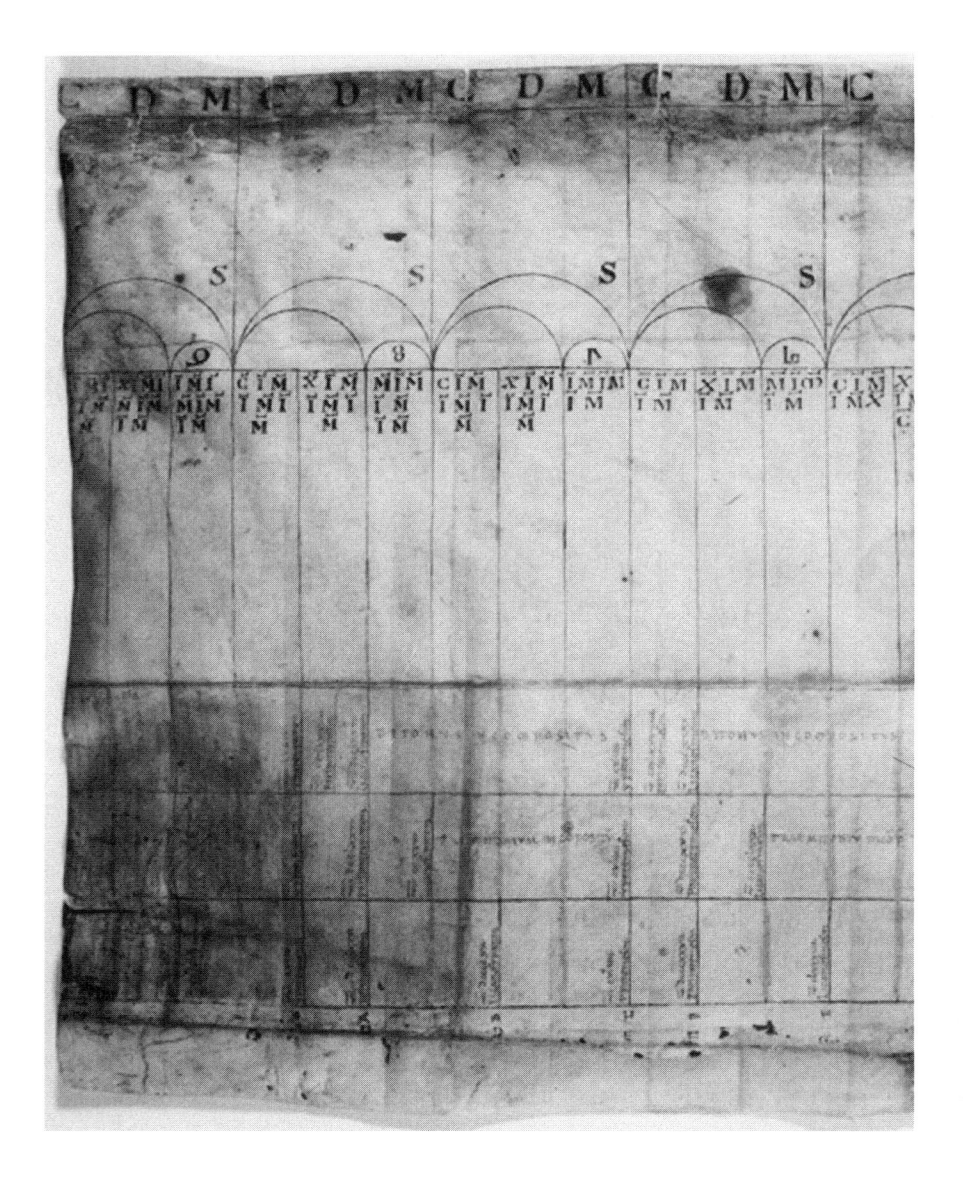

Plate 1. The Echternach abacus board (MS Bibliothèque nationale de Luxembourg, 770; reproduced with permission)

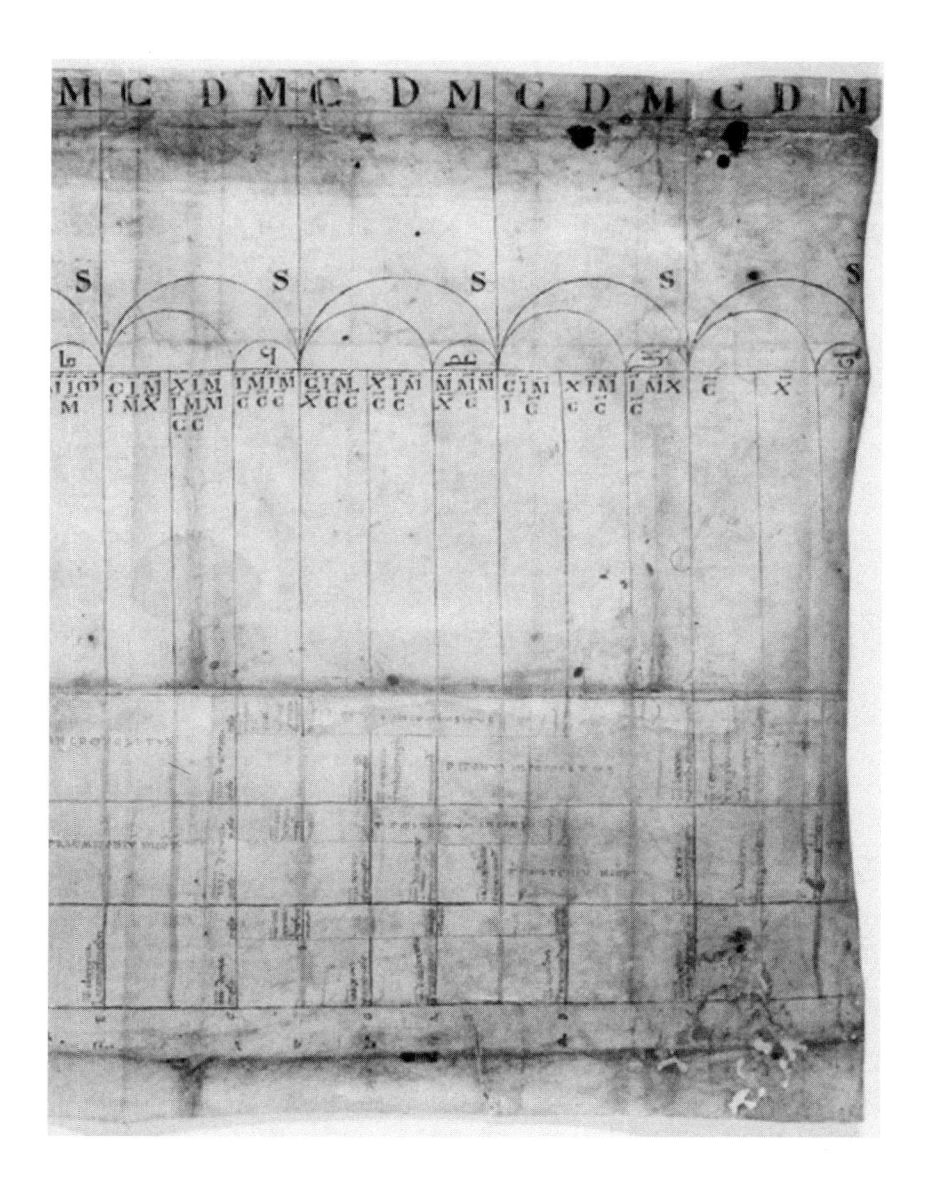

text to the other items is also suggested by the two folds still visible on the sheet. The latitudinal fold has been made to separate off the *mensura monochordi*. The longitudinal fold, however, crosses through the middle of an abacus column, damaging the letters in the column, whereas the scribe of the *mensura monochordi* has scrupulously avoided writing where the fold occurs.

3 The Echternach manuscript sheet as an example of Gerbert's own abacus board

We may compare the Echternach example, point by point, with the earliest descriptions of abacus boards: (a) the description of Gerbert d'Aurillac's own abacus by his pupil Richer in his *Historiae*, III, ch. 54 (= Rich.), [8] and (b) a fuller description of the abacus in the earliest set of instructions for calculation that contains such a description, namely the *Liber abaci* of Bernelinus of Paris, who attributes his knowledge to Gerbert's writings (= Bern.).[9]

  i) Rich.: 'Abacum, id est tabulam dimensionibus aptam, opere scutarii effecit' ('He <Gerbert> employed the offices of a shield-maker to make an abacus, i.e. a board fitted with divisions of measurement'). The usual medieval shield was made of wood covered with leather.[10] The Echternach table consists of only one sheet of parchment and hence is not as sturdy as the example described by Richer. It does, however, have the length and width of a breastplate.

  ii) Rich: 'cuius longitudini in .xxvii. partibus diductae novem numero notas omnem numerum significantes disposuit' ('He divided its length into 27 columns of which he marked nine with a numeral, these signifying all possible numbers'[11]); Bern.: 'Tabula ... diligenter undique prius polita, per .xxx. dividatur lineas, quarum tres primas unciarum minutiarumque dispositioni reservamus'

---

[8]Richerus, *Historiarum libri quattuor*, ed. G. Waitz, Monumenta Germaniae Historica, Scriptores rerum germanicarum in usum scholarum, Hannover, 1877.

[9]Bernelin, élève de Gerbert d'Aurillac, *Libre d'Abaque*, ed. B. Bakhouche, comm. J. Cassinet, Collection Istoría Matematíca Occítana 3, Montpellier, 2000, p. 21: 'abaci rationes persequar diligenter, negligentia quidem apud nos iam paene demersas, sed a domino papa Gerberto quasi quaedam seminaria breviter et subtilissime seminatas' ('I shall carefully pursue the rules of the abacus, now almost sunk beyond trace amongst us because of negligence, but, as it were, sown again in nurseries (seminaries) briefly and most subtly by our lord Pope Gerbert'). One may infer from how Bernelinus refers to Gerbert in this text that it was written during Gerbert's papacy (999–1003).

[10]C. Blair, *European Armour*, London, 1958, p. 18.

[11]Grammatically, this translation seems to be more correct than the usual interpretation: '... on which he arranged the nine numerical symbols which signify all possible numbers'. I have taken 'notas' as a past participle passive referring to 'partes', rather than as a noun referring to the

('A board, which has first been thoroughly polished, should be carefully divided into 30 columns, of which we reserve the first three for fractions'). Only 24 columns of the Echternach table survive, but, since the first column (marked with the Arabic numeral '2') is for the thousands (10^3), it is evident that originally it consisted of at least three more columns (for 10^0, 10^1, and 10^2), of which the first would have been marked with an Arabic '1'. The missing columns are due to the sheet being cut down to fit the binding of the Riesen-bibel. That it did not include the three columns for fractions mentioned by Bernelinus is indicated by the following facts. (1) Arabic numerals,marking every third column, fit a 27-column abacus, but not a 30-column abacus,[12] and indeed are found on extant early depictions of 27-column abacuses,[13] but not in Bernelinus's description, which is confirmed by a depiction of a tenth-century 30-column abacus.[14] (2) From the evidence of the longitudinal fold, if we assume the table was folded in half, there is exactly the right space for three columns only.[15]

iii) Bern.: '... reliquarum vero .xxvii., per ternas et ternas, haec certa mensurandi proveniat regula: primam de tribus lineam circinus in hemisperii modum teneat amplexam; maior autem circinus duas reliquas amplectatur;sed easdem tres maximus complectatur pariter ... per omnem abaci tabulam omnes eaedem praescribantur litterae, id est C, D, M, S, hoc modo: C super centenum, D super decenum, M super unitatem, sive ut monos designet, sive ut mille significet; cui supponatur S singulare significans' ('the other 27 columns should be grouped in threes, linked in the following way: a semicircle should embrace the first of the three, a larger semicircle the other two, but an even larger semicircle should embrace all three.... The same letters should be repeated through the columns: namely, C, D, M, S, in this way: C over the hundreds, D over the tens, M over the units, whether because it indicates "monos" or "mille". Below it is placed an 'S' indicating "singulare"'). These details, which are not mentioned by Richer, are all present on the Echternach table.

iv) Bern.: 'Prima linea...unitate signetur ita: I, secunda, X, id est deceno; tertia

---

numeral form (for which 'c(h)aracter' or 'figura' is the usual term in the abacus literature).

[12] The Arabic numerals at the heads of columns (unlike those on the abacus counters) do not have an arithmetical function, but, presumably, are added to show symbolically what a huge range of numbers they can represent. Moreover, as can be seen from paragraph vi below, they give examples of the symbols to be written on the counters.

[13] MSS **P** and **J** (see n. 5 above).

[14] MS **B** (see n. 5 above). This abacus agrees in almost every respect with Bernelinus's description. MS **Q**, however, which otherwise is identical to MS **B**, does include the numerals, and the 'S' below the 'M'.

[15] See n. 7 above. Further confirmation will be given below.

C, id est centeno; quarta elemento I, superaddito titulo, id est milleno; quinta elemento X, superaddito titulo, id est deceno milleno; sexta elemento C, superaddito titulo, id est centeno milleno; septimam vero hæc elementa praesignent, additis titulis, M.I., id est mille milia. Octavam hæc X.MI, additis tantum titulis super M et I, id est decies mille milia. Idque in omnibus

column number and value	Bernelinus	Echternach table
1. $1 \ (10^0)$	I	<I>
2. $10 \ (10^1)$	X	<X>
3. $100 \ (10^2)$	C	<C>[16]
4. $1{,}000 \ (10^3)$	$\overline{\text{I}}$	$\overline{\text{I}}$
5. $10{,}000 \ (10^4)$	$\overline{\text{X}}$	$\overline{\text{X}}$
6. $100{,}000 \ (10^5)$	$\overline{\text{C}}$	$\overline{\text{C}}$
7. $1{,}000{,}000 \ (10^6)$	M.I	IM
8. $10{,}000{,}000 \ (10^7)$	X.$\overline{\text{MI}}$	X$\overline{\text{IM}}$
9. $100{,}000{,}000 \ (10^8)$	C.$\overline{\text{MI}}$	C$\overline{\text{IM}}$
10. $1{,}000{,}000{,}000 \ (10^9)$	M.MI	MMM
11. $10{,}000{,}000{,}000 \ (10^{10})$	X.$\overline{\text{MMI}}$	X$\overline{\text{IM}}$
12. $100{,}000{,}000{,}000 \ (10^{11})$	C.$\overline{\text{MMI}}$	C$\overline{\text{IM}}$
13. $1{,}000{,}000{,}000{,}000 \ (10^{12})$	M.$\overline{\text{MMI}}$	$\overline{\text{IMIM}}$
14. $10{,}000{,}000{,}000{,}000 \ (10^{13})$	X.$\overline{\text{MMMI}}$	X$\overline{\text{IMIM}}$
15. $100{,}000{,}000{,}000{,}000 \ (10^{14})$	C.$\overline{\text{MMMI}}$	C$\overline{\text{IMIM}}$[17]
16. $1{,}000{,}000{,}000{,}000{,}000 \ (10^{15})$	M.$\overline{\text{MMMI}}$	M$\overline{\text{IMIM}}$
17. $10{,}000{,}000{,}000{,}000{,}000 \ (10^{16})$	X.$\overline{\text{MMMMI}}$	X$\overline{\text{IMIM}}$
18. $100{,}000{,}000{,}000{,}000{,}000 \ (10^{17})$	C.$\overline{\text{MMMMI}}$	C$\overline{\text{IMIM}}$
19. $1{,}000{,}000{,}000{,}000{,}000{,}000 \ (10^{18})$	M.$\overline{\text{MMMMI}}$	$\overline{\text{IMIMIM}}$
20. $10{,}000{,}000{,}000{,}000{,}000{,}000 \ (10^{19})$	X.$\overline{\text{MMMMMI}}$	X$\overline{\text{IMIMIM}}$
21. $100{,}000{,}000{,}000{,}000{,}000{,}000 \ (10^{20})$	C.$\overline{\text{MMMMMI}}$	C$\overline{\text{IMIMIM}}$
22. $1{,}000{,}000{,}000{,}000{,}000{,}000{,}000 \ (10^{21})$	M.$\overline{\text{MMMMMI}}$	M$\overline{\text{IMIMIM}}$
23. $10{,}000{,}000{,}000{,}000{,}000{,}000{,}000 \ (10^{22})$	X.$\overline{\text{MMMMMMI}}$	X$\overline{\text{IMIMIM}}$
24. $100{,}000{,}000{,}000{,}000{,}000{,}000{,}000 \ (10^{23})$	C.$\overline{\text{MMMMMMI}}$	C$\overline{\text{IMIMIM}}$
25. $1{,}000{,}000{,}000{,}000{,}000{,}000{,}000{,}000 \ (10^{24})$	M.$\overline{\text{MMMMMMI}}$	$\overline{\text{IMIMIMIM}}$
26. $10{,}000{,}000{,}000{,}000{,}000{,}000{,}000{,}000 \ (10^{25})$	X.$\overline{\text{MMMMMMMI}}$	X$\overline{\text{IMIMIMIM}}$
27. $100{,}000{,}000{,}000{,}000{,}000{,}000{,}000{,}000 \ (10^{26})$	C.$\overline{\text{MMMMMMMI}}$	C$\overline{\text{IMIMIMIM}}$

cavendum volumus ut, his elementis X et C <exceptis>, posthac superaddantur tituli omnibus ceteris[18] ...' ('The first line should be marked in the following way: I; the second, X, i.e. a ten; the third, C, i.e. a hundred; the fourth, the letter I with a tilde ontop, i.e. a thousand; the fifth, the letter X with a tilde on top, i.e. a ten thousand; the seventh is marked with the following letters, with tildes added: $\overline{\text{M}}.\overline{\text{I}}$., i.e. 1,000,000; the eighth, with these: X.$\overline{\text{M}}\overline{\text{I}}$, with tildes over the M and the I only, i.e. 10,000,000. We should warn the reader that in every case after this there are no tildes on the letters X and C, but tildes on all the other letters'). These indications of the powers of ten are followed in the Echternach table, except that, in the latter, (a) the two ways of indicating '1,000' — I with a tilde and M with a tilde[19] — alternate, and (b) '$\overline{\text{XM}}$' and 'CM' are always used instead of 'X$\overline{\text{MM}}$' and 'C$\overline{\text{MM}}$'.

According to the editor, there are errors in all the manuscripts of Bernelinus.[20] In the Echternach table, however, there is not a single mistake here.

v) Bern.: 'Quibus praesignatis, a prima priori linea usque ad primae vicesimam septimam quatuor trahantur lineae aequali spatio differentes inter se, quarum prima primus trames, ultima vero quartus nuncupabitur, duarum autem mediarum secunda secundus, tertia tertius nominabitur. Quod quare statuerim, diligenter intuentem non latebit cum ad divisiones venerit' ('When this has been done, four lines equally distant from each other should be drawn from the line before the first <column> to the line 27th in respect to the first <column>, of which the first is called the "first cross-line", the last, the "fourth cross-line", and the two middle ones respectively "the second" and "the third cross-lines". The careful observer will see why I have arranged them in this way when it comes to describing how to do division.'[21]). Here the Echternach sheet contains

---

[16] The first three columns on the Echternach sheet, now missing, can be presumed to have had these headings.

[17] A tilde on 'C' has been erased.

[18] Bakhouche's edition (n. 9 above, p. 24) gives 'his elementis X et C posthac superaddantur tituli omnibus ceteris generaliter superadditis', which appears to be corrupt. I have emended the text to make it conform to the examples that Bernelinus goes on to give.

[19] It seems irrational to add a tilde to M, when it already means '1,000', but Bernelinus states that this is done, and Bernelinus's prescription is confirmed both by the Echternach manuscript sheet and by most early abacus depictions. An exception is MS **V** (see n. 5 above) in which M is always written without a tilde (In MS **J** M without the tilde is used for '1,000' in the fourth column, but M with a tilde is used in all other cases).

[20] Bernelin, *Libre d'Abaque*, ed. B. Bakhouche (n. 9 above), p. 24, note 7.

[21] As far as I can see, Bernelinus never returns to describing the purpose of these cross-lines, which can only be inferred from MS **B** and **Q** (see n. 5 above) in which the second cross-line contains

a serious mistake. Numbers that should have been on the second cross-line have been added to the ends of the numbers on the first one, as follows:[22]

column and numerals in the first cross-line	numerals from the second cross-line
7. $\overline{\text{IM}}$	$\overline{\text{XC}}$ i.e. 10 x 100,000
8. $\overline{\text{XIM}}$	$\overline{\text{CC}}$ i.e. 100 x 100,000
9. $\overline{\text{CIM}}$	$\overline{\text{IC}}$ i.e. 1,000 x 100,000
10. $\overline{\text{MMM}}$	$\overline{\text{XC}}$ i.e. 10,000 x 100,000
11. $\overline{\text{XIM}}$	$\overline{\text{CC}}$ i.e. 100,000 x 100,000
12. $\overline{\text{CIM}}$	$\overline{\text{XCC}}$ i.e. 10 x 100,000 x 100,000
13. $\overline{\text{IMIM}}$	$\overline{\text{CCC}}$ i.e. 100 x 100,000 x 100,000
14. $\overline{\text{XIMIM}}$	$\overline{\text{MCC}}$ i.e. 1,000 x 100,000 x 100,000
15. $\overline{\text{CIMIM}}$	$\overline{\text{X}}$ (incomplete)

vi) Rich.: 'Ad quarum etiam similitudinem mille corneos effecit caracteres, qui per .xxvii. abaci partes mutuati, cuiusque numeri multiplicationem sive divisionem designarent' ('In the likeness of which <i.e. the columns marked with Arabic numerals> he (Gerbert) made 1,000 counters of horn, which, by being moved around through the 27 columns of the abacus, could show the multiplication or division of any number you like'); Bern.: 'His igitur expeditis, ad ipsos caraceres veniamus et quibus figuris praenotentur adscribere properemus. Unitas, quae primus caracter dicitur, sic figuratur : 1, sive per Graecum A alpha. Binarius...' ('Having done this, let us turn to the characters themselves, and let us hurry to describe the shapes by which they are represented. The one, which is called the first character, is the following shape: 1 or represented by the Greek A = alpha. The two...'). Both Richer and Bernelinus are describing the counters that are marked with each of the nine digits. Richer does not give the forms of

---

numbers which by being added (in columns 1 to 6 — disregarding the first three columns of fractions) and multiplied (in columns 7–27) produce the number at the head of the column; the third cross-line contains the halves of those numbers; and the fourth cross-line lists the names, symbols and equivalents in roman numerals, for fractions. The fact that no numerals are included in the first three fraction columns of MSS **B** and **Q** suggests that these columns were added at a later stage than the four cross-lines.

[22] The values from the second cross-line are exactly those in MSS **B** and **Q**, and it may be significant that the scribe has started to copy them from the point where they change from being additive to multiplicative (see previous note).

the Arabic numerals; Bernelinus gives the forms of all of them, as well as the alternative Greek letters that can be used. In the Echternach table the counters are, of course, not depicted. One must presume that they existed separately and were moved over the table for the purpose of making calculations. The columns are wide enough to accommodate such counters. We can infer from Richer's passage that they would have had, written on them, the same forms of the Arabic numerals as have been written over every third column in the Echternach table. These shapes are remarkably similar to those in the text of Bernelinus, but are common to other early abacus texts.[23] Bernelinus gives Latin names to these counters: 'unitas', 'binarius', 'ternarius' etc. ('the one', 'the two', 'the three', etc.), but other abacus texts have the distinctive names (some of Arabic origin): 'igin, andras, ormis, arbas, quimas, calctis, zenis, temenias' and 'celentis'.[24]

The Echternach manuscript sheet, then, has no features that are not mentioned in Richer or Bernelinus's descriptions of the abacus board, nor do either of these descriptions include features absent from the Echternach table, with the exception of the four cross-lines. However, the fact that some of the numbers which belong to the second of these cross-lines have also been copied onto the Echternach sheet (but in the wrong place) indicates that the cross-lines had been present in a previous version. What we appear to have, then, on the Echternach sheet, is the abacus board as devised by Gerbert, but before the modification of Bernelinus, which consisted in extending the number of columns from 27 to 30, and dropping the Arabic numerals as headings to columns.

4 A second Echternach abacus

Virtually a facsimile of the abacus on the Echternach manuscript sheet can be found in another Echternach manuscript, namely Trier, Stadtbibliothek, 1093/1694. The depiction of the abacus fills fol. 197r (see Plate 2). At 510 x 345 mm it approaches the size of the Echternach sheet more closely than any other known abacus table, and it is written in the same bold capital letters. Moreover it is complete, and so confirms the assumption that the Echternach sheet, before being cut down in size, consisted of 27 columns and not 30. The first three columns of the Trier abacus,

---

[23] A table of forms is given in M. Folkerts, *"Boethius" Geometrie II, ein mathematisches Lehrbuch des Mittelalters*, Wiesbaden, 1970.

[24] The most complete account of the occurrence of these names is in M. Folkerts, 'Frühe westliche Benennungen der indisch-arabischen Ziffern und ihr Vorkommen' in *Sic itur ad astra: Studien zur Geschichte der Mathematik und Naturwissenschaften. Festschrift für den Arabisten Paul Kunitzsch zum 70. Geburtstag*, ed. M. Folkerts and R. Lorch, Wiesbaden, 2000, 216–33.

Plate 2. The second Echternach abacus (MS Trier, Stadtbibliothek, 1093/1694, fol. 197r; reproduced with permission)

as one would have expected, are devoted to the units, tens and hundreds, but there are two features which could not have been guessed from the Echternach sheet: i) a blank is left where one would have expected the upper-case roman 'I' for the units; instead, the Arabic 'I' (which is of course identical in shape) in the small semicircle above the first column serves as the only heading; ii) again, above the first column, the large 'S', which in every other case is placed over the first of each set of three columns to the right of the large semicircle, is doubled in size and put at the top right-hand corner of the sheet. Both these features enhance the visual aspect of the table, and show the reader that the table begins from the right. In the parts of the table that the Trier manuscript and the Echternach manuscript sheet share in common, the only differences to be observed are the following:

i) the letters M D C, repeated eight times at the heads of the columns, are missing in the Trier version, though two horizontal lines have been drawn (absent in the Echternach sheet) which may have been expected to contain them. The space between these lines is occupied only by the above-mentioned large 'S' over the first column. The depictions of the abacus in MSS **P** and **J** (see n. 5 above) show that such an S replaces the M in the first column, while S and M appear together in all other cases.

ii) In the case of the Arabic numerals in the Trier version the lower loop of the '3' is written as a separate stroke rather than as a continuation of a curved line passing through the central horizontal stroke, and the horizontal stroke of the '6' leaves the vertical stroke at the mid-point rather than at the base. In the latter case the numeral is more similar to its equivalent in Arabic script.

iii) While the same numbers from the second cross-line follow immediately after the numbers of the first cross-line in both abacus tables, in the Trier version they always begin on a new line, except in the case of the last, incomplete, number ('X') which is tagged onto the end of the first number. Here the Trier version could be said to be more accurate, but the inclusion of only a few numbers from the second cross-line, and their placement immediately after those of the first cross-line, would still seem to be a mistake, shared by both versions.

iv) There are trivial differences in the line-arrangement of the higher numbers.

The Trier table, then, is marginally more correct and may be closer to an archetype. The absence of the headings M D C, however, show that it is not the exemplar from which the Echternach sheet was copied.

But the Trier version has another feature which is not present on the Echternach sheet: namely, it adds some miscellaneous texts on calculation, fractions and Arabic numerals. These texts do not reveal anything further on the construction of the abacus, but they are clearly meant to be complementary to the depiction of the instrument. They consist of the following:

1) Fol. 197v. Rules for multiplication and division. These begin with instructions on how to determine in which columns the product of a multiplication falls ('Singularis multiplicator quemcumque multiplicaverit, in eundem pone digitum quem multiplicat, in ulteriorem articulum ...'). These resemble most closely the rules for the abacus given by Heriger of Lobbes.[25] They are followed by miscellaneous rules for multiplying and dividing numbers ('Si multiplicare vis quemlibet numerum et semis ...'; 'Omnis divisor sive simplex sive compositus denominationem (?) semper in prima ponit linea et quot lineas rediens divisor descenderit, tot subposita denominatio ascendet ...'; 'Si compositus est divisor solus, qui dux vocatur ...'; 'Dividendus autem, sive sit simplex sive compositus, semper erit per denominationes et divisores, facta multiplicatione, reformandus ...'; 'In simplici divisori simplici dividendo et in simplici divisori composito dividendo "quotiens" dicendum est semper ...'). Here 'linea' is used for decimal place.

2) Fol. 197v. Rules for 'equating', apparently in regard to finding the quotients in long division ('De eo aequandis haec prima servatur ratio. Primus superior semper debet maior esse reliqui minores inferioribus ...').

3) Fol. 197v-198r. The problem of what to do with the remainder in a division leads without a break into a discussion of Roman fractions ('Omne quod remanet de maioribus indivisum per uncias primum erit dividendum. As enim .xii. habet uncias et ideo per duodenarium habes invenire quot uncias remanentia maiorum possint reddere. Unciis autem ita inventis, per maiorum divisores iterum dividere curabis. Si quid autem de unciis remanserit, per scripulos dividendum erit...'). The scribe describes the divisions of the Roman *as* in detail, and includes a table of the 22 different divisions, with their symbols and their values in *uncie* and *scripuli*. The table is very similar to that associated with the 'Incertus abacista saeculi decimi' edited by Bubnov,[26] while the textual account of the relationship between the various fractions is similar to that found in the *Calculus* of Victorius in the version used by Bernelinus[27] and with the commentary by Abbo of Fleury.[28]

4) Fol. 198r. A detailed account of the etymology of each of the Roman fractions ('Libra vel as sive assis xii untiae. Libra dicta a librando quasi libera, eo quod liberaliter pondera adaequat ...'). This does not correspond to any

[25] Bubnov, *Gerberti opera* (n. 4 above), pp. 224-5.

[26] Ibid., pp. 228-9.

[27] Bernelinus, *Liber abaci* (n. 9 above), Book IV.

[28] See Abbo of Fleury and Ramsey, *Commentary on the Calculus of Victorius*, ed. A. Peden, London, 2002, Appendix 8.

text that I know of, but quotes 'Comminianus',[29] Augustine,[30] Priscian,[31] and Ambrosius.[32]

5) A poem on the names of the nine Arabic numerals and the zero ('Ordine primigeno nomen iam possidet ygin').[33] One line is devoted to each numeral and each line is followed by the relevant roman and Arabic form of that numeral (See illustration).

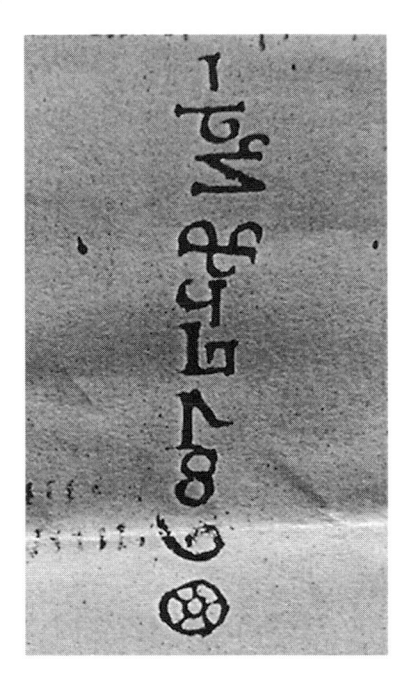

The forms differ a little from those on the abacus table on the previous folio: the '3' is upside down, the '4' has an extra loop below the horizontal line, the

---

[29]'Assis secundum antiquos nominativus fuit, teste Comminiano, sed hodie genitivus'. Comminianus is a fourth-century Latin grammarian.

[30]'Augustinus: scripulus per diminutionem, a lapillo breve qui scrupus vocatus est, scripulus translato de littera nomine appellatur ...'.

[31]'Unde Priscianus dicit: semina sex alii siliquis latitantia curvis'.

[32]'Ambrosius quoque in expositione Lucae dicit siliqua est humillimum genus ligni nascens in littoribus maris'.

[33]This poem has been edited by Menso Folkerts in 'Frühe westliche Benennungen' (n. 24 above), but without the use of this manuscript. The readings of the manuscript conform to those of the edition.

horizontal stroke of the '6' leaves the vertical stroke at the base (agreeing here rather with the Echternach manuscript sheet), and the '9' gives the appearance of lying on its back. The zero is a real wheel (the 'rota' in the poem), composed of two concentric circles linked to each other by five short lines.

6) The beginning of another account of the divisions of the *as* ('Assis enim libra est .xii. untiarum, ex quibus si subtracta fuerit una .xi., quae remanent deunx vocatur ...'). This breaks off in the middle of a sentence.

7) The etymology of 'digit' ('Digitos ideo appellari puto quod sicut digiti corporales a brevibus iuncturis incipientes in longiores extenduntur, sic et innumerabiles a finita[te] unitatis quantitate inchoantes in infinitum'[34]).

8) A depiction of the portable sundial (labelled as the 'horologium viatorum').

These miscellaneous notes do not correspond consistently with any known source. They appear to be written by a single scholar, at different times, who has made frequent erasures and corrections.

Conclusions

MS Trier, Stadtbibliothek, 1093/1694 belongs to a group of fine manuscripts of a large format, copied in Echternach towards the end of the tenth century, by a hand (known as 'hand B') which also completed a copy of a manuscript dedicated to the monastery of Echternach by Leofsin.[35] The English-born Leofsin had fled to Echternach in 993 from Mettlach, where he had been abbot. Thus the Trier manuscript, including the abacus and notes on calculation, is likely to have been written soon after this date, and therefore is probably contemporary with the Echternach manuscript sheet.[36] In Leofsin's time at Mettlach learning greatly flourished, due, it seems, to a previous abbot, Ruotwic (941–ca. 975) who had sent two of his monks to Gerbert as students. One of these students, Nizo or Nithard, succeeded Ruotwic as abbot, and kept in touch with Gerbert by letter. The Trier manuscript includes a poem by Gerbert concerning Boethius and texts that reflect the teaching curriculum established by Gerbert in Reims. Thus the information provided by the Trier

---

[34]The sentence is incomplete; the following words have been erased.

[35]J. Schroeder, *Bibliothek und Schule der Abtei Echternach um die Jahrtausendwende*, Inaugural-Dissertation, Freiburg i. B., 1975, pp. 39–43 and 78–81. The historical information in this paragraph is summarised from Schroeder's book. The detailed investigation of the Echternach manuscripts currently being carried out by Thomas Falmagne may necessitate a slight revision of the argument.

[36]A later hand ('X') also appears in the manuscript and was responsible for adding the calendar which has a starting date of 1049. But even if the abacus and notes on calculation were added by hand X, this would not necessarily place them later than the writing of the Echternach manuscript sheet.

manuscript corroborates the connection with Gerbert d'Aurillac which we deduced in the first half of this article from the similarity of the abacus of the Echternach manuscript sheet to the instrument devised by Gerbert himself. Bernelinus of Paris, while ascribing the description of this kind of abacus to Gerbert (as has already been noted[37]), singled out the Lotharingians as those contemporaries of his who were experts in the use of the instrument.[38] It may not be by chance, then, that a depiction of an abacus, and what is probably an abacus board itself, surviving from Bernelinus's time, should come from the monastery of Echternach, which lies in the heart of medieval Lotharingia.

---

[37] See n. 9 above.

[38] Bernelinus, *Liber abaci* (n. 9 above), p. 16: 'Quod si tibi taedium non esset, harum fervore, Lotharienses expetere, quos in his, ut cum (= quam?) maxime expertus sum, florere...' ('If, because of your passion for these (rules) it were not too difficult for you to seek out the Lotharingians, whom, as I have especially experienced, excel in these matters, <I would not have written this treatise>').

II

ABBON DE FLEURY *ABACI DOCTOR**

Abbon a exprimé son soulagement de mettre un point final à son commentaire sur le *Calculus* de Victorius sous la forme de trois vers :

> Apportez maintenant de l'eau pour les mains car nous avons assez mangé
> Celui qui se déverse en larmes reçoit ici les joies de la moisson
> Ici Abbon, l'abbé, le maître dans l'enseignement de l'abaque (*abaci doctor*) se retire pour se reposer[1].

En citant ces vers, le grand historien russe des mathématiques, Nicolai Bubnov, faisait remarquer, de façon peu indulgente, que « bien qu'Abbon se glorifie d'être *abaci doctor*, il ne semble pas avoir été très doué en cet art »[2]. Bubnov basait son jugement sur la pauvreté de l'exposé d'Abbon sur les règles de l'abaque dans ses œuvres conservées. L'objectif de cette contribution est de réexaminer les éléments dont nous disposons sur l'enseignement et les écrits

---

* Traduction par David Juste.

[1] N. Bubnov, *Gerberti postea Silvestri II papae Opera mathematica (972-1003)*, Berlin, 1899, p. 203 : « Nunc manibus fer aquas, quia sat iam sumpsimus escas. / Qui serit in lacrimis, recipit hic gaudia messis. / Hic Abbas abaci doctor dat se Abbo quieti ».Voir maintenant A. Peden éd., *Abbo of Fleury and Ramsey : Commentary of the Calculus of Victorius of Aquitaine*, « Auctores britannici medii aevi, XV », Oxford, 2003, p. XXXVIII. Peden donne la leçon « his » pour « hic ». Ces vers ne figurent que dans Bruxelles, Bibliothèque royale, ms. 10078, s. XI. De ce fait, Peden estime qu'il s'agit d'un ajout dû au scribe de ce manuscrit.

[2] *Ibid.*: « Quamquam Abbo se abaci doctorem esse gloriatur, hujus artis tamen non satis peritus exstitisse videtur ».

d'Abbon concernant l'abaque, et de voir dans quelle mesure nous pouvons évaluer plus positivement la contribution d'Abbon.

Un excellent point de départ est le fameux manuscrit du début du XIIe siècle de l'abbaye de Ramsey (Oxford, St. John's College, 17), qui comprend, parmi de nombreuses matières de mathématiques et de comput, une table de nombres présentée sous la forme d'un diptyque et intitulée comme suit : « Dans cette figure un nombre infini est décrit, car il commence avec le nombre 1 et finit avec le nombre 900.000 »[3] (**fig. 1**). Viennent ensuite quelques règles de calcul intitulées *Ratio Abbonis supra praefatum numerum* (« Règle d'Abbon sur le nombre indiqué plus haut » – c'est-à-dire le nombre infini). Pratiquement la même table et les mêmes règles apparaissent dans l'*Enchiridion* composé en 1011 par Byrhtferth de Ramsey, l'élève d'Abbon[4] (**fig. 2**). Ici, cependant, le titre est différent : « Le nombre 1000 est parfait, comme le montrent amplement le nombre et la figure qui suivent »[5]. Byrhtferth poursuit en disant « Il nous plaît d'apprécier la signification[6] qu'Abbon, de digne mémoire, prêtait à ce nombre. Les miracles qui ont suivi sa mort révèlent avec quel éclat sa dignité brillait dans cette vie. Car il était rompu aux savoirs scientifiques, et accompli en philosophie. Il avait l'habitude de dire... »[7] (suivent les règles de calcul).

On ne peut donc guère douter que ce texte fut présenté par Abbon à ses étudiants de l'abbaye de Ramsey, où il séjourna durant les années 985-987. Nous pouvons nommer les éléments communs du ms. d'Oxford et de

---

* [3] Éd. N. Bubnov, p. 203 : « In hac figura descriptus est numerus infinitus: incipit enim ab uno pervenitque usque ad nongentesimum millesimum ». Pour un commentaire détaillé sur cette table, voir F. Wallis, *Oxford St. John's College 17 : A Mediaeval MS in its Context*, thèse non éditée, Toronto, 1985, II, p. 393-394.

[4] Le fait qu'il y ait une plus grande cohérence, comme nous le verrons plus loin, entre le titre et le contenu dans le ms. St. John's 17 suggère que la version de ce ms., bien qu'ayant été copiée à une date plus tardive, est antérieure à l'*Enchiridion*.

[5] Byrhtferth, *Enchiridion*, IV.1., fig. 35, éd. et trad. P. S. Baker et M. Lapidge, Oxford, 1995, p. 228 : « Millenarius perfectus est, sicut liquido numerus demonstrat sequens et figura ».

[6] Dans la traduction, il est difficile de conserver la similarité entre les mots du titre qui introduisent les règles dans le ms. St. John's 17 (*ratio [...] supra praefatum numerum*) et ceux utilisés par Byrhtferth (*ratio huius numeri [...]*).

[7] *Ibid.*: « Ratio huius numeri, quam digne memorie Abbo super hunc invexit, libet libari. Iste vero quante dignitatis refulsit in vita ostendunt post mortem miracula. Erat enim in doctrinali scientia peritus et in philosophia perfectus. Aiebat enim [...] ».

l'*Enchiridion* de Byrhtferth « la table et les règles de Ramsey ». Il n'est pas vraiment approprié de qualifier ces éléments de « texte », car ils consistent, d'une part, en une représentation visuelle de la relation entre les nombres et, d'autre part, en une litanie de courtes règles qui, ensemble, couvrent moins de deux folios dans un manuscrit. L'élément visuel est la table, composée de six colonnes et de neuf lignes. Les colonnes indiquent les puissances de 10, en chiffres romains, de 1 à 100 000 ; les neuf lignes fournissent l'équivalent de ces puissances de 10 pour chacun des neuf chiffres. Les règles de calcul expliquent dans quelle colonne (c'est-à-dire dans quelle puissance de 10) il faut mettre le produit de la multiplication. La terminologie habituelle est utilisée, c'est-à-dire celle qui dérive du « calcul digital », du calcul sur les doigts (le mode de calcul que les étudiants d'Abbon ont pu encore utiliser) : un *digitus* (c'est-à-dire un « doigt » ou, plus précisément, un doigt levé) est un nombre appartenant à la plus petite position décimale (de 1 à 9) ; un *articulus* (c'est-à-dire un doigt replié) est un nombre de la position décimale suivante (de 10 à 90). Suit le groupe de formules :

> « Si tu veux multiplier une unité par un 10, assigne 10 à chaque *digitus* et 100 à chaque *articulus* ».
> « Si tu veux multiplier un 10 par une unité, assigne 10 à chaque *digitus* et 100 à chaque *articulus* ».
> « Si tu veux multiplier un 10 par un 10, assigne 100 à chaque *digitus* et 1000 à chaque *articulus* », etc.

Il s'agit ici de formules mnémotechniques plutôt que d'instructions, car il n'y a aucun exemple. Il ne sera donc peut-être pas inutile à ce stade d'en donner un. Supposons que l'unité soit 4 et la dizaine 30. 4x3 = 12, dont 2 est le *digitus* et 1 l'*articulus*. En vertu de la règle n°1, qui dit : « assigne 10 à chaque *digitus* et 100 à chaque *articulus* », on obtient 2 = 20 et 1 = 100, et le résultat est 120. Pour ce faire, on peut aussi utiliser la table : on regarde dans la colonne des « dizaines » jusqu'à ce qu'on trouve la ligne du « 2 », où nous lisons xx = 20. Ceci peut sembler évident, mais je voudrais montrer qu'il ne s'agit pas ici simplement d'arithmétique. Dans le ms. St. John's 17 comme dans l'*Enchiridion* de Byrhtferth, l'accent est mis sur le symbolisme et le mystère du nombre lui-même.

Dans le ms. St. John's 17, la table est censée montrer, si on s'en réfère au titre, le « nombre infini », c'est-à-dire le potentiel qu'ont les nombres de progresser à l'infini. En réalité, ce n'est pas le cas ; la table montre seulement les nombres jusqu'à la cinquième puissance de 10, comme le titre l'indique explicitement : « car il commence avec 1 et finit avec 900 000 ». Ce qui est important, néanmoins, c'est que la table est une *figura*, une « image » (construite comme le triptyque d'un autel médiéval, où le caractère merveilleux du nombre est représenté). Le fait qu'il y ait six colonnes et que 6, comme Abbon l'aura puisé dans le *De institutione arithmetica* de Boèce, est le premier nombre parfait (correspondant à la somme de ses facteurs : 1 + 2 + 3 = 1 x 2 x 3), est probablement aussi significatif, comme l'est le groupement de 3 par 3 au moyen de deux demi-cercles. Cependant, l'absence de contexte ne nous permet pas de savoir si Abbon voulait attirer l'attention sur ce point.

Dans l'*Enchiridion* de Byrhtferth, en revanche, nous trouvons un contexte. La table et les règles représentent l'aboutissement d'une série de chapitres consacrés à la signification mystique de chaque nombre. Ils ont été choisis, comme nous l'avons vu, pour illustrer le mystère du nombre 1 000. Ainsi, le « nombre infini » du ms. St. John's 17 a été remplacé par 1 000, qui est explicitement qualifié de « parfait »[8]. Ceci a résulté en une modification légère, mais significative, dans la table elle-même : après la colonne avec le 900 000 (qui correspondait au « nombre infini » d'Abbon) Byrhtferth a ajouté « ı » (= 1 000) dans une septième colonne où devrait plutôt se trouver 1 000 000. Le titre de Byrhtferth fait écho à un texte bien connu des *Moralia in Job* de saint Grégoire, paraphrasé par Bède, qui explique la perfection du nombre 1000 : « Il est appelé parfait parce qu'il représente le cube solide du nombre 10 »[9]. La perfection du nombre 1000, poursuit Byrhtferth, convient à la perfection d'Abbon lui-même (... *in philosophia perfectus*), et les règles de calcul sont présentées comme s'il s'agissait d'une litanie (c'est-à-dire une prière consistant en une formule répétée) récitée par Abbon lui-même : *aiebat enim...*

Qu'il faille interpréter ce court morceau dans un sens symbolique ou arithmologique, et non comme un calcul ayant une utilité pratique, est

---

[8] La question est de savoir s'il existe un lien entre la convention médiévale représentant le nombre 1.000 au moyen de deux cercles juxtaposés (« ∞ ») et l'utilisation mathématique du même symbole pour l'infini, comme me l'a fait remarquer David Juste, reste à explorer.

[9] Cf. saint Grégoire, *Moralia in Iob*, xxxv. 16.42 (CCSL, cxliiiB. 1802) et Bède, *De templo*, ii (CCSL, cxixA. 211). Le passage en question est cité dans le commentaire de P. S. Baker et M. Lapidge à l'édition de l'*Enchiridion* (p. 366), où l'on trouvera d'autres références sur la signification arithmologique du nombre 1 000.

également suggéré par une erreur dans la dernière règle, que Byrhtferth n'a pas corrigée : « Si tu veux multiplier 10 000 par 10 000, assigne à chaque *digitus* 100 fois 100 000 »[10], formule qui aurait dû être formulée comme suit : « ... assigne à chaque *digitus* 1 000 fois 100 000 ».

Ce texte faisait partie de l'enseignement d'Abbon, comme le montre la présence du même matériel arithmétique, avec la même spéculation symbolique, dans son commentaire au *Calculus* de Victorius, rédigé quelques années avant son séjour à Ramsey. On y retrouve la même table[11]. Le titre, cependant, est différent, l'accent étant plutôt mis sur le potentiel qu'a l'unité d'engendrer tous les nombres :

> Nous devons montrer comment toute multitude de nombres est multipliée au moyen d'un système pratique où l'on peut observer quelle nature seconde est dans les « dizaines », quelle nature troisième est dans les « centaines », quelle nature quatrième est dans les « milliers » , tous <ces types de nombres> étant indiqués dans une seule figure [soit la figure][12].

Le langage d'Abbon a de quoi surprendre. Pour lui, l'unité (le caractère de ce qui est unique) consiste en différentes « natures » : les unités, les dizaines, les centaines, les milliers, etc. Tous les nombres d'une même colonne appartiennent à l'une de ces « natures ». Il faut donc regarder dans la colonne de la « nature » appropriée pour trouver le nombre recherché (comme nous l'avons vu avec l'exemple du produit de 4 x 30).

Abbon a ajouté à la table elle-même les mentions *latitudo* (« largeur ») et *longitudo* (« longueur »). Il explique ensuite le contenu de la table de façon relativement détaillée, en nommant les lignes « vers », comme s'il s'agissait des vers d'un poème. Je voudrais maintenant citer un passage assez long qui nous donnera un aperçu du ton de son exposé[13]

---

[10] « Si multiplicaveris decenum millenum per decenum millenum, dabis unicuique digito centies centum milia... », éd. N. Bubnov, p. 203 ; éd. P. S. Baker et M. Lapidge, p. 228.

[11] Voir Abbon de Fleury, *Commentary on the* Calculus *of Victorius of Aquitaine*, chap. 60, éd. A. M. Peden.

[12] *Ibid.*, chap. 59 : « Doceamus qualiter omnis numerorum multitudo multiplicetur quodam argumenti conpendio, ubi animadverti poterit quae sit in singularibus prima unitatis natura, quae in decenis secunda, quae in centenis tertia, quae in millenis quarta, quorum subscriptio stlais sit uno ordine disposita ».

[13] *Ibid.*, chap. 61. La langue d'Abbon étant particulièrement alambiquée, il est impossible de donner une traduction littérale.

Puisque neuf « vers » appartiennent à la longueur, mais six à la largeur < de la table >, on doit considérer chacun d'entre eux à la fois en eux-mêmes et en rapport avec les autres, de sorte qu'ils puissent enseigner les uns aux autres toute formule de multiplication. À partir des « singuliers », qui sont premiers, nous obtenons une règle qui peut être appliquée dans le cas des autres nombres. Nous les nommons « singuliers » parce qu'ils découlent des « unités », de même que les « dizaines » sont ainsi nommées parce qu'elles croissent à partir de 10, les « centaines » à partir de 100, les « milliers » à partir de 1 000. Chaque puissance[14] de l'unité est la même sur toute la longueur, sauf que le début de la première unité n'a pas de multiplication, mais le début de la seconde augmente par elle-même jusqu'à 100. La fin de la longueur consiste toujours en un 9 ou en un nombre nommé d'après la quantité du 9. Chaque ligne de la largeur est marquée par des nombres qui tirent leur nom des « singuliers » placés devant eux, sauf la première qui se distingue par des principes différents, tout en étant néanmoins sujette à la nature de l'unité[15]. Car à partir de 2 sont nommés 20 (lequel est aussi nommé *biginti*, c'est-à-dire « 10 deux fois né »), 200, 2 000, 20 000, 200 000, etc. Dans le second vers de la largeur[16], se trouve l'origine du mot dont tu peux observer l'utilité dans les autres < colonnes > : 2 multiplié par lui-même donne 4 ; 20 se délecte aussi de lui-même avec sa progéniture de 400 ; 200 aussi, multiplié par lui-même, grandit en 40 000. De la même façon, trois groupes de 3 font 9, 30 x 30 donnent 900, et 300 x 300 donnent 90 000. Vois-tu comment les simples lignes de vers se suivent dans la figure, avec leurs petites sommes vues sur ce côté-ci et ce côté-là, lorsque, à partir de 4, dérivent 400 et 40 000, dont l'un vient de 20 et l'autre de 200 ? À partir du nom de 9 sont produits 900 et 90 000, dont l'un croît à partir de 30 et l'autre à partir de 300. Ainsi, il se produit que la seconde unité regarde la première unité, la troisième la seconde et la quatrième la troisième.

Ceci peut sembler compliqué (et la traduction est considérablement simplifiée par rapport à l'original latin). La raison en est qu'Abbon cherche à montrer la coïncidence entre le dérivé grammatical (c'est-à-dire *duo milia*, *ducenti*, *viginti*, tous dérivant de la racine *duo*) et le dérivé arithmétique (2 000, 200, 20 provenant du nombre 2) et utilise la métaphore d'accroissement et de

---

[14] Remarquons qu'Abbon utilise ici *potentia* plutôt que *natura*.

[15] C'est-à-dire que les nombres *decem*, *centum*, *mille* ne peuvent être dits dériver de la racine *unus*.

[16] C'est-à-dire dans la seconde colonne.

génération naturels pour englober les deux. La phrase qui suit rappelle, justement, le titre de l'*Enchiridion* de Byrhtferth :

> Cette quatrième (unité) < c'est-à-dire 1000 > est considérée comme particulière par les arithméticiens, parce que c'est un grand cube produit par deux angles égaux[17].

Nous voyons ici que Byrhferth est justifié lorsqu'il dit qu'Abbon prêtait une signification importante au nombre 1 000. Nous pouvons aussi y voir une autre référence (bien qu'elle soit formulée de façon obscure) à l'explication de saint Grégoire à propos de la perfection du nombre 1 000.

Il s'avère donc que l'explication d'Abbon n'a pas pour objectif de montrer à l'étudiant comment multiplier les nombres, mais plutôt de mettre l'accent sur la merveilleuse façon dont les nombres peuvent se produire et se relier entre eux. On retrouve le même ton dans la section suivante, dans laquelle Abbon montre, à nouveau au moyen d'une figure faite de demi-cercles reliant les nombres, les rapports entre les cinq premiers nombres. Ensuite, Abbon expose plus en détail, et dans une prose élaborée, les règles de calcul qui font suite à la table dans « la table et les règles de Ramsey ». Voici encore une fois un aperçu de son langage :

> Nous pouvons conclure que les nombres dont nous avons parlé prennent, jusqu'à un certain point, une similarité ou participation dans la multiplication en ceci qu'ils partagent le même nom. Le calcul des plus grands nombres sera plus facile si tu commences par multiplier chaque « singulier » pour voir ce qui en résulte. Par exemple, si tu ne connais pas le produit de 20 x 20 ou de 20 x 30, vois comment 2 x 2 font 4 et 2 x 3 font 6, et, puisque tu as à t'occuper d'un nombre 10 multiplié par 10, attribue 100 au résultat de la multiplication des petits nombres, et tu trouveras ainsi que 20 x 20 font 400 et que 20 x 30, sans aucun doute, font 600.

Abbon fournit ici exactement la même information que la troisième formule de « la table et des règles de Ramsey », qui dit : « si tu veux multiplier un 10 par un 10, assigne 100 à chaque *digitus*... », mais, comme on le voit, les règles de Ramsey sont beaucoup plus faciles à mémoriser que la prose rhétorique du commentaire sur le *Calculus*. Abbon poursuit avec des exemples

---

[17] *Ibid.*, chap. 61: « Quae videlicet quarta ob id ab arithmeticis seponitur, quod maiori quadratae figurae duo anguli aequales habentur ».

utilisant les nombres des différentes puissances de 10, et qui couvrent à peu près toutes les formules figurant dans les règles de Ramsey. Il conclut finalement cette section du *Commentaire* en disant : « Vu que ces matières relèvent davantage des règles sur l'abaque, elles appartiennent à une autre tâche et à une autre discussion »[18].

C'est la première fois, dans les textes dont nous avons discuté jusqu'ici, qu'apparaît le mot « abaque ». Et il est important de noter qu'Abbon fait la distinction entre cette discussion et celle d'un traité sur l'abaque. Il ne sera pas inutile de dire un mot concernant l'abaque dont il est question. Il ne s'agit pas, bien entendu, d'un boulier tel qu'on en rencontre aujourd'hui, mais plutôt d'un tableau, ou d'une feuille de parchemin, divisé en colonnes pour chaque puissance de 10. Des jetons marqués des nombres 1 à 9 sont placés dans les colonnes appropriées. Par exemple, pour représenter « 92 », un jeton « 9 » est placé dans la colonne des dizaines et un jeton « 2 » dans la colonne des unités. Si le nombre comprend un zéro, la colonne est simplement laissée vide. Il n'y a aucune raison de supposer que ce type d'abaque est une invention arabe. Ce qui semble par contre bien être le cas, c'est que Gerbert d'Aurillac, qui fut en contact avec les mathématiques arabes en Catalogne en 967, a introduit la pratique qui consiste à marquer les jetons avec les chiffres arabes plutôt qu'avec les chiffres alphabétiques romains ou grecs[19]. Les règles d'utilisation de cet abaque expliquent comment disposer les jetons dans le tableau, ainsi que les substitutions à faire dans les opérations de calcul.

Abbon ne fait qu'emprunter les règles propres aux textes sur l'abaque et les utilise à ses propres fins. Vu que la discussion du commentaire sur le *Calculus* de Victorius est un exposé détaillé des matières qui seront plus tard intégrées dans « la table et les règles de Ramsey », ces matières ne doivent pas non plus être comprises comme formant un texte sur l'abaque. Qu'Abbon ait

---

[18] Abbon de Fleury, *Commentary on the* Calculus *of Victorius*, chap. 64 : « Sed quoniam haec pertinent ad rationem abaci, alterius sunt disputationis ac negotii ».

[19] C'est ce qu'on peut déduire de l'hexamètre joint à deux des plus anciennes représentations de l'abaque (Berne, BB, Cod. 250, fol. 1r, s. x^{ex} et Vatican, BAV, Vat. lat. 644, fol. 77v-78r, s. xi^{in}) : *Gerbertus Latio numeros abacique figuras* (« Gerbert <a offert> à la culture latine les nombres et les formes de l'abaque »), voir M. Folkerts, « Frühe Darstellungen des Gerbertschen Abakus », dans R. Franci, P. Pagli et L. Toti Rigatelli (éds), *Itinera mathematica: Studi in onore di Gino Arrighi per il suo 90a compleanno*, Sienne, 1996, p. 23-43.

puisé ses règles dans des écrits sur l'abaque est confirmé par les textes contemporains qui traitent expressément de cet instrument. Bubnov avait déjà remarqué que les règles de Ramsey étaient étroitement liées à celles de Gerbert d'Aurillac[20]. Des règles très semblables apparaissent dans les traités sur l'abaque d'Hériger de Lobbes, qui est l'exact contemporain de Gerbert[21], ainsi que dans un texte sur l'abaque datant du Xe siècle et provenant de l'abbaye d'Echternach[22]. Mais, alors qu'il reprend ces règles, Abbon évite toute allusion au fait qu'elles s'appliquent à un abaque. Dans les versions d'Hériger et d'Echternach, on lit que « si un 10 est multiplié par une unité, place les *digiti* dans la même < colonne >, mais les *articuli* dans la < colonne > suivante »[23], et ainsi de suite pour chaque puissance de 10. Ici, nous lisons le verbe « placer » (*ponere*) : les instructions parlent bien de « placer » dans le tableau de l'abaque des jetons portant le chiffre approprié. Abbon dit simplement « assigne (*dabis*) 10 à chaque *digitus* et 100 à chaque *articulus* », expressions qui, comme je l'ai rappelé, peuvent aussi bien s'appliquer au calcul digital. Plus significatif est que la table d'Abbon, en dépit de son caractère vivant, n'est pas la représentation d'un tableau d'abaque. Elle a en commun avec l'abaque la division en colonnes, une pour chaque puissance de 10, mais toutes les représentations de l'abaque dont nous disposons, ainsi que les règles d'utilisation, montrent que les plus petites puissances de 10 sont disposées sur la droite et les plus grandes sur la gauche. Cette disposition de droite à gauche est tout à fait naturelle lorsqu'il s'agit d'arithmétique pratique. En effet, lorsque les jetons sont placés sur le tableau, ils représentent les nombres composés de la même façon qu'ils s'écrivent sur une page, quelle que soit la forme des chiffres utilisés (p. ex. « cxlviii » en chiffres romains ou en chiffres arabes : les centaines sont à gauche, les unités à droites. Cette disposition correspond à la façon dont nous prononçons les nombres, que ce soit en latin, en anglais ou en français : les plus grands nombres précèdent les plus petits

---

[20] N. Bubnov, *Gerberti opera*, p. 203.

[21] *Ibid.*, p. 224.

[22] Voir Ch. Burnett, « The Abacus at Echternach in ca. 1000 A.D. », *Sciamus*, 3, 2002, p. 91-108.

[23] Hériger (éd. N. Bubnov, p. 224) : « Singularis quemcunque multiplicet, in eodem digitos ponet, in ulteriore articulos… Decenus quemcunque multiplicet, in secundo ab illo digitos, in ulteriore articulos » ; ms. Trèves, Stadtbibliothek, 1093/1694 (originaire d'Echternach), fol. 197v : « Decenus quemcumque multipli<caverit>, in secundo ab eo quem multiplic<cat> pone digitum, in ult<eriorem> articul<um> ».

(« cent quarante-huit »). Abbon renverse l'ordre, et il le fait délibérément, car son but est de montrer comment les nombres grandissent à partir de l'unité. En conséquence, l'unité doit être à gauche.

En résumé, ni dans « la table et les règles de Ramsey », ni dans le commentaire sur le *Calculus* de Victorius, Abbon ne donne d'instructions pour effectuer des opérations de calcul sur l'abaque, et il spécifie bien au lecteur, dans le commentaire sur le *Calculus*, qu'il n'est pas en train d'écrire un traité sur l'abaque. En revanche, il utilise des éléments issus des traités sur l'abaque qui circulent à son époque (tout comme il utilise le texte de Victorius), pour mettre en évidence le pouvoir merveilleux du nombre[24]. Lorsque son biographe, Aimoin de Fleury (m. 1008), énumère les réalisations d'Abbon dans les arts libéraux, il ne mentionne pas non plus l'abaque, mais il indique qu'Abbon « appréciait grandement la multiplicité des nombres géométriques »[25], c'est-à-dire qu'il était fasciné par les nombreuses façons dont les nombres sont liés entre eux[26]. La remarque désobligeante de Bubnov à propos de l'incompétence d'Abbon en matière d'abaque n'est dès lors pas justifiée. Il pensait que « la table et les règles de Ramsey », ainsi que la discussion du commentaire sur Victorius, constituaient l'exposé d'Abbon sur l'abaque, et il les critique parce qu'elles omettent le matériel (comme les règles de division) que l'on s'attend à trouver dans les traités sur l'abaque.

Il reste la question de savoir ce qu'Abbon a voulu dire en se qualifiant lui-même d'*abaci doctor*. Il est possible, mais peu probable, qu'Abbon ait écrit un traité sur l'abaque qui n'aurait pas été conservé. Il est plus vraisemblable qu'*abacus*, dans ce contexte, signifie « calcul » en général. Cette signification est largement attestée plus tard au Moyen Âge. Lorsque Léonard de Pise (Fibonacci) écrit son *Liber abaci*, qui n'a rien à voir avec l'abaque à proprement parler, « école de l'abaque » devint en Italie l'expression standard

---

[24] Ce point constitue un leitmotiv dans les écrits d'Abbon, comme l'a mis en évidence E.-M. Engelen, *Zeit, Zahl und Bild: Studien zur Verbindung von Philosophie und Wissenschaf bei Abbo von Fleury*, Berlin / New York, 1993.

[25] Aimoin de Fleury, *Vita s. Abbonis*, cité par N. Bubnov, p. 197 : « geometricorum multiplicitatem numerorum non mediocriter agnovit ».

[26] N. Bubnov (*loc. cit.*) pense qu'Aimoin se réfère ici à l'abaque, parce que l'instrument était considéré comme relevant de la géométrie plutôt que de l'arithmétique. Cependant, dès lors que les nombres sont envisagés sous l'angle de leurs relations mutuelles, on peut parler de « géométrie » (tout comme le cinquième livre des *Éléments* d'Euclide est consacré au rapport et à la proportion).

pour désigner les institutions où l'arithmétique des affaires était enseignée. Mais, avant cela, la fréquente utilisation du terme *abacista* pour désigner les mathématiciens des XIe et XIIe siècles[27] montre qu'il avait une acception plus générale que « expert dans l'utilisation de l'abaque ». Le contexte dans lequel Abbon se nomme lui-même *abaci doctor*, c'est-à-dire en tant qu'auteur d'un commentaire sur l'ensemble le plus complet de tables de calcul du haut Moyen Âge, est suffisant en lui-même pour justifier le titre de « maître dans l'enseignement du calcul ».

---

[27] Le mot est utilisé par Robert de Losinga (m. 1095) et Walcher de Malvern (m. 1135), voir Ch. Burnett, *The Introduction of Arabic Learning into England*, Londres, 1997, p. 16.

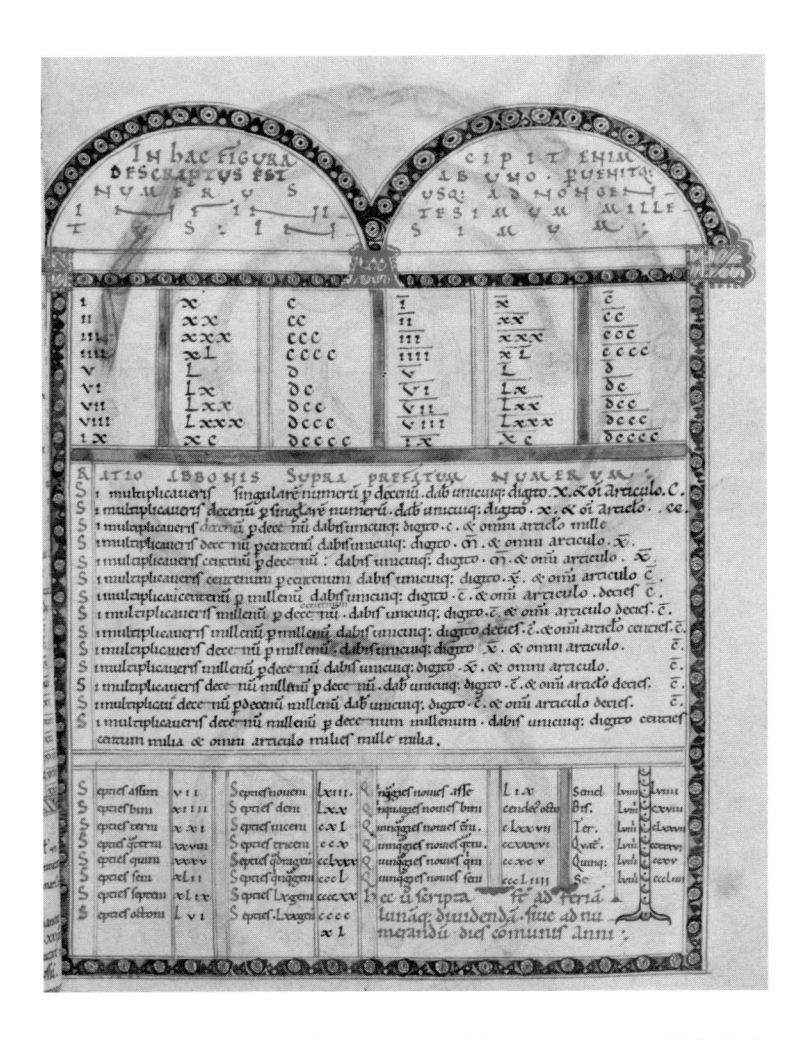

Fig. 1. Oxford, St John's College, ms. 17, fol. 35r. Thorney, vers 1110–1111. L'abaque d'Abbon. By permission of the President and Scholars of Saint John Baptist College in the University of Oxford.

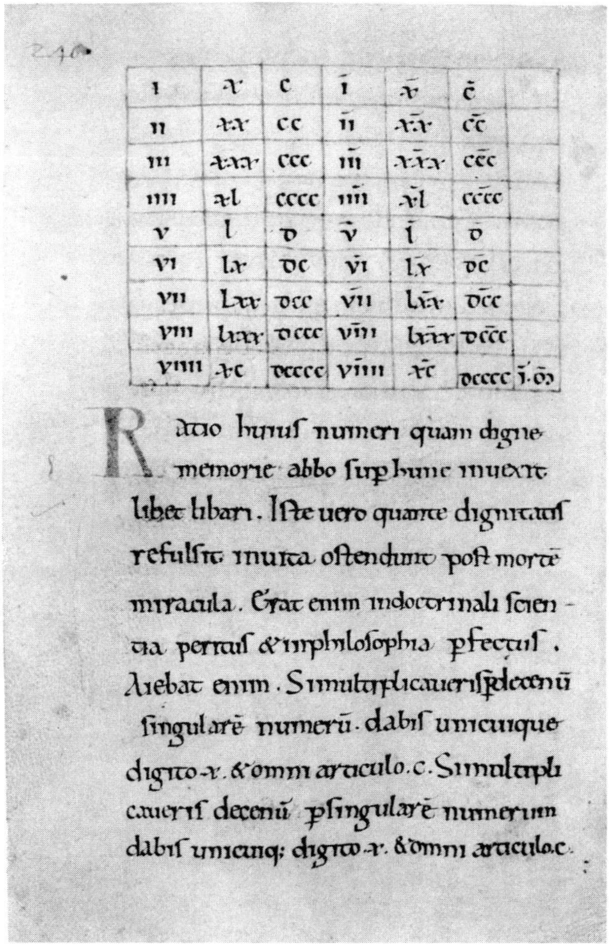

i	x	c	i	x	c	
ii	xx	cc	ii	xx	cc	
iii	xxx	ccc	iii	xxx	ccc	
iiii	xl	cccc	iiii	xl	cccc	
v	l	d	v	l	d	
vi	lx	dc	vi	lx	dc	
vii	lxx	dcc	vii	lxx	dcc	
viii	lxxx	dccc	viii	lxxx	dccc	
viiii	xc	dcccc	viiii	xc	dcccc	i. m

Ratio huius numeri quam digne memorie abbo sup hunc inuenit liber libari. Iste uero quante dignitatis refulsit multa ostendunt post morte miracula. Erat enim indoctrinali scientia peritus & in philosophia pfectus. Aiebat enim. Si multiplicaueris p decenu singulare numeru. dabis unicuique digito .x. & omni articulo .c. Si multiplicaueris decenu p singulare numerum dabis unicuiq; digito .x. & omni articulo .c.

Fig. 2. Oxford, Bodleian Library, ms. Ashmole 328, p. 240, xi[e] s. L'abaque d'Abbon, Byhrtferth, *Enchiridion*. By permission of the Bodleian Library. Cf. p. 130.

III

Algorismi vel helcep decentior est diligentia: the Arithmetic of Adelard of Bath and his Circle[1]

A mid-twelfth-century manuscript of arithmetical works whose earliest known provenance is among the Grey Friars of Coventry, begins with an introduction to arithmetic in which we read:

> Concerning the practical side of arithmetic there are what is called a 'book on how arithmetic is practised' (*operatio*), and texts on the aba-

---

1 I am very grateful to Marvin Colker, Luc Deitz, Menso Folkerts, the late Margaret Gibson, Jan Hogendijk, Wilbur Knorr, Johannes Thomann, Rodney Thomson, the Niel Ker Fund and all the participants of the Wolfenbüttel Colloquium for helpful criticisms and suggestions. The following abbreviations will be used throughout this article:
Adelard of Bath = Adelard of Bath: An English Scientist and Arabist of the Early Twelfth Century, ed. C. Burnett, Warburg Institute Surveys and Texts, XIV (1987)
B = MS British Library, Royal 15 A.XXVII.
C = MS Cashel (Tipperary), G. P. A. Bolton Library, Medieval MS 1
O = MS Oxford, Bodleian Library, Auct. F.1.9
P = MS Paris, Bibliothèque nationale, lat. 6626
R = MS Cambridge, Trinity College, R.15.16 ('Coventry manuscript')
T = MS Oxford, Trinity College, 47
When this article was about to go to press M. André Allard very kindly sent me the page-proofs of his book *Mohammad ibn Mūsā al-Khwārizmī, Le Calcul indien: édition critique, traduction et commentaire des versions latines remaniées du xiie siècle* (Paris and Namur, 1992). Allard gives some very convincing arguments concerning the inter-relationship of the three versions of the *Liber ysagogarum Alchorismi* (which he has identified) and the *Liber Alchorismi de practica arismetice*, as well as concerning their possible authorship. There has not been time to follow up in detail the implications of M. Allard's arguments, but they are essentially compatible with, and complementary to, those that I have arrived at independently in this article. The findings of Menso Folkerts and H. L. L. Busard in their recent edition of Version II of Euclid's *Elements – Robert of Chester's (?) Redaction of Euclid's* Elements, *The So-Called Adelard II Version* (Basel etc., 1992) – should also be taken into account in examining the arguments put forward in this article.

Originally published in *Mathematische Probleme im Mittelalter: Der lateinische und arabische Sprachbereich*, ed. M. Folkerts (*Wolfenbütteler Mittelalter-Studien* 10). Wiesbaden: Harrassowitz, 1996, pp. 221–331 cus and rhythmomachy. But the study of the algorism or 'helcep' (i.e., 'numeration') is more fitting (*decentior*) than these.[2]

In the next paragraph, which is about the texts on geometry, the unknown author refers to 'modern scholars' who have become famous in the subject: 'Alardus, Iohannes, Willelmus.'[3]

It is difficult to avoid identifying the first of these three with Adelard of Bath, whose name often appears in the French or Anglo-Norman form 'Aelardus' or 'Alardus';[4] and this identification becomes almost certain when we come across, in the section of the Coventry manuscript devoted to practical arithmetic, paragraphs which correspond to a text addressed by a certain 'Ocreatus' to 'Aelardus Batensis magister suus'. This text describes arithmetical procedures, applying to them the very term 'helcep' which is identified with the algorism in the Coventry manuscript.[5]

This article investigates the kind of arithmetic practised by Adelard of Bath, his colleagues and his immediate successors. This will lead us to re-examine the introduction of the algorism into Europe, and, incidentally, to make some comments on the terminology for, and use of, the zero, and the authorship of the Latin versions of Euclid's *Elements* known as Version I and Version II. The key texts are Adelard's passage on arithmetic in his *De eodem*

2 MS *R*, fol. 3r: Est autem et circa huius artis practicam qui dicitur liber operationis, et tractatus super abacum et Rimachiam. Algorismi vero vel helcep – id est numerandi – ipsis decentior est diligentia. This is a corrected passage from the edition of the 'Coventry introduction to arithmetic' in my 'Innovations in the Classification of the Sciences in the Twelfth Century', in *Knowledge and the Sciences in Medieval Philosophy. The Proceedings of the Eighth International Congress of Medieval Philosophy (S.I.E.P.M.)*, ed. S. Knuuttila et al. (Helsinki, 1990), II, pp. 25–42. See Plate I below.

3 MS *R*, fol. 3r: Sicut et quidam modernorum in eadem claruerunt, ut Alardus, Iohannes, Willelmus.

4 The form 'Alardus' is used in the MSS of Adelard's *De opere astrolapsus* and the abridged form of Adelard's treatise on falconry (I owe this information to Baudouin Van den Abeele), and was known to Roger Bacon who used an 'editio specialis [of Euclid's *Elements*] Alardi Bathoniensis': see M. Clagett, 'The Medieval Latin Translations from the Arabic of the *Elements* of Euclid, with Special Emphasis on the Versions of Adelard of Bath', *Isis*, 44 (1953), pp. 16–42 (23).

5 Ocreatus's text was first edited by C. Henry, 'Prologus N. Ocreati in Helceph ad Adelardum Batensem magistrum suum', *Abhandlungen zur Geschichte der Mathematik*, 3 (1880), pp. 129–39. A new edition of this text and of the parallel passages in the Coventry manuscript can be found in Appendices I and II below.

et diverso,[6] his *Regulae abaci*,[7] the versions of Euclid's *Elements* associated with the name of Adelard of Bath,[8] glosses to Boethius's *Music* which mention Adelard,[9] glosses to Boethius's *Arithmetic* in the same manuscript as those to Boethius's *Music*,[10] the *Helcep Sarracenicum* of H. Ocreatus, and the contents of the Coventry manuscript.

This investigation breaks new ground since little scholarly work has been done on the *Regulae abaci* and the *Helcep Sarracenicum*, partly because of the poor quality and inaccessibility of their editions. The Coventry manuscript has been neglected by historians of mathematics and the Cashel manuscript of the *Helcep* has been identified only recently. This field is interesting, not only because these works shed light on the mathematical concerns of an inter-related group of scholars, working, I imagine, in the West Midlands of England in the mid-twelfth century, but also because they provide some names and approximate dates for the introduction of the algorism into Europe. We now know much more than before concern-

6 Adelard of Bath, *De eodem et diverso*, ed. H. Willner, Beiträge zur Geschichte der Philosophie des Mittelalters, 4,1 (Münster, 1903).
7 Adelard of Bath, *Regulae abaci*, ed. B. Boncompagni, in *Bullettino di bibliografia e di storia delle scienze matematiche e fisiche*, 14 (1881), pp. 1–134.
8 The three versions which are attributed in manuscripts to Adelard were first named as Version I, II and III, and described by Marshall Clagett, 'The Medieval Translations' (see n. 4 above). Clagett's conclusions have been modified and refined, and many further details have been added in M. Folkerts, *Euclid in Medieval Europe*, Questio de rerum natura, 2 (Winnipeg, 1989) and substantially repeated in the introduction to Busard and Folkerts' edition of Version II (see n. 1 above). Version I has been edited by H. L. L. Busard, *The First Latin Translation of Euclid's 'Elements', Commonly Ascribed to Adelard of Bath*, Pontifical Institute of Mediaeval Studies, Studies and Texts, 64 (Toronto, 1983). The distinct status of the arithmetical books of the *Elements* (i. e., books 7–9) was generally recognized in the twelfth century, as is shown by the facts that (1) they are included separately from the remaining books of the *Elements* in the arithmetical section of the *Heptateuchon* of Thierry of Chartres (Chartres, Bibliothèque municipale, 498 (now destroyed), fols. 122r–124v), the earliest MS of Version II; and (2) they are described as being 'demonstrative arithmetic' in MS *R* (see below p. 246).
9 These glosses, in MS *T*, are edited in *Adelard of Bath*, pp. 82–3.
10 These glosses have not been edited. The text of Boethius's *Arithmetic* and *Music* will usually be cited as it is found in MSS *T* and *R*, but reference will also be made to the edition of the two texts by G. Friedlein, *Boetii De institutione arithmetica. De institutione musica* (Leipzig, 1867).

224

ing the earliest texts which introduced into Europe al-Khwārizmī's
On Indian Calculation – the text which gave the algorism its name
– thanks especially to the work of Kurt Vogel[11] and André Allard,[12]
but we know almost nothing about the authorship and the date of
composition of these texts. The principal Latin texts are the *Dixit
Algorizmi* – apparently a literal translation of an Arabic text (with
Latin additions), which only represents in a partial way al-
Khwārizmī's *On Indian Calculation*;[13] the *Liber ysagogarum Alchorismi in totum quadrivium*, of which the first three books give a
version of the algorism independent of *Dixit Algorizmi*;[14] and two
texts deriving from the same Latin source: the *Liber pulveris*[15] and
the *Liber Alchorismi de practica arismetice*.[16] The last-named is the
most complete and advanced of all these texts. The only date associated with any of these algorisms is the year 1143, which is mentioned in a Viennese manuscript which includes a fragment of the
Liber ysagogarum Alchorismi; but even this date is not necessarily
the year in which the manuscript was written.[17] The only names
associated with these texts are a 'magister A' in one manuscript of
the *Liber ysagogarum* (Paris, Bibliothèque nationale, lat. 16208),

11 K. Vogel, *Mohammed ibn Musa Alchwarizmi's Algorismus: Das früheste
Lehrbuch zum Rechnen mit indischen Ziffern* (Aalen, 1963).
12 A. Allard, *Le Calcul indien* (n. 1 above). See also idem, 'La Formation
du vocabulaire latin de l'arithmétique médiévale', in *Méthodes et instruments du travail intellectuel au moyen âge*, ed. O. Weijers (Turnhout,
1990), pp. 137–81, for a full list of the arithmetical terms used in the
early algorisms.
13 Ed. Vogel, *Mohammed ibn Musa* (n. 11 above), and Allard, *Le Calcul
indien* (n. 1 above), pp. 1–22, both from Cambridge, University Library,
Ii.6.5; a second manuscript containing a fuller text has been discovered
by Menso Folkerts (New York, Hispanic Society of America MS 297/
726): see M. Folkerts, 'Eine neue Handschrift der Arithmetik von al-
Ḥwārizni' in *Cosmographia et Geographica: Festschrift M. Nobis*
(Munich, 1994), pp. 181–93.
14 Edited in Allard, *Le Calcul indien* (n. 1 above), pp. 23–61, and in B. G.
Dickey, 'Adelard of Bath: An Examination Based on Heretofore Un-
examined Manuscripts', unpublished Ph.D dissertation, University of
Toronto, 1982, pp. 77–111.
15 Ed. Allard, *Le Calcul indien* (n. 1 above), pp. 62–224 (left hand
columns).
16 Edited ibid., pp. 62–224 (right hand columns).
17 H. von Fichtenau, 'Wolfger von Prüfening', *Mitteilungen des österreichischen Instituts für Geschichtsforschung*, 51 (1937), pp. 313–57 (320),
considers A. D. 1143 as a *terminus post quem* for MS Wien, Österreichische Nationalbibliothek, cod. 275 which includes on fol. 27r a summary of some operations described in *Liber ysagogarum Alchorismi*.

and a 'magister Iohannes' in several manuscripts of the *Liber Alchorismi* (one manuscript gives *Iohannes Hispalensis*); the identities of both these authors will be discussed later on. In *Helcep Sarracenicum*, however, we have both an approximate date and the name of an author: the text was written within the lifetime of Adelard and by an author called H. Ocreatus.[18]

I

Adelard's name is attested in four documents written at Bath between 1100 and 1122, in a Pipe Roll (listing him as belonging to the County of Wiltshire) in 1130, and in a charter of King Stephen drawn up between 1135 and 1139. His *De opere astrolapsus* is dedicated to 'Henricus regis nepos' (the future Henry II) and was probably written between 1149 and 1150. If we can confidently attribute to him a set of political horoscopes described by John North, he was still alive in 1151.[19] He dedicates his *Natural Questions* to Richard, bishop of Bayeux, and the literary introduction to the seven arts which he calls *De eodem et diverso* to William, bishop of Syracuse. From these works we gather that he has studied in Tours, taught at Laon, and travelled to Salerno, 'Graecia maior', Sicily and the Norman principality of Antioch.[20] His translations from Arabic, however, connect him more closely with scholars working north of Bath, and further up the valley of the river Severn. These scholars are Petrus Alfonsi from Huesca in Aragon, and Walcher from Lorraine, who died in 1135 as prior of the Abbey of Great Malvern,[21]

18 The designation of the author as *N.* Ocreatus is false (as can be readily see from the reproduction of the superscript in Plate II), and is apparently due to a mistake on the part of its first editor, Charles Henry, since Moritz Cantor (*Vorlesungen über Geschichte der Mathematik*, third edition (Leipzig, 1907), I, p. 906), had correctly read 'H. Ocreatus'. Amable Jourdain, *Recherches sur les anciennes traductions latines d'Aristote*, second edition (Paris, 1843), p. 99, had read 'O'Creati', which he gallicized to O'Créat, and this pseudo-Irish form of the name found its way into secondary sources, such as D. E. Smith and L. C. Karpinski, *The Hindu-Arabic Numerals* (Boston and London, 1911), pp. 119–20, L. C. Karpinski, 'Two Twelfth Century Algorisms', *Isis*, 3 (1921), p. 409, and G. Sarton, *A History of Science: Hellenistic Science and Culture in the Last Three Centuries B. C.* (Cambridge, Mass. 1959), p. 50.

19 See *Adelard of Bath*, pp. 47–61.

20 See Burnett, 'Introduction', and M. Gibson, 'Adelard of Bath', in *Adelard of Bath*, pp. 3–4, 7–16.

21 See C. H. Haskins, *Studies in the History of Mediaeval Science* (second edition, Cambridge, Mass. 1927), pp. 113–120.

and their work is best represented in a manuscript written between 1120 and 1140 at the Priory of Worcester Cathedral, in whose diocese Great Malvern is situated. This manuscript, now Oxford, Bodleian Library, Auct. F.1.9 (our MS *O*), not only includes two works by Walcher, one of which is the 'opinion of Petrus the Jew, called Alphonso, on the lunar nodes, which Walcher, prior of the church of Great Malvern translated into the Latin language',[22] but also Adelard's own version of the astronomical tables of al-Khwārizmī; this version can be understood as a revision with further reference to an Arabic text, of a translation made originally by the same Petrus Alfonsi.[23] Earlier mathematical texts in this Worcester manuscript originate from the Continent, and include very fine copies of works on the computus and the abacus by Bernelinus, an anonymous commentary on Gerbert d'Aurillac's *Regulae abaci*, and a work on rhythmomachy by Odo of Tournai. The significance of these texts on the abacus and rhythmomachy will be brought out later.

But first to Adelard's *De eodem et diverso*. This work survives in only one manuscript, where it accompanies the same author's *Natural Questions*. The section on geometry in the *De eodem et diverso* is devoted to the practical geometry of the Roman land-measurers and shows no knowledge of Euclid's *Elements*. On the other hand, the section on arithmetic does not mention the practical side of the subject at all, and there is no reference either to the abacus or to the algorism here. The maiden personifying arithmetic discourages the practical aspect, telling the reader that arithmetic's 'rules demand the practitioner's entire attention and that he should forget all external things'.[24] Adelard summarizes definitions from Boethius's *Arithmetic*, I.3–19 and 28 and arranges the material in a logical way so that the 'beautiful diagram of Nicomachus' for the

22 Sententia Petri Ebrei, cognomento Anphus, de dracone, quam dominus Walcerus prior Malvernensis ecclesie in latinam transtulit linguam. For the dating of this MS and the grouping of its scribes see E. Macintyre, 'Early Twelfth-Century Worcester Cathedral Priory, with Special Reference to the MSS Written There', unpublished D. Phil dissertation, University of Oxford, 1978, pp. 28, 34, 45, and 214.

23 See Mercier, 'Astronomical Tables in the Twelfth Century', in *Adelard of Bath*, pp. 87–119 (97–100). For a slightly different conclusion see G. J. Toomer's review of O. Neugebauer, *The Astronomical Tables of al-Khwārizmī*, Det Kongelige Danske Videnskabernes Selskab, histor.-filosof. Skrifter, 4, 2 (Copenhagen, 1962), in *Centaurus*, 10 (1964), pp. 203–12 (211–2).

24 *De eodem et diverso*, ed. Willner, p. 25.9–11: Hoc autem unum scio: eius precepta artificem totum sibi vacantem exterioraque omnia obliviscentem exposcere.

even numbers matches the sieve of Eratosthenes for the odd numbers.[25] Only in a section on proving the compatability of the philosophies of Plato and Aristotle does Adelard mention the abacus – as an instrument on which the results of a multiplication are tested by division.[26] Although Adelard praises his episcopal dedicatee for his prowess in mathematics,[27] the *De eodem et diverso* is evidently meant to entertain and as an encouragement to the study of the arts, rather than as a textbook for teaching the subjects of the trivium and quadrivium.

It is as a sweetener to make the way to the quadrivium easier that Adelard introduces his *Regulae abaci*. The text is addressed to a certain 'H.' There is no way of telling whether this is the 'H. Ocreatus' who composed the *Helcep Sarracenicum*, 'Henricus regis nepos', the dedicatee of Adelard's *De opere astrolapsus*, or another 'H'. Once again it shows no knowledge of Euclid's *Elements*, and cites only Boethius's *Arithmetic* and *Music*. Verbal similarities, however, suggest that Adelard also used the pseudo-Boethian *Geometry II* which includes a section on the abacus. The *Geometry II* is the earliest work we know in which the abacus numerals have an 'Arabic' form.[28] Adelard claims to be following the method of Gerbert d'Aurillac and 'a certain pupil of his whom they call Guichardus.'[29] The abacus-texts in MS *O* belong to this tradition and might

25 *Ibid.*, ed. Willner, p. 24: in ea figura...quam Nicomachus eiusque sequaces admirabilem vocaverunt. The reference is to Boethius, *Arithmetic*, I.28, which elicits a similar response from the glossator of MS *R*, fol. 13r: in hac figura quam admirabilem dicunt ... The sieve of Eratosthenes is described in Boethius, *Arithmetic*, I.17.

26 *Ibid.*, p. 11: Amat enim et compositio divisionem et divisio compositionem. Unde si quid in digitis et articulis abaci numeralibus ex multiplicatione crescerit, id utrum recte processerit, divisione eiusdem summae probatur.

27 *De eodem et diverso*, p. 3.20–21: Tibi, igitur, Willelmo, Syracusie praesul omnium mathematicarum artium eruditissime, hanc orationem direxi.

28 M. Folkerts, 'Adelard's Versions of Euclid's *Elements*', in *Adelard of Bath*, pp. 55–68 (60–61), and idem, *„Boethius" Geometrie II, ein mathematisches Lehrbuch des Mittelalters* (Wiesbaden, 1970), pp. 139–44, and Plates 1–21. Werner Bergmann shows that there is no positive evidence for believing that Gerbert and his earliest followers used 'Arabic' (or *ghubār*) numerals; see his *Innovationen im Quadrivium des 10. und 11. Jahrhunderts* (Stuttgart, 1985), p. 212.

29 *Regulae abaci*, ed. Boncompagni, p. 100.21–25: Tamen quia super his tractavit Girbertus philosophus vir subtilis ingenii diligenter et compendiose, quidam eciam quem discipulum eius predicant quem Guichardum (v. l. Guicarilum, Huicarilum) nominant diligenter et prolixe.

include the work of this mysterious 'Guichard'; they should be compared with the *Regulae abaci*. The Gerbertian abacus whose use is described in these texts, requires a separate counter for each of the nine digits. These counters are marked with symbols – *caracteres* – some of which are recognizable as derivative from Arabic numerals, and are called by names which appear to be debased forms of the names of their Arabic counterparts:[30]

1	2	3	4	5	6	7	8	9
igin	andras	hormis	arbas	quimas	caletis	zenis	temenias	celentis

In performing calculations these counters are placed in the relevant columns – or *arcus* – of the abacus board and columns are left blank when necessary. For example, to represent 4,000 one places the counter called 'arbas' in the thousands column, while the hundreds, tens and units columns are left empty. Thus, in the original form of the Gerbertian abacus there was no counter for zero. In the diagrams in Pseudo-Boethius's *Geometry II* and in at least two English manuscripts – Oxford, St. John's College 17 (written at Thorney Abbey between 1102 and 1110), and Hereford Cathedral, MS O.i.vi (written in Cirencester between 1131 and 1147)[31] – a tenth counter is shown, marked with a circle and called 'sipos'. The origin of this term is debated. Adelard makes specific mention of this counter in his *Regulae abaci* (in fact, he only mentions one other counter by name).[32] Strangely enough, as Gillian Evans has pointed out, Adelard has run together the names of zero and nine, and calls the counter 'sipos-celentis'.[33] This might indicate that the use of the counter for zero was recent, or at least still unfamiliar to Adelard or his scribes, at the time of the writing of *Regulae abaci*. To give an example from the text: Adelard says that, to show the number 90,707 one should

30 These numerals tend to be written in a more elaborate way than the Arabic numerals of the algorism, and the latter (with the exception of zero) are rarely called by their Arabic names in the texts describing the algorism. An exception is the *Ars algorismi* of Frankenthal (s.xiie–s.xiii^in), edited by A. Allard, 'A propos d'un algorisme latin de Frankenthal: une méthode de recherche', *Janus*, 65 (1978), pp. 118–41 (129), where the names are given in the first chapter, but not used thereafter.

31 See F. E. Wallis, 'MS Oxford, St. John's College, 17: A Medieval Manuscript in its Context', unpublished Ph.D dissertation, Pontifical Institute of Medieval Studies, Toronto, 1985 and G. R. Evans, 'Schools and Scholars: the Study of the Abacus in English Schools, c. 980–c. 1150', *English Historical Review*, 94 (1979), pp. 71–89 (85–6).

32 *Regulae abaci*, pp. 99–100: zenis vero caracterem...

33 Evans, 'Schools and Scholars' (n. 31 above), p. 83.

'add the character *siposcelentis* to the empty (columns of the) thousands and the tens'.[34] One can see how close in appearance a calculation using the Gerbertian abacus counters plus the zero would be to a calculation using Arabic numerals written on a sand-board or on parchment. The 'golden' method of division (or 'division' without *differentiae*) is chronologically later than the 'iron', and is particularly close to the method of division used in the algorism. However, in none of his works does Adelard use the Arabic numerals of the algorism.

A new stage in the history of Western mathematics is heralded by the translations from Arabic of the full text of Euclid's *Elements.* These provided a much greater challenge and inspiration to mathematicians than the bare listing of definitions, postulates, axioms, and propositions from hardly more than the first three books, in the old 'Boethian' version. Recent scholarship has intimated that Adelard's authorship of the three versions of Euclid's *Elements* described by Clagett in 1953 is far from certain. Nevertheless Version I and Version II, at least, must be closely associated with Adelard's circle. Version I seems to be a direct translation from the Arabic made by Adelard himself (probably with the help of an arabophone); for its style and vocabulary match that of other translations by Adelard, as will be shown elsewhere. Version II consists of two elements: (a) the definitions, postulates, axioms and propositions, which are compiled from earlier and possibily fresh, translations, and (b) directions for proof which appear to be subsequent to the first element and may have been composed by several scholars over a period of time; there are considerable variations in both these elements among the manuscripts.[35] The earliest manuscript of Version I – Oxford, Trinity College 47 (our MS *T*) – also contains Version II; it could have been written within the lifetime of Adelard, and it firmly attributes both versions to the English scholar, using characteristically Anglo-Saxon forms of the name.[36] Some texts of

34 *Regulae abaci*, ed. Boncompagni, p.99.6–7: Adde...deceno et milleno vacantibus siposcelentis caracterem. For explanations of this and the following procedures see F.A. Yeldham, *The Story of Reckoning in the Middle Ages* (London, 1926), pp.36–45; G.Flegg, *Numbers: Their History and Meaning* (London, 1983), pp.153–7, and Bergmann, *Innovationen* (n.27 above), pp.187–90.

35 See Folkerts, *Euclid in Medieval Europe* (n.8 above).

36 Version I, MS *T*, fol.171r: ad ⊦ elardū baToniensē; Version II, *ibid.*, fol.104v: adelardū batHoniensē. The character ⊦ for 'h' deriving from the Greek rough breathing, is also used in the 'Coventry' introduction to arithmetic (see Burnett, 'Innovations' (n.2 above), p.34) and in the Paris MS of the *Helcep Sarracenicum.* The date of the writing of MS *T*

Version II include references which imply that Adelard is still alive and available for consultation.[37] Moreover, the same MS *T* includes two glosses to Boethius's *Music* which record the opinions of Adelard. The first of these happens to be the most substantial gloss to the text, and is in a hand which has glossed only this passage of Boethius. The text is Boethius's discussion of Archytas's proof that no number can be a mean between two numbers which are in superparticular ratio (i. e., $(n+1):n$), which represents a rare example of early Greek mathematical theory.[38] The gloss refers to the opinion

is still under review. Most scholars have proposed a mid-twelfth-century date: see G. Lacombe et al., *Aristoteles Latinus*, codices (Paris and Brussels, 1939), I, no. 379 ('saec. XII med.'), and L. Minio-Paluello, *Opuscula: The Latin Aristotle* (Amsterdam, 1971), p.358 ('circa 1130–40'), but recently Bernhard Bischoff in a private communication to Menso Folkerts has proposed a more recent date ('not very long before 1200'). However, the mid-twelfth-century date has been asserted again by Professor Rodney Thomson in a personal communication.

37 See below, p.231.

38 MS *T*, fol.87v, gloss to Boethius, *Music*, III.11: (*Boethius's text*) Superparticularis proportio scindi in equa medio proportionaliter interposito numero non potest...C E...sunt superparticulares, E numerus C numerum parte una sua eiusque transcendit. Sit hec D [(*gloss*) hoc enim est in omni superparticulari proporcione, quod differentia et se et inferiorem et superiorem metitur]. Dico quoniam D non erit numerus sed unitas. Si enim est numerus D et pars eius est, qui est E, metitur D numerus D numerum (*supra*: se ipsum); quocirca et E numerum (*supra*: C est pars) metietur, quo fit ut C quoque metiatur. Utrumque igitur C et E numeros metietur D numerus, quod est inpossibile [(*gloss*) *Quod est impossibile*, scilicet quod sit numerus, cum sit differentia limitum secundum Manegaldum. Sed dico quod si sic legatur, ad nullum inconveniens ducit adversarium, quod tamen intendit cum dicit *Si enim est numerus* et etiam nichil per hoc idem probaret. Dicatur igitur sic: *quod est impossibile*, scilicet ut differentia metiatur et se et inferiorem et superiorem, et accipe 'metiri' per tale quid quod sit numerus, scilicet binarium, ternarium et cetera, non communem mensuram omnium numerorum, que est unitas. Hec enim mensura non reputatur pro mensura quandoque secundum Adelardum]. For discussions of this problem see T. L. Heath, *The Thirteen Books of Euclid's Elements*, second edition (Dover reprint, London, 1956), II, p.295, W. R. Knorr, *The Evolution of the Euclidean Elements* (Dordrecht, 1975), pp.212–25, and Boethius, *Fundamentals of Music*, edited and translated by C. M. Bower and C. V. Palisca (New Haven, 1989), pp.103–5. For 'one' not being a number see also Boethius, *Arithmetic*, II.28, ed. Friedlein, p.117.5–7 ('Binarius autem, numerus primus') and the even clearer statement in Pseudo-Boethius, *Geometry II*, ed. Folkerts, lines 467–8: unitas enim, ut in arithmeticis est dictum, numerus non est, sed fons et origo numerorum.

of a 'Manegaldus', but ends with the sentence: 'Thus (the unit as the common measure of all numbers) cannot be considered to be a measure in any context, according to Adelard'. The words 'secundum Adelardum' at the end of the gloss may refer to the whole gloss rather than to the last sentence alone. The second gloss to mention Adelard refers to his way of annotating music and implies that a previous gloss was written by Adelard himself. Both the latter glosses are in the hand of the principal scribe (hand 'A').[39] This hand is responsible for pointing out the equivalent Arabic terms for Boethius's *pariter par* and *impariter par* in a gloss to Boethius's *Arithmetic* in the same manuscript. These terms – *zauj al-zauj* and *zauj al-fard* – are used in the Arabic versions of Book VII of the *Elements*, but I have found no Latin translation which keeps these words in their original Arabic form. The explanation of these terms in the gloss corresponds to that of Version I.[40]

The most significant references to Adelard within the directions for proof in Version II itself are the following two comments: 'What has been proposed will be clear to the Bathonian's intelligence',[41] and 'What has been proposed will be clear to Adelard's perspicacity'.[42] Other insertions seem to cast some light on the cheerful banter of the classroom. We read the enigmatic phrases: 'This proposition of ours is proved to be true without the help of John'; and 'From the wallet (?) of Reginald. Whoever does not know that he should

---

MS *T*, fol. 87v is illustrated in A. White (Peden), 'Boethius in the Medieval Quadrivium', in *Boethius: His Life, Thought and Influence*, ed. M. Gibson, Oxford, 1981, pp. 162–205; White is the first person to have drawn attention to this gloss (pp. 182 and 202).

39 See *Adelard of Bath*, pp. 81–83. The significant words are 'Adelardus id inprobat per superiorem notam'. The note which immediately precedes is on the same subject and has no attribution, so implying that it is a statement by Adelard himself.

40 MS *T*, fols. 50r–51r, to Boethius, *Arithmetic*, I.9 and I.10: zaug el zaug; unde Euclides: pariter par est cuius parium eum numerantium vices pares // zaug el fart; unde Euclides: parium eum numerantium vices impares. Compare Euclid, *Elements*, Version I, book VII, defs. v and vi, ed. Busard, p. 196: ⟨v⟩ Numerus pariter par est cuius omnium parium eum numerantium vices pares. ⟨vi⟩ Numerus pariter impar est cuius parium eum numerantium vices impares. For the Arabic texts see the article of Sonja Brentjes in this volume.

41 *Elements* II.7, MS *T*, fol. 108r: quod propositum est Bathoniensis patebit ingenio.

42 *Elements* II.8, MS *T*, fol. 108v: Adelardi pro⟨p⟩ositum patebit acumini. For other internal references to Adelard see Folkerts in *Adelard of Bath*, p. 63.

reply to you in this way should give you a white cow!'[43] Other scholars mentioned are 'Radulfus' and 'Eggebericus'.[44] Most puzzling are the several references to the occurrence or non-occurrence of some passages of directions for proof in an 'ocrea' of John, or Reginald. The phrases are 'ex ocrea Johannis'; 'ex ocrea Regineri'; 'in ocrea Johannis nondum reperitur'.[45] The most common meaning of 'ocreae' in England at this time was 'out-door footware'. It is unlikely that John or Reginald kept their notes in their boots, though it is possible that 'ocreae' could refer to some kind of boot-shaped leather pouch. In any case, since both John and Reginald evidently possessed 'ocreae' (whatever these might have been), they would have been called 'ocreati', and, as we have seen, an 'ocreatus' appears as the author of the *Helcep Sarracenicum*.[46]

43 Gloss after *Elements*, X.31: vera ergo probatur esse proposicio nostra absque Johannis industria. Gloss at the beginning of *Elements*, X: Vale Reginere. Quicumque nesciret tibi sic respondere utinam daret tibi vaccam albam. Compare 'ex walle Regineri', at the end of the directions for proof for *Elements*, IX.15 (British Library, Additional 34018, fol. 36r; Oxford, Bodleian Library, Auct. F.5.28, fol. xxiii^r has the spelling 'valle'), for which other manuscripts give 'ex ocrea Regineri' (*Adelard of Bath*, p. 64). The etymology of 'wallet' is disputed, but if it can be derived from a Teutonic root *wall- (Oxford English Dictionary, s. v. 'wallet') a Latinized form 'wallis' is plausible. The 'vacca alba' may be a piece of virgin parchment on which the correct answer can be written. In the same position in MS Oxford, Bodleian, Auct. F.5.28 (English, before 1260), fol. xxv^r, a variation of the same phrase, with the name of a different scholar, is found: Vale, Radulfe. Quicumque nescierit tibi sic respondere et magistrum se facit, utinam daret tibi pellicium agninum (I owe this reference to Wilbur Knorr).
44 *Adelard of Bath*, p. 64.
45 'Ex ocrea Johannis' is found appended to directions for proof to *Elements* V, def. 16 and X.15; 'in ocrea Johannis nondum reperitur' appears after *ibid.*, X.25 and X.31 (see *Adelard of Bath*, p. 64). For 'ex ocrea Regineri' see n. 43 above. A comment in the text of Version II in MS Cambridge, Gonville and Caius College 504, fol. 54v reads: Et ex ocrea Iohannis contra, sic: Si fuerit totum commensurale uni parti... (to *Elements*, X.8).
46 One may note that 'ocreae' was also used as a synonym for 'hosae' ('trousers'): King Henry I's uncle Robert Courthose, duke of Normandy, was known as 'Robelinus curta ocrea'; William of Malmesbury, *Gesta regum*, ed. T. D. Hardy (London, 1840), II, p. 607. Both 'ocrea' and 'hosa' were used to describe containers for wine (presumably leather bottles: see DuCange, *Glossarium*, s. v. ocrea (1), and H. Brunner, *Deutschen Rechtsgeschichte* (Berlin, 1958), II, p. 530, n. 11: ocrea id est hosa vini). By the fourteenth century 'ocrearius' and 'hosarius' designated a

What is interesting is that there are two other contexts in which 'Ocreatus' occurs as a proper name. In a medieval catalogue of the manuscripts of Lanthony Secunda (Augustinian Canons; near Gloucester) there is mentioned 'Glose magistri Iohannis Ocreati de geometria'.[47] At the end of the seventeenth century Edward Bernard saw a manuscript in the Royal Library (now part of the British Library) headed *Euclidis Elementa ex Arab. in Lat. vers. per Joan. Ocreatum.*[48] This manuscript has been identified as British Library, Royal 15 A.XXVII (our MS *B*), which consists of nothing but the propositions of the *Elements* in Version II with directions for proof as far as *Elements*, I.32. Unfortunately, the first folio is now lost and so the title as read by Bernard is no longer there. However, the propositions in this manuscript are close in their wording to those in MS *T*, which has the directions for proof mentioning Adelard, and there is a similarity of hands between the two manuscripts.[49] Moreover the lay-out of both MSS is the same: the propositions of the *Elements* have been written in the centre of the page, and the

'cellarman' (*Dictionary of Medieval Latin from British Sources*, s.v. 'hosarius'). 'Ocrea' was also used for other (leather) containers; e.g., the pouch into which a certain quantity of gold was sewn: see *Die Regensburger Schottenlegende – Libellus de fundacione ecclesie Consecrati Petri*, ed. P.A. Breatnach (Munich, 1977), p.239: aperierunt archam regiam et duas ocreas latas et longas fecit impleri de electo auro; quas...ocreas fecit impleri...et consui...et exterius signari cum suo sigillo. In this case the apparent interchangeability of 'ocrea' and 'wall(is)' in Version II (see n.43 above) is understandable: they would be alternative words for a leather wallet. Perhaps, then, 'ocreatus' was a man carrying such a wallet or satchel; i.e., a scholar? I suggest below (n.56) an additional, jocular, interpretation of 'ocreatus'. For the possibility that 'Ocreatus' is a variant (encouraged by these puns) of the family name 'Hosatus', which is that of the most prominent family in Bath in the time of Adelard, see my 'Ocreatus' in *Vestigia mathematica: Studies in Medieval and Early Modern Mathematics in Honour of H.L.L.Busard*, ed. M.Folkerts and J.P.Hogendijk, (Amsterdam, 1993), pp.69–77. I am grateful to Thomas Peiss of the *Mittellateinisches Wörterbuch*, David Howlett and Richard Sharpe of the *Dictionary of Medieval Latin from British Sources* and Kieran Devine of *Dictionary of Medieval Latin from Celtic Sources* for providing me with examples of the use of 'ocrea'.

47 I am grateful to Richard Sharpe, as general editor of the Corpus of British Medieval Library Catalogues, for this information.

48 E.Bernard, *Catalogi librorum manuscriptorum Angliae et Hiberniae in unum Collecti* (Oxford, 1697), n.8639 (in the *Aedes Jacobaeae*).

49 See Plates III and IV for specimens of the hands in these two manuscripts.

directions for proof have been added (as far as they go) in the margins.

So, is this 'booted John' the same as the 'John' from whose 'ocrea' comments were taken in several manuscripts of Euclid's *Elements*? Is he the same as the John mentioned alongside Adelard and William as 'modern geometers' in the Coventry manuscript? Finally, is he the same as, or a colleague of, the H. Ocreatus whom Adelard asked to explain 'Saracen calculation' - i.e. the author of the *Helcep Sarracenicum*? Let us turn to this text.

II

The *Helcep Sarracenicum* up to now has been known only from one manuscript, Paris, Bibliothèque nationale, lat. 6626 (our MS *P*), where it was discovered by Amable Jourdain, and from which Charles Henry (on the suggestion of Moritz Cantor) made an edition. The existence of a second manuscript – Cashel (Tipperary), G. P. A. Bolton Library, Medieval MS 1 (*C*) – was first signalled by Marvin Colker in an unpublished description of the manuscript. Both manuscripts are full of errors indicating that the copiers were neither accurate scribes nor good mathematicians (They may be exonerated to a certain extent because of the complexity and novelty of the subject-matter). The scribes of both manuscripts omit phrases, or wrongly incorporate what were probably originally glosses. However, since the copies are clearly independent of each other, they both help to establish a text which is closer to an archetype.

The relevant portion of the Cashel manuscript appears to be of the late twelfth century and includes much mathematical material which needs to be explored; amongst this material are two further algorisms using Arabic numerals. The calendar on pp. 71-6 is derived from the calendar of the Benedictine abbey of Tewkesbury, which is not far from Gloucester and Lanthony Secunda.[50] The booklet containig the *Helcep* begins with one of the oldest copies of Quṣṭa b. Lūqā's *De differentia spiritus et animae* (the oldest copy itself being part of the collection of manuscripts left by the doctor 'Herbertus' to Durham Cathedral library in the third quarter of the twelfth century; now Edinburgh, Advocates, 18.6.11), and the beginning of an otherwise unattested commentary on Porphyry's *Isagoge*. In the Paris manuscript the *Helcep* has been copied along-

---

50 Nigel Morgan, who is editing the English Benedictine calendars, kindly provided me with this information. More detailed descriptions of this and the Paris MS are given in Appendix I below.

side Seneca's *De clementia* and *De beneficiis*.[51] The works of Seneca
tell us little since they were frequently copied in the twelfth century.
But the orthography shows English, and in particular Anglo-Norman, features.

Ocreatus starts with a short but effusive prologue. His reference
to his master's 'ingenium' (3 'cuius quidem compendium ingenio
vestro placiturum non ambigo') recalls the comment in Version II
that a proposition will be clear to 'Adelard's *ingenium*'.[52] Ocreatus
says that Adelard asked him to explain the Saracen way of calculation. This implies that he knew Arabic better than Adelard; he
may have been employed by Adelard to help him interpret Arabic
science. MS *B* once stated that 'Joan. Ocreatus *translated* Euclid's
Elements', and this would imply that he did more than write the
sketchy comments and directions for proof to part of the first book.[53]
However, as will become clear, the *Helcep Sarracenicum* is not a
direct translation from Arabic. It is rather Ocreatus's description of
some arithmetical procedures which employ the algorism. Only three
Arabic words are used. The first is *helcep* itself. I am now convinced
by Derek Latham's interpretation of this as al-ḥisāb – 'arithmetic,
reckoning, computation'.[54] This would make sense of the interpreta-

51 For the popularity of this text in the twelfth century and its presence
    in England see L. D. Reynolds and N. G. Wilson, *Texts and Transmis-
    sion* (Oxford, 1983), pp. 363–5. Reynolds points out that king Henry II
    (1155–89), who received the dedication of Adelard's *De opere astrolap-
    sus* before his coronation, is said to have always kept a copy of Seneca's
    De clementia at his side: Gerald of Wales, *Topographia Hibernica*, 48,
    in *Giraldi Cambrensis Opera*, ed. J. F. Dimock (London, 1867), V, p. 191.
52 See the addition to *Elements*, II.7 (n. 41 above). Since *Elements*, II.7 is
    the geometrical restatement of the binomial theorem which also con-
    cludes the *Helcep* (see below), the reference to the Master's *ingenium*
    in both cases may be significant.
53 This is the second of the three alternatives put forward by Clagett,
    'Medieval Translations of Euclid' (n. 4 above), in a very perceptive
    analysis of Ocreatus's possible role in the composition of Version II
    (pp. 21–2).
54 *Dictionary of Medieval Latin from British Sources*, s. v. 'helcep'. Ex-
    amples of the unvoicing of the final consonant occur in MS *T*: e. g.,
    fol. 109v, 'dhenep atoz' (hand 'A') for *dhanab al-ṭāwūs* (repeated in
    another hand on fol. 167v), and 'zaug elfart' in hand 'A' for *zauj al-fard*
    (n. 40 above). Either the first syllable could have succumbed to metathe-
    sis, encouraged by the ambiguous significance of 'h' in Medieval Latin,
    or 'hel-' may be written instead of 'el-' as has happened in the case of
    'helmuahym' (= *al-mu'ayyin*) used for 'rhombus' in Version II in Ox-
    ford, Bodleian Library, Auct. F. 5.28 (English, before 1260); see Burnett,
    'The Use of Geometrical Terms in Medieval Music: elmuahim and

tion given in the Coventry manuscript: 'helcep – id est numerandi', and that given here: 'helcep sarracenicum de multiplicandi scilicet numerorum et divisione' (2).

The second word (or rather, phrase) is *sinaphihi* (**66**), which, inexplicably, is used where one would expect the Latin 'in/per semetipsos' ('by themselves'), and is apparently a miscopying of the equivalent Arabic expression 'fī nafsihi' ('by itself') of 'fī nafsihimā' ('by themselves [dual]'), through confusion of the 'long-s' and 'f'. The latter Arabic phrase occurs (in the form 'feeffehem' < 'fē eſſeheme'?) as a marginal note in a thirteenth-century copy of Version I in Bruges, Stadsbibliotheek, 529, fol. 22r, opposite the text 'ex ductu *ha* et *ad* in seipsas' (within Euclid. *Elements*, III.35; lines are being multiplied by themselves).

The third Arabic word Ocreatus uses is *cifre* for zero. This is a transliteration of the Arabic *ṣifr*, which appears for the first time in Latin texts in Petrus Alfonsi's and Adelard of Bath's versions of the *Tables* of al-Khwārizmī,[55] and the *Liber ysagogarum Alchorismi*.[56]

elmuarifa and the Anonymous IV', *Sudhoffs Archiv*, 70 (1986), pp. 198–205 (203).

55 H. Suter, et al., *Die astronomischen Tafeln des Muhammed ibn Mūsā al-Khwārizmī in der Bearbeitung des Maslama ibn Aḥmed al-Madjrītī und der latein. Übersetzung des Athelhard von Bath*, Det Kongelige Danske Videnskabernes Selskab, Skrifter, 7 Række, histor.-filos. Afd., 3 (Copenhagen, 1914), c. 7: 'figuram cifrae t', and c. 8: 'cifre ... id est nullus'.

56 *Liber ysagogarum*, ed. Allard, *Le Calcul indien* (n. 1 above), p. 25: Utuntur etiam ciffre hoc modo 0 vel hoc t; *ibid.*, p. 33: si in aliqua nil remanserit, ciffre ponatur. In every other context, however, 'circulus' is used for zero; see Allard, 'La formation du vocabulaire' (n. 12 above). Note also the gloss in Munich, Clm. 18927 (another MS of the *Liber ysagogarum*), fol. 1r: 0 et t sunt cifre (R. Lemay, 'The Hispanic Origin of Our Present Numeral Forms', *Viator*, 8 (1977), pp. 435–62, Fig. I). If it is true, then, that the word *cifre* was first introduced into Latin in the circle which included Adelard and H. Ocreatus, it is tempting to suggest an additional significance for the term 'ocreatus' which would be in keeping with the light-hearted banter characteristic of this circle. From Classical times onwards 'ocrea' was often written instead of oc⟨h⟩ra ('burnt ochre', 'yellow'), and this spelling, for example, is found in the version of the *Mappae clavicula* with additions made from Arabic and Anglo-Saxon sources possibly by Adelard himself (see *Adelard of Bath*, pp. 29–32). From the same Arabic three-letter root ṣ-f-r are derived *ṣifr* (=zero) and *ṣufra* ('yellow'). To an untrained ear Arabic 'i' and 'u' sound virtually the same after 'ṣ'. The fact that these scholars adopted a form *cifra* or *cifre* (rather than, say, 'cefer'), with a final vowel which has no equivalent in the Arabic for zero, also suggests an

The signs used for *cifre* in the manuscripts of the *Helcep* are either a simple lower-case 't' (MS *C*) or a 't' with transverse strokes added to each end of the cross-stroke (MS *P*): similar signs are used in Adelard's version of the astronomical tables[57] and the *Liber ysagogarum Alchorismi*.[58] The *Liber ysagogarum* is the only early algorism in which the word *ciffre* is used for *zero*: here the circle form, which the abacists had already popularized, is used throughout for the algorism, but the 't'-form is used in a table of eras in the astronomical book.[59] There is clearly a close connection between the *Liber ysagogarum Alchorismi* and the *Helcep Sarracenicum*, and that connection might have something to do with Petrus Alfonsi, since the table of eras (for 1 October 1116) in the astronomical section of the *Liber ysagogarum* forms part of the calendrical tables of Petrus's version of the *Tables* of al-Khwārizmī.[60] It is curious, however, that Ocreatus uses the astronomical numerals for his algorism and not Arabic numerals. Can we see here another example of the 'overcompensation' in introducing Arabic science to a *Latin* audience which is apparent in the clumsy attempt to calculate the tables of mean motion for the Latin calendar instead of the *Arabic* calendar in Petrus Alfonsi's version of al-Khwārizmī's tables? That Roman numerals were the original intention of Ocreatus and not a substitution due to a scribe is shown by the fact that Ocreatus has to insist

assimilation of the word to the Arabic for the abstract form for 'yellow' (*ṣufra*) which would yield the transliteration *sufra*, or the feminine adjective *ṣafrā'*, which would yield *sufre*. This could also explain the alternative form for zero found uniquely in MS *R* (see below, p. 251): 'solfra'.

57 In MS *O* the forms for zero vary between ⊖ ♂ Ⴀ (see Plate V); a variant of the last form is used in the two MSS of Petrus Alfonsi's version of the tables: Oxford, Corpus Christi College, 283 and London, Lambeth Palace, 67. The use of 't', or a sign resembling 't', for zero in astronomical tables goes back to Classical times, and appears in Greek papyri; see *Adelard of Bath*, p. 42, and R. A. K. Irani, 'Arabic Numeral Forms', *Centaurus*, 4 (1955-6), pp. 1–12 (4–5, 11).

58 See n. 56 above.

59 See Plate VI.

60 Petrus, the converted Jew, may also have been responsible for other Hebrew calendrical material in the *Liber ysagogarum*. That he is responsible for any of the extant versions of the text is more debateable. 'Magister A', to whom the text is attributed in the Paris MS, is an unlikely abbreviation for Petrus Alfonsi: one would expect 'magister P'. See Allard, *Le Calcul indien* (n. 12 above), pp. viii–xxiii, for a very fine evaluation of the evidence linking Petrus Alfonsi with the *Liber ysagogarum Alchorismi*, and the suggestion that the versions of the text were compiled by other scholars.

238

that his reader treats the numerals, in whatever column they occur, as 'digits', whether they are units, tens or hundreds. In the examples of calculation he uses the Roman numerals as Arabic digits: writing .iii. iii. for .xxxiii. etc.[61]

Ocreatus starts by describing the principles of place value in a way common to the abacists and the algorists. He adds the comment that God created nine orders of celestial spirits because of the nine different numbers. Then he describes a way of calculating the square of a single digit (16–25). For the squares of numbers between 6 and 10 he applies a formula which he correctly attributes to Nicomachus: it is found, with that attribution, in Boethius's *Arithmetic*, II.43, where it is given as a characteristic of an arithmetical mean.[62] We can express it algebraically as:

$$a^2 = (a-b)(a+b) + b^2$$

This formula is applied by making 'b' the difference between the number and 10.

Ocreatus then (27–42) considers the squares of *articuli* (i.e., a number followed by one or more zeros), and discovers that they can be arrived at through using the formula of a geometrical mean:

If $a:b = c:d$, then $b \times c = a \times d$
When b and c are the same number, $a \times d$ gives the square of that number.
E.g., if you want to find the square of 20 you use the formula: $4:20 = 20:100$, and therefore $20 \times 20 = 4 \times 100$.

Ocreatus could have found a perfectly appropriate description of the geometrical mean in Boethius's *Arithmetic*. However, instead, he cites Euclid, *Elements*, VII.19: 'If four numbers are in proportion, what results from the multiplication of the first and the fourth is just as what results from the multiplication of the second and the third'.[63] He does this probably for the sake of symmetry: as he re-

---

61 *Helcep Sarracenicum* **60**. Ocreatus might have characterized the procedure as 'Sarracenicum' and not 'Indicum' because he was *not* using the Arabic numerals, which are called 'Indian' in Arabic and in the earliest Latin algorisms.

62 Boethius, *Arithmetic*, II.43, ed. Friedlein, p. 143: Illud quoque subtilius, quod multi huius disciplinae periti nisi Nicomachus nunquam antea perspexerunt, quod in omni dispositione vel continua vel disiuncta, quod continetur sub duabus extremitatibus minus est eo numero qui ex medietate conficitur, tantum quantum possunt duae sub se differentiae continere, quae inter ipsos sunt terminos constitutae.

63 Euclid, *Elements*, VII.19, Version II, MS *B*, fol. 21v: Si fuerint quatuor numeri proportionales, quod ex ductu primi in ultimum producetur

sorted to a work on arithmetic for the arithmetical mean, so he goes
to a work on geometry for the geometrical mean. But he might also
have cited Euclid out of respect for his teacher. For, since this is
beyond the point where the 'Boethian' version of Euclid's *Elements*
breaks off, Ocreatus must know one of the versions from the Arabic,
though from the evidence of this passage it is not possible to say
whether he used Version I or Version II which are very similar at
this point. It is interesting to note that he uses *theorema* for *propo-
sitio*, something that we find only in Martianus Capella and the
'mixed' versions blending Version II and the 'Boethian' version, in
which we find a predilection for Greek terms.[64]

The next paragraph (43–45) gives two rules for multiplying any
two numbers which are different from each other. The first rule can
be expressed algebraically as:

$a \times b = 10b - b(10 - a)$
where a and b are under 10 and $a > b$.
$a \times b = 100b - b(100 - a)$
where a and b are under 100 and $a > b$, etc.

The second rule (46–47) can be expressed:

$a \times b = a^2 + a(b - a)$
where b is larger than a.

The next rule (48–54) applies the rule of three to discovering the
result of multiplying two numbers. One must find which number
has the same ratio to the lower multiplicand as the higher multiplicand has to the first number in the higher decimal place. I. e., when
you multiply 5 times 6, you must find which number has the same
ratio to 5 as has 6 to 10 (which is the first number in the decimal
place higher than the units). The answer is 3. Since $3 : 5 = 6 : 10$,
then 5 times 6 is equivalent to 3 times 10, i. e., 30.

equum erit ei quod ex ductu secundi in tercium. Cf. Version I, ed.
Busard, p. 209: Si fuerint quattuor numeri proportionales, erit quod ex
ductu primi in quartum, sicut quod ex ductu secundi in tertium.
64 Martianus Capella, *De nuptiis Philologiae et Mercurii*, ed. J. Willis
(Leipzig, 1983), p. 258.17 (VI.724). The 'mixed' versions are described
in M. Folkerts, 'Anonyme lateinische Euklidbearbeitungen aus dem 12.
Jahrhundert,' *Österreichische Akademie der Wissenschaften, math.-nat.
Kl., Denkschriften*, 116, 1 (Vienna, 1971). The use of the word *theorema*
in respect to Euclid's *Elements* was familiar to the later English scholar,
Alexander Nequam (1157–1217), who included in a curriculum of study
'theoremata geometrie que ordine artificiosissimo disponit Euclides in
suo libro' (quoted by Haskins, *Studies* (n. 21 above), p. 374).

To this section should be added the last two sentences of Ocreatus's work (127-8) which set out the binomial theorem, expressed algebraically:

$$(a + b)^2 = a^2 + b^2 + 2ab.$$

These rules for multiplication do not occur in the original form of the algorism or in its earliest adaptations, with one significant exception: two manuscripts of the *Liber ysagogarum Alchorismi* include the binomial theorem (one MS as a gloss).[65] A similar set of rules, however, is given in the English algorism in MS British Library, Egerton 2261,[66] and was included by Ocreatus's fellow-countryman, John of Sacrobosco, in the chapter on multiplication in his *Algorismus vulgaris* (ca.1230).[67] John of Sacrobosco, too, adopts and popularizes the word *cifra* for zero. The sources of the *Algorismus vulgaris* have not been studied in depth, but Ocreatus's text gives the earliest example we know of these multiplication rules, and should be taken into account when studying Sacrobosco's sources.

The algorism proper is the subject-matter of the remaining portion of Ocreatus's work (55-126). As Ocreatus promised in his preface, he describes how to multiply numbers together and, in testing the result, demonstrates the procedure for division. He also promised in his preface to include a section about the multiplication of fractions ('de multiplicatione proportionum'). In fact this turns out to be an explanation of how to make a fraction out of the remainder,

65 Paris, BN 16208, fol. 67r (in the margin): Omnis numerus ductus in se tantum efficit quantum utraque pars eius ducta in se et altera in alteram bis; cf. ed. Allard, *Le Calcul indien* (n. 1 above), p. 55 (an addition in Milan, Ambrosiana, A.3.sup.): Omni namque numero in duas partes diviso idem fit ab unaquaque parte in se et una in alteram bis quod a duabus simul iunctis et in se ductis. The binomial theorem is described in geometrical terms in Euclid, *Elements*, II.4; cf. Version II, MS *B*, fol. 6v: Si fuerit linea in duo divisa, illud quod ex ductu totius in se ipsam equum erit his que ex ductu utriusque in seipsam et alterius in alteram bis; Version I, ed. Busard, p. 74, is very similar; the 'Boethian' version, ed. Folkerts, „*Boethius*" *Geometrie II* (n. 28 above), pp. 200-2 is markedly different. The versions in *Helcep Sarracenicum* and the marginal addition in the Paris MS of *Liber ysagogarum Alchorismi* are closer to each other (both using the correlatives 'tantum(dem)/quantum') than to the translations of Euclid.

66 See p. 305 below. The first rule also appears in the French algorism edited by E. G. R. Water in 'A Thirteenth Century Algorism in French Verse', *Isis*, 35 (1928), pp. 45-84 (46).

67 See p. 305 below.

when the divisor cannot be divided into the dividend an exact number of times. The procedures he uses are those of the algorism, except that, as already noted, he uses Roman numerals instead of Arabic numerals. The numerical examples do not correspond to those found in any of the early algorisms, and, as before, Ocreatus uses as his main examples the multiplication of numbers by themselves – i.e., squares. Another peculiarity of this algorism is that zeros are sometimes added to columns which are higher than the column which has the highest digit, producing configurations such as t. 1. 8. 9. (83). Again this is characteristic of the tables of al-Khwārizmī, and particularly of the fifth book of the *Liber ysagogarum Alchorismi* in which 't.' is placed in the thousands column for year-values less than 1,000.[68]

What can one conclude concerning the *Helcep Sarracenicum*? It is not a great work of mathematics, and covers only a small part of the algorism. The author knows the method of the algorism, but is reluctant to use Arabic numerals. As in the algorism he starts with multiplication (the abacists tended to start with division). However, his terminology is still largely taken from earlier abacus-works and Boethius's *Arithmetic*. He uses the forms 'quotus/totus' and 'secundare' which are familiar from abacus texts. On the other hand, in addition to using the verb *ponere* for 'placing the numbers' (which is the verb proper to the abacus), he uses the verb *scribere* which can only be used in performing the algorism on a sand-board or parchment. He employs Boethius's word for the product of a multiplication: i.e., 'quod continetur sub'. The use of the term 'helcep Sarracenicum' for the algorism suggests that the work was written before Arabic calculation had become generally known by the name of the Arabic mathematician who wrote about it. *Helcep Sarracenicum* would seem to be the more appropriate term, but it evidently failed to catch on.[69]

Ocreatus's work shows several affinities to the *Liber ysagogarum Alchorismi*, especially to the version which is attributed to 'magister A'. This version states in its title that the work is an introduction to

68 See Plates V and VI.
69 It is always possible that descriptions of the method of the algorism were transmitted orally, or by example, and independently of al-Khwārizmī's work. Evidence for the practice of arithmetic in the maghreb using *ghubār* numerals and the sand board is given in the *talqīḥ al-afkār fī ʿamal rasm al-ghubār* of Ibn Yāsamīn, a mathematician of Berber origin who died ca. 1204; see R. Köbert, 'Zum Prinzip der ġurāb[sic!]-Zahlen und damit unseres Zahlensystems', *Orientalia*, 44 (1975), pp. 108–12.

astronomy ('in artem astronomiam'); a purpose made clear in the preface contained in all the versions of the *Liber ysagogarum*. The use of the astronomical forms of numerals in the *Helcep Sarracenicum* suggests that it, too, was written in the context of astronomical texts. Multiplication and division is required for calculating the rising times of the signs in a particular latitude from tables of right ascension, in the absence of tables of oblique ascension. One notable feature about the tables of al-Khwārizmī in their Latin versions is that they do not include tables of oblique ascension, so that anyone using them on a latitude other than the equator, would have had to multiply and divide as well as add and substract.[70] This is perhaps the significance of a passage in the introduction to arithmetic in the Coventry manuscript, which states that the part of arithmetic which is the 'scientia multiplicandi, proporcionandi et dividendi' – i.e. the algorism or *helcep* – 'comes to help problems in astronomy, e.g. by determining the hours, the signs of the zodiac and the parallels'.[71] Why the examples in the *Helcep Sarracenicum* are squares is more puzzling: only one calculation in the tables of al-Khwārizmī requires the astronomer to find the square of a number, and that is an alternative method for finding the height of the Sun from the length of the shadow cast by a gnomon (c. 28a). The greatest difference between the calculations of the *Helcep Sarracenicum* and those in the astronomical tables is, of course, that the *Helcep Sarracenicum* is based on a decimal system, whereas the astronomical tables employ a sexagesimal system.

The parallels between Ocreatus's *Helcep Sarracenicum* and the *Liber ysagogarum Alchorismi* are obvious and suggest that the two works were composed in the same context. However, they do not go as far as verbal correspondences. Given its apparently early form, the *Helcep Sarracenicum* might be expected to have some relation to the most primitive form of the Latin algorism – *Dixit Algorizmi*.

70 A table of right ascensions is referred to in c. 26a, but is missing in all the Latin MSS; see Neugebauer, *The Astronomical Tables* (n. 23 above), p. 54. We read the encouraging words in the preface (ed. Suter, p. 1): ab hac radice [i.e., the prime meridian of Arin] per regulas geometricales et arithmeticas ceteras regiones et tempora determinare non sit difficile.

71 MS *T*, fol. 2r: Sciencia multiplicandi, proporcionandi et dividendi astronomice difficultati subservit, ut in distinctionibus horarum, signorum et parallelorum. 'Iohannes Hispanus/Hispalensis' in the *Sentencie de diversis libris excerpte*, advises that whoever wishes to approach astronomy should devote his whole effort to learning arithmetic, and above all the 'kind of numeration which is done with Indian letters, used especially by the Saracens, and invented with wonderful skill by the honourable al-Khwārizmī' (see n. 73 below).

After all, the oldest manuscript of this work came from the Abbey of Bury St Edmunds and appears to have been written within the twelfth century. However, it is with another algorism that the *Helcep Sarracenicum* shows a closer affinity. For the opening of Ocreatus's text (**5-8**) is word for word the same as the beginning of a paragraph within the *Liber Alchorismi de practica arismetice.*[72] Until recently this text had been attributed to John of Seville on the basis of an attribution in one manuscript of the fourteenth century (Paris, Bibliothèque nationale, lat.7359). The identity of John of Seville and the establishment of his oeuvre are much debated subjects. No manuscript of the *Liber Alchorismi* is older than the thirteenth century, and other manuscripts attribute the work only to a 'magister Iohannes'.[73] The passage in question in *Liber Alchorismi* has no equi-

72 *Liber Alchorismi*, ed. Allard, *Le Calcul indien* (n. 1 above), p. 67–8: Ordines vero (*v. l.* igitur) sive limites numerorum a primis numeris qui digiti vocantur et sunt 9 per decuplos (*v. l.* decuplationes) in infinitum (*v. l. omit.* in infinitum) procedunt, unde in unoquoque limite numerorum sunt 9 termini (*v. l.* termini 9), nec plures excogitari possunt. Omnes autem qui sunt in ceteris limitibus preter primum articuli solent appellari, ut sit primus limes ab uno usque ad 10, secundus a 10 per decuplos primorum digitorum usque ad C, tercius vero a C per decuplos secundorum usque ad mille, et sic quartus per decuplos terciorum, et sic deinceps in aliis (*v. l.* ceteris). This correspondence between the two texts had been pointed out by S. R. Benedict, *Comparative Study of Early Treatises Introducing into Europe the Hindu Art of Reckoning* (Concord, New Hampshire, 1914), but she believed Ocreatus to be the debtor.

73 The traditional arguments for attributing the *Liber Alchorismi* to John of Seville are set out in Allard, *Le Calcul indien* (n. 1 above), pp. xiv–xv. The most significant of these is that the 'Iohannes Hispanus' or 'Iohannes Hispalensis' who wrote the *Sententie de diversis libris excerpte*, in which the disagreements between different astronomical tables is discussed, promises to deal with 'illud genus numeri quod fit per litteras Indorum quo Sarraceni maxime utuntur quod probus Alchoarismi (Millás erroneously corrects this to Alchoarismus) mirabili racione invenit'; *Sententie de diversis libris excerpte*, ed. J. M. Millás Vallicrosa in idem, *Estudios sobre historia de la ciencia española* (Barcelona, 1949), pp. 273–88 (274). Allard rejects the attribution to John of Seville, who translated mainly astrological texts in the second quarter of the twelfth century, and makes the plausible suggestion that 'magister Iohannes' is the scholar of that name who collaborated with Dominicus Gundissalinus in Toledo in the third quarter of the same century. This does not preclude the possibility that the author of the *Liber Alchorismi* is the same as the author of the *Sentencie de diversis libris excerpte*, i. e. that 'magister Iohannes' and 'Iohannes Hispanus' are the same, but further research is needed.

valent in the *Liber pulveris*. The sentences shared by *Helcep Sarracenicum* and *Liber Alchorismi* sound like the beginning of a treatise. The presence of a connecting word 'igitur' is not an argument against this; in fact, Adelard himself uses 'igitur' at the beginning of his text of the *De opere astrolapsus*, and straight after his dedication, in exactly the same way as does Ocreatus.[74] Thus I would be tempted to think of the *Liber Alchorismi* as the debtor.

Whichever way the influence spread, the verbal correspondence is significant. If the *Liber Alchorismi* is a source for Ocreatus then it must predate the *Helcep Sarracenicum*: this would give the first indication of the date of the text that we have up to now. If, on the other hand, we are to say that Ocreatus's text is the source of the *Liber Alchorismi* we must see the *Helcep Sarracenicum* as one stage in the production of what was to become the most comprehensive and coherent account of the algorism before the *Liber abaci* of Fibonacci.[75] Moreover, we are bound to ask whether 'magister Johannes', instead of being John of Seville, might not rather be 'John the geometer' or 'Johannes Ocreatus'.[76]

III

Further light can be shed on the context of Ocreatus's *Helcep Sarracenicum* by looking at the Coventry manuscript of arithmetical texts. This manuscript consists solely of arithmetical texts and calculations.[77] If one follows M. R. James in his description of the manuscript in the catalogue of the manuscripts of Trinity College, Cambridge, one is led to imagine that an early thirteenth-century scholar

---

74 See edition in Dickey, 'Adelard of Bath: An Examination' (n. 14 above), p. 112.

75 Cf. Allard, *Le Calcul indien* (n. 1 above), p. xxxv.

76 It is unlikely, however, that the author of *Helcep Sarracenicum* is the same man as the author of *Liber Alchorismi*, since the latter calls the zero 'circulus', and does not use the terms 'totus/quotus' or 'secundare'; i.e., he excises the terms which H. Ocreatus took from the abacus treatises, and regularly uses 'scribere' and 'delere' for writing and erasing the numerals.

77 See Appendix II. One can point to one connection between Coventry and Worcester (where MS *O* was written). For in Coventry there was a special *studium* for the friaries of the custody of Worcester according to D. Knowles and R. N. Hadcock, *Medieval Religious Houses: England and Wales* (London, 1953), p. 224.

found a twelfth-century manuscript of Boethius's *Arithmetic* and filled it out with his own additions. However, Rodney Thomson has confirmed for me that the hand of James's 'early thirteenth-century scholar' is rather that of an English scholar writing in the mid-twelfth century. To me the manuscript seems to have been planned from the beginning by a scholar (hand 'A') who employed a scribe (hand 'B') to copy Boethius's *Arithmetic* and then took over the preparation of the rest of the manuscript himself. But he never completed his project. The scholar added glosses to Boethius's text and then started to write a preface to the art of arithmetic which tails off with a couple of notes or *aides-memoire* which were to remind him how he intended to continue. The preface states that arithmetic is divided into a theoretical and practical part and was clearly intended to be followed by texts illustrating both these aspects. After the theoretical (or speculative) arithmetic of Boethius the scholar wrote the rubric 'Now begins the second part of the art, which is the practical part of it, according to the Greeks, Arabs and Indians'.[78] This text occupies the last half page of a gathering and ends with a *custos* indicating the first words of the next gathering. Either this gathering is now lost, or it was never written. Tantalizingly, the text ends with the word *caracteribus* (in this context, the 'Arabic' abacus numerals). Instead of the lost gathering there is a bifolium covered with the scholar's own notes on practical arithmetic, which exhibit a certain degree of disorder (as befits notes). Here we find the rules of rhythmomachy, discussions of means, and, most significantly, the rules for multiplication corresponding to those in the *Helcep Sarracenicum*. The scholar has added further notes to the fly-leaves at both ends of the manuscript.

Unlike in the case of the manuscripts of the *Helcep Sarracenicum* in which scribes have negligently transcribed a text perhaps two or three removes from the pen of the original mathematician, here we can see the mathematician at work: we have his autograph. A note in the top margin of the first page of the introduction to arithmetic – 'ex cerebro testardi' – would suggest the classroom banter characteristic of Adelard's circle – 'from a bighead's noddle' (*testardus* meaning 'stubborn' in Medieval Latin) – rather than that the author was called 'Testardus'.[79] More significant are the erased and partially

---

78 MS *R*, fol. 59v: Incipit secunda pars artis que est practica eiusdem, secundum Grecos, Arabes, et Indos. See Plate VII.

79 The possibility that Testardus is Robert Grosseteste, the 'bighead' who was part of the bishop's household in Hereford in the last years of the twelfth century cannot be ruled out if the manuscript is late enough. It appears that no authentic writings on arithmetic by Grosseteste have

obliterated names of two early owners on fol. 3v, to which we shall return. But first we should give an idea of the nature of the arithmetic discussed in this manuscript, pointing out how it relates to the work of Adelard and Ocreatus.

The introduction to arithmetic takes the familiar form of an *accessus ad artem* describing, in respect to arithmetic, what it is, what is its genus, its matter, its species, its parts, its role, its practitioner, its instrument, its aim, the reason for its name, and the order in which it should be taught and learnt. Arithmetic, like the other arts, is divided into *scientia* and *operatio*, which the Greeks call 'theorica' and 'practica'. Sometimes the theoretical division is more worthy than the practical; sometimes it is the other way round.[80] There are two species: one is demonstrative, which is what Euclid teaches us in the seventh, eighth and ninth books of his art of geometry; the other is 'probable', which Nicomachus has dealt with.[81] Arithmetic is further divided into four parts, which themselves correspond to the four subjects of the quadrivium: 'Number in itself' is arithmetic proper, 'related number' corresponds to music, 'shaped number' corresponds to geometry, and, as we have seen, the multiplication and division of numbers corresponds to astronomy.[82] There are two

been found (see S. Harrison Thomson, *The Writings of Robert Grosseteste*, Cambridge, 1940, p. 234), but this is a subject that would have interested him, especially early in his career.

80 MS *R*, fol. 2r: Scientia enim numerorum quam dicimus arithmeticam, sicut et alie artes, bifariam dividitur, scilicet in scientiam et operationem, quam discretionem Graii dicunt theoricam et practicam. In utraque autem tres inspectiones occurrunt sine quibus neutra sciri potest, scilicet de proprietatibus numerorum, de proportionibus eorumdem, et de caracteribus ipsorum. His ergo doctus et docens, ab horum distat operatore. Quandoque vero hoc dignius, quandoque illud.

81 MS *R*, fol. 2r: Species eius due sunt. Est enim quedam demonstrativa quam docet Euclides in .vii. et .viii. et ix. volumine artis geometrice, alia est probabilis, quam digessit Nichomacus. The author of this introduction considers Boethius's *Arithmetic* (aside from the two 'prefaces') as a direct translation of Nicomachus's *Introduction to arithmetic*, as one can see from his gloss to *Arithmetic* I.2 (MS *R*, fol. 7r): Incipit Nicomachus.

82 MS *R*, fol. 2r: Partes artis huius sunt .iiii.: de numero per se, de numero relativo, de numero figurali, de proportionalitate et medietate numerorum. Quatuor autem hee partes secundum quatuor disciplinas matheseos assignantur. Scientia numerandi per se quantum ad suam propriam conceptionem. Scientia de numeris ad aliud relatis, quantum ad armonice subtilem disquisitionem. Scientia numerandi figuraliter, quod dicunt numerum geometricum, fit ad figurarum geometricarum ordinatam genituram. ... Scientia multiplicandi, proporcionandi et divi-

kinds of practitioner, the one teaches the art, the other operates with it. The first is called the *arithmeticus*, the second the *computator*. Both must know the properties of numbers, their relations to each other, and the way they are written ('caracteres eorum').[83]

Particularly interesting is the way the author speaks of the abacus:

> The instrument of arithmetic according to some is the abacus...But we say that the instrument of this art is not the abacus,...for the abacists divide the unity, which arithmetic refuses to do.[84] The abacus is the instrument of geometry since it speculates on quantity rather than number. For by creating different configurations of the figures in the different squares of the counting board, one considers how large or how small something is.

The author claims that chess, too, is an instrument of geometry.[85] The true instrument of arithmetic, on the other hand, is rhythmomachy, presumably because one does not make calculations by moving the counters about and then 'reading off' the result from the configuration of the counters, but rather one makes Boethian arithmetical calculations from the values marked on the pieces on each side of the board respectively. In the gathering at the end of the manuscript the scholar has filled one page with a diagram of the rhythmomachy board and the pieces arranged on it. This is followed by brief instructions which are amplifications of the rhythmomachy of Odo of Tournai, so establishing a connection between MS *R* and

dendi astronomice difficultati subservit, ut in distinctionibus horarum, signorum et paralellorum.

83 See n. 80 above.

84 The author is referring to the section of the abacus-treatises on the divisions of the *as* which follows the section on integers.

85 MS *R*, fol. 2r: Instrumentum huius artis secundum quosdam abacus,...Sed dicimus nec huius instrumentum esse abacum,...Nam abaciste unitatem etiam dividunt, quod recusat arithmetice...At abacus geometrie est instrumentum, quoniam quantitatis pocius quam numeri habet speculationem. Rerum enim quantitatem investigans, dum per figure posituram per campeolos arte vagatur, quam multum quamve parvum aliquid sit consideratur. Troianum etiam vel Thebanum diludium quod a captura Scaccarium dicunt teste diligentia geometrica pollet industria quidque etiam figurarum geometrie non paucas in solido dant admirationes. Compare the statement in the late 12th-century rhythmomachy of Werinher edited by M. Folkerts, in *Vestigia Mathematica* (n. 46 above), pp. 107–42 (130): Non enim aliter arismetice opus rithmachia representat quam musica in cytharis et organis, et geometria in abaci opere et astronomia in h⟨or⟩oscopis et astrolabii sollertia consistit.

MS *O* and other manuscripts containing works of Adelard of Bath.[86] These amplifications include a reference to the algorism.[87]

The editorial hand of the scholar can be seen in the copy of Boethius's *Arithmetic* which follows the introduction to arithmetic. The first book is headed 'About the equality, inequality and ratios of numbers',[88] the second 'About the conversion and shaping and finding of the means of numbers'.[89] The scholar has added glosses explaining grammatical, historical and philosophical points. He comments that the *quadrivium* is an introduction to *physica*.[90] He admits that the subject-matter of arithmetic and geometry is close, and makes the following comment on how to differentiate between them:

86 The version in MS *R* is different from any of the several versions edited in the comprehensive volume of A. Borst, *Das mittelalterliche Zahlenkampfspiel* (Heidelberg, 1986). It consists of (1) a glossed version of the rhythmomachy of Odo of Tournai (Borst's text IV, pp. 344–55), including cross-references to the copy of Boethius's *Arithmetic* in MS *R*, and comparisons with chess; (2) the first addition to Odo of Tournai's text (Borst's text Va, pp. 356–7); (3) a description of the arithmetical, geometrical and harmonic means and their application to the playing of rhythmomachy (including some of the words you might say to your opponent in the course of a game): this has no equivalent in Borst; (4) a two-sentence summary of the rules (Borst, text Vd, p. 371); (5) on fol. 60r there is an illustration of the rhythmomachy board (Borst, item Vb, pp. 358–60). (1) and (2) occur (in their original form) in MSS *O*, Hereford Cathedral, MS O.i.vi, and the late twelfth-century Victorine MS, Paris, Bibliothèque Nationale, lat. 15119, which is the earliest of the three MSS to contain Adelard's *Regulae abaci*. (1) (2) (4) and (5) only occur together elsewhere in MS Avranches, Bibliothèque municipale, 235 of the twelfth century, where they were once immediately followed by Adelard's translations of Abū Maʿshar's *Isagoge minor* and Thābit ibn Qurra's *Liber prestigiorum*, which were subsequently excised from the MS. This is significant because MS Avranches 235 is the earliest manuscript in which these two rarely-copied texts are known to have existed; see *Adelard of Bath*, p. 135.

87 MS *R*, fol. 61v: Fiant tractus ex alterutra parte alternatim, tum abaci vel algorismi, tum propriis et arithmeticis vel geometricis rationibus. 'Fiant...alternatim' is Odo of Tournai's text (ed. Borst (previous note), p. 346, lines 7–8); 'tum abaci...rationibus' is the addition in MS *R*.

88 MS *R*, fol. 27v: Explicit liber primus de equalitate et inequalitate et proportione numerorum (hand 'B').

89 MS *R*, fol. 59v: De conversione et figuratione medietateque numerorum (hand 'A' or 'B').

90 MS *R*, fol. 5r: Hoc enim quadrifarie introducimur ad fisicam, velud trivio ad ethicam, ut secundum sui naturam animus trivio, corpus quadruvio curaretur.

It is the property of arithmetic to say 'what number of/such a number of' (quotus/totus), but of geometry to say 'how great/so great' (quantus/tantus). Proportion belongs to geometry, but difference to arithmetic.[91]

This conspicuous use of the correlatives 'quotus/totus' is not *
of course 'correct' Latin, in which 'totus' can only mean 'the whole', but is found in Adelard's *Regulae abaci*,[92] Version II of the *Elements* (but not in Version I, or the mixed 'grecizing' version in Paris, BN 10257),[93] the *Liber ysagogarum Alchorismi*[94] and the *Helcep Sarracenicum* (e. g., **13–14, 55, 61**). The terms apply in particular to the position of counters in the abacus columns, or the decimal position of numerals. Thus, in the *Helcep*, 'quotus' on its own can imply 'in what decimal place' as is seen from the manuscript variation between 'quotus + ordo' and 'quotus' without 'ordo' in **55**.

The text on practical arithmetic is tantalizingly brief and is not found in any other manuscript. The few words that remain are difficult to understand. The author seems to be saying:

---

91 MS *R*, fol. 19r: Nota arismeticam geometricas vicinitate usurpare dignitates. Arismetice enim proprium est dicere 'quotus totus', geometrie 'quantus tantus'. Ipsius est proporcio, arismetice vero differentia.
92 Ed. Boncompagni (n. 7 above), p. 94.28: secundus articulus a secunda parte, tercius a tercia, quartus a quarta, et deinceps, a tota parte requiruntur denominaciones quoti (*read*: quota?) ipsi recesserunt a denario. Adelard is probably indebted to earlier abacus treatises for this use of 'quotus/totus', for it appears in Turchillus, *Reguncule super Abacum*, ed. E. Narducci, *Bullettino di bibliografia e di storia delle scienze matematiche e fisiche*, 15 (1882), p. 138, and the anonymous commentary to Gerbert's *Regulae abaci* in MS *O*, fols. 33v–41v and Cambridge, Trinity O.7.41 (incipit: Numeri ex qualibet multiplicatione...).
93 E. g., *Elements*, VII.6, MS *B*, fol. 20r-v, MS *T*, fol. 117r: Si fuerint quattuor numeri quorum primus tote partes secundi quote tercius quarti, erunt primus et tercius pariter accepti tote partes secundi et quarti pariter acceptorum quote primus secundi. The author of the version in Paris, BN 10257 has replaced 'tote' in this context with the more Classical 'tot et tante'; G. D. Goldat, 'The Early Medieval Traditions of Euclid's *Elements*', unpublished Ph. D dissertation, University of Wisconsin, 1957, p. 288. On the other hand, the version of the *Elements* attributed to Hermann of Carinthia retains the 'quotus/totus': *The Translation of the Elements of Euclid from the Arabic into Latin by Hermann of Carinthia (?), Books VII–XII*, ed. H. L. L. Busard (Amsterdam, 1977), p. 26.
94 See ed. Allard, *Le Calcul indien* (n. 1 above), pp. 40 (Allard's MS O) and 42 (Allard's MSS M Cu AP).

There are two parts of the art of numbering. The first – which Boethius took over from a Greek invention – being open to speculation alone, is used as an introduction to geometry and music, and goes with the practice of rhythmomachy. The other one, which we have got hold of only from scattered texts and translations, is a procedure less involved with speculation, but more necessary for geometry and astronomy. It was unacknowledged[95] by the Greeks, but reveals the mysteries of the Indians and the Arabs, excepting the rules of the abacus which are (already) exchanged amongst us.

The incomplete last sentence goes on to mention both *caracteres* and a book of axioms.[96]

This 'book of axioms' may be a reference to one of the texts included in the notes on the gathering which comes next. Here we read: 'Incipiunt anxiomata artis arismetice'.[97] There follow statements about number put in an axiomatic form; then definitions and rules. The Euclidean model in the exposition is unmistakable. Where the eighth axiom of the first book of Euclid reads 'Every whole is greater than its part', the Coventry manuscript reads 'Every number is either even or odd', 'every number is either simple or composite', 'every number is either in the first decimal place or the second or a higher decimal place', 'every number is prime to that number into which it cannot be divided' etc.; where Euclid defines the line, the surface, the point, the Coventry manuscript defines multiplication, division, *denominatio* and *differentie*. Some of the rules for multiplication which follow are put in the infinitive just like a proposition of Euclid: 'The problem is to find which number has the same ratio to the smaller number as the larger number has to the first number of the higher decimal place'. There is a *probatio* and a *porisma*. Although some the definitions can also be found in Euclid, *Elements* VII, the author's arrangement is his own, and represents a crude,

95 Literally 'ungreeted'; the word occurs only once in Classical Latin, in Virgil's *Aeneid*, 9.288, but is used by St Jerome, Ennodius, Sigebert de Gembloux and Bernard of Clairvaux.

96 MS *R*, fol. 59v: Artis numerandi due sunt partes. Altera quam Boecius a Greca transferens inventione soli vacans speculationi geometrie musiceque habet introductioni, et rithmachie prosequitur industriam. Altera quam apprehendimus sparsim edita et translata, opus magis (*read*: minus?) exercens speculatione, magisque geometrie et astronomie necessaria, Grecis insalutata Indorum Arabumque revelat misteria, exceptis hic datis vel acceptis abaci regulis. Nichil autem Libri anxiomatum exceptis cara(cteribus dicimus) [the last words in brackets are a custos for the next folio or quire, which is missing]. See Plate VII.

97 See Appendix II.

but, I think, important attempt to apply the axiomatic method to arithmetic.[98] It is precisely the rules for multiplication from this text that recur in Ocreatus's *Helcep Sarracenicum* where they are described as being 'like certain axioms' (**54**).

The technique of the algorism itself is only adumbrated in the Coventry manuscript in the form of brief notes which could equally apply to the abacus (see p. 298 below). There are, however, two examples of the use of Arabic numerals on the front and back fly-leaves respectively.

On the inside of the front fly-leaf there are two lines of verse giving the 'Arabic' names of the abacus *caracteres*. Underneath each name is given the equivalent Roman and Arabic numeral. But the Arabic numerals are no longer in the forms in which we find them in the abacus-treatises. They are the Arabic numerals of the algorism. Moreover, two forms are given for the numbers 2, 4, 7, and 8. Two of these numbers — 4 and 7 — are the very numbers given as examples of numbers which can be written in two or more different ways in the *Liber Alchorismi* of 'magister Iohannes'.[99] The word for zero is not included in the two lines of verse but is added to the left of them. Indeed several names are given, starting with that adopted by the author of the *Liber ysagogarum* and Ocreatus and including the abacists' 'sipos', and the two symbols given for zero are the two alternatives for zero mentioned by the author of the *Liber ysagogarum*: t and 0, with t coming first (fol. Av):

cifra vel solfra vel nichil
t. 0. cim(er)a sipos[100]

---

98 The axiomatic method adapted to arithmetic in works by Jordanus de Nemore and Gernardus is rather different from that in the Coventry manuscript, and apparently has developed quite independently; see G. Eneström, 'Über eine dem Jordanus Nemorarius zugeschriebene kurze Algorismusschrift', *Bibliotheca Mathematica*, 3 Folge, 8 (1908), pp. 135–53, and idem, 'Der „Algorismus de integris" des Meisters Gernardus', (*ibid.*), 13 (1913), pp. 289–332.

99 Allard, *Le Calcul indien* (n. 1 above), p. 152.

100 See Plate VIII. For 'cifra' and 'solfra' see above, n. 56; 'cim(er)a' – i.e., chimaera, 'the imaginary beast' or 'figment of the imagination' – is a curious word for 'nothing', but cf. Walter Anglicus, *Fabulae*, 60.13: fallax ergo decor atque chimera nihil (quoted in *Dictionary of Medieval Latin from British Sources*, s. v. chimera). Also note the jotting on fol. 63r: nescio quia nichil (est), 'I do not know (*scil.* the zero?) because it is nothing'.

252

On the inside of the last fly-leaf we see examples of the algorism being used in calculation. Here there is a mixture of Roman and Arabic numerals used in sums involving marks, *denarii, libre* and *solidi*.[101] The number forms correspond to those on the first fly-leaf.

IV

This article has raised more questions than it has answered. Only tentative conclusions can be drawn, which are provisional on further research.

1. The compiler of the Coventry manuscript and the author of *Helcep Sarracenicum* share the same interests and the same sources. They use the same terminology, to a striking degree. Both use *limes* and *ordo* for decimal place (instead of, for example, *differentia*), and hesitate about whether to call the right-most column the 'first' or the 'last'.[102] Both have a predilection for *responsiones* (marked by an ℞ in MS *R*, and ℞ and *Respondeo* in Ocreatus). Both give scant attention to addition and subtraction. Moreover, of the two erased owners' names in MS *R*, fol. 3v, one can be read, under ultraviolet light, as 'Radulfus' – the same name as that of the scholar from whose 'wallis' a paragraph at the beginning of the *Elements*, Version II, Book X, was taken,[103] – whilst the surname of the second appears to be '(h)osatus', which, as we have seen, may be a synonym for 'ocreatus'.[104] However, it seems that the compiler of the Coventry manuscript is writing later than H. Ocreatus, for he now knows the term 'algorismi' and uses the Arabic numerals of the algorism with facility. That he should mention 'helcep' at all (and his is the only mention of this term outside Ocreatus's work) shows that he belongs to H. Ocreatus's group, and the verbal parallels between the *Anxiomata artis aritmetice* and the *Helcep Sarracenicum* suggest that either he knew H. Ocreatus's text, or that they were both depending on a third source. In any case, the Coventry arithmetician plunders this source to write out a series of notes, whereas H. Ocreatus attempts to put his material into a literary form. The poor state of the manuscripts that preserve the *Helcep Sarracenicum* may lead us into

101 The use of these monetary values is compatible with a twelfth-century dating of the MS, and its presence in England; see P. Spufford, *Handbook of Medieval Exchange* (London, 1986), p. 198.
102 MS *R*, fol. Av: In addendo a prima (*supra*: ultima), id est dextra; cf. *Helcep Sarracenicum*, **91**: Duco ergo primum – id est ultimum.
103 See n. 43 above.
104 See n. 46 above.

appreciating Ocreatus's endeavour less than it merits. Other material common to the Coventry manuscript and the *Helcep* is found in the Egerton MS.[105] The evidence this gives is ambiguous since the manuscript is later than MS *R*, but it is worth noting that the author of the Egerton MS uses *ut si* to introduce examples, which is characteristic of H. Ocreatus, whereas the Coventry arithmetician has substituted *verbi gratia* (see conclusion no. 3 below).

2. Can the evidence put forward in this article shed any more light on the authorship of the translations of Euclid's *Elements* known as Version I and Version II? We have seen that hand 'A' of Boethius's *Arithmetic* in MS *T* who gives the equivalent Arabic terms for Boethius's words, cites Version I. The author of the gloss is clearly aware of the original Arabic text of the *Elements*; Version I appears to be a direct translation of an Arabic text. It is possible, then, that the author of the gloss and the translator of Version I are the same person. This would be Adelard or his arabophone collaborator.

3. A distinctive vocabulary is shared by the gloss on fol. 87v of MS *T*, some of the directions for proof in Version II and the *Helcep Sarracenicum*. Both the gloss and these directions for proof use the term 'adversarius' for the proponent of the false argument.[106] The use of *ut* (*si*), instead of *verbi gratia* or *exempli gratia* to introduce an example or a proof is common to a large part of the directions for proof of Book V of Version II and the *Helcep*, as is the exclamation 'Ecce' to highlight the result.[107]

To identify H. Ocreatus with the 'Joan. Ocreatum' who is credited with the translation of Version II in MS *B* would be pressing the evidence too far. It is worth noticing, however, that the Arabic transliterations in this manuscript are remarkably accurate, and that, while the short glosses and directions for proof are not the same as those found in the text of the majority of manuscripts, some statements recur in glosses in other manuscripts.[108] Moreover, on several

105   See p. 305 below.
106   See n. 38 above. See the article by Wilbur Knorr in this volume.
107   'Ut si' is found in the directions for proof for Elements V.4, 5, 6, 7, 8, 9, 10, 14, 15, 16, 19, 20, see also IX.1, 2, 3, 4, and the 'B' proofs of VII.15, 16, 17, and 20; in *Helcep Sarracenicum*, **8** (ut sit), **57**, and **107**. For 'Ecce' see *Elements* V.8 and *Helcep*, **10** and **21**. Note also the frequent use of first person singular forms and the conjunction 'ergo', especially 'age ergo' (*Elements*, VII.20, *Helcep*, **76**), and the use of 'quotus/totus' (see n. 93 above). The fact that 'verbi gratia' (or occasionally 'exempli gratia') is more common in other parts of the directions of proof of Version II (e. g., book VIII) suggests that more than one scholar was involved in composing these directions.
108   For example, the gloss on MS *B*, fol 1v: 'Elmuharifa vel trapezia,

folios, there are vestiges of erased directions for proof, of the kind found both in glosses to Version I and in the directions for proof of Version II.[109] My provisional conclusion is that we are dealing with several scholars who worked closely together (or in succession), of whom the earliest were colleagues and pupils of Adelard. It is probably these pupils whose names appears in the certain manuscripts of Version II: Johannes, Reginerus, Radulfus, Eggebericus, etc. They may have adopted the nickname 'ocreati'.

4. At some stage the *Liber ysagogarum Alchorismi* came into the orbit of this group. Not only are four propositions in the geometrical book of the *Liber ysagogarum* identical with those of Version II and with no other version of the *Elements*,[110] but also some of the arithmetical additions and the treatment of the zero are significantly close to what we find in the *Helcep Sarracenicum*.

5. We have seen that the spread of the technique of the algorism did not necessarily entail the use of Arabic numerals. The works securely attributed to Adelard himself (*De eodem et diverso*, *Regulae abaci*, and *De opere astrolapsus*) show no evidence of the use either of the Arabic numerals or of the technique of the algorism. However, these works (with the possible exception of the *Regulae abaci*) were written for a noble, essentially amateur, audience. The versions of Euclid's *Elements*, the glosses to Boethius' *Arithmetic* and *Music*, and the *Helcep Sarracenicum* can be seen as representing more accurately the interests and activities of 'professional' mathematicians who were interested in new techniques and speculations. Since the *Helcep* is addressed to Adelard as teacher, he must be regarded as the central figure and inspiration for this group of mathematicians.

Addendum

Wilbur Knorr has kindly pointed out to me that the rubric of *Helcep Sarracenicum* in the Paris manuscript (the Cashel manuscript lacks a rubric) appears to have been added to the manuscript after the

idem quod irregularis. Elmuharifa enim arabice, trapezia grece, irregularis latine', recurs in the English MS Oxford, Bodleian Library, Auct. F.5.28 (written before 1260), fol. 1r: (helmuarifa sunt) trapezie et sunt figure irregulares; see Burnett, 'The Use of Geometrical Terms in Medieval Music' (n. 54 above), p. 203.

109 The directions for proof, introduced by 'per (descriptionem)' or 'a' (fol. 21v: a xiiii.^ta), can be found on fols 13v, 14r, 14v, 15r, 21v, 24r, 25r, 25v, 39v, and 40r.

110 See Folkerts in *Adelard of Bath*, pp. 62–63.

text had been written, and, since only in the rubric are the names 'H. Ocreatus' and Adelard mentioned, one must be cautious about accepting these attributions. Since this has implications for the dating of the *Helcep* and, consequently, of other references to 'ocreati', it is worth re-evaluating the evidence of the rubric.

The rubric has clearly been written after the text (see Plate I). This, however, was the general practice in writing rubrics in this period, and does not itself imply that the rubric was written much later, or added by a scribe who was unaware of the archetype of the text. Palaeographically the rubric appears to be of the same period, if not of the same hand, as the text. The ungainly way in which the rubric has been added is due to the fact that the lay-out of the text has been badly planned, as can be seen from the haphazard, and often erroneous, layout of the tables, which no one would suggest are not integral with the text. Similar ungainly rubrics can be observed in MS *O*. Even so, since the rubric has been written after the text, one should take a cautious attitude towards the information it contains, as one should in the case of any rubric. It is best, therefore, to examine the rubric and the text independently from each other, to see what can be gleaned concerning dates and authorship:

1) The text. This makes no reference to Ocreatus or to Adelard, though its subject is clearly *ars helcep* (see sentences **2** and **68**). *Helcep*, as we have seen, is a very rare term. In spite of the absence of references to Ocreatus and Adelard, a date and context for the *Helcep* may be indicated by the following facts: a) the text appears to have been written before *Liber Alchorismi de practica arismetice* of 'magister Iohannes' (see p. 243 above). If one follows the plausible arguments of André Allard, this 'magister Iohannes' could well be the scholar who collaborated with Dominicus Gundissalinus on translations of works by Algazel and Avencebrol in the third quarter of the twelfth century.[111] b) The *Helcep* shows no evidence of knowledge of the use of the term *algorismi* for the algorism. c) The bizarre use of Roman numerals in place of Arabic numerals would seem to situate it early in the transmission of the Latin algorism. d) The coincidence of a large part of the *Helcep* with notes in MS *R* written in the mid-twelfth century (this being the only other manuscript in which the term *helcep* is used), would suggest that it is close in date to that manuscript.

2) The rubric. It is theoretically possible that an author and a dedicatee might be attached to a mid-twelfth-century text by an early thirteenth-century scribe. But it is difficult to explain why that scribe

111 Allard, *Le Calcul indien* (n. 1 above), pp. xix–xxi.

would have chosen these particular names. If he had wished to enhance the authority of the text he would have attributed it to Adelard
directly rather than to an otherwise unattested 'H. Ocreatus'. One
suspects a rubric precisely when it gives information which has clearly been extracted from the text itself. The very fact that the rubric
includes information *not* in the text (as well as the word *helcep*
which *is* in the text) would seem to be an argument in favour of its
authenticity. There remains the possibility that a 'H. Ocreatus' might
have wished to enhance his own authority by pretending to be a
pupil of Adelard, or at least to belong to Adelard's academic lineage.
Pending further evidence, however, this would seem unlikely.

Appendix I.
An Edition and Translation of the *Helcep Sarracenicum* of H. Ocreatus

THE MANUSCRIPTS

C   Cashel (Tipperary), G. P. A. Bolton Library (formerly Cathedral Library), Medieval MS 1, comprises three octavo-size manuscripts bound together, of which the first one (A; pp. 1-120) consists of several booklets on arithmetic, music, astronomy, divination and the calendar.[1] The booklet containing the *Helcep* appears to be in late-twelfth-century hands and consists of three works:

1. Pp. 105-11. Qusṭa b. Lūqā, *De differentia spiritus et animae*, translated by John of Seville and Limia.
2. Pp. 111-17. *Helcep Sarracenicum*.
3. Pp. 118-20. An unidentified commentary on the *Isagoge* of Porphyry, beginning: *Cum sit necessarium.* Cum in reliquis dialecticis libris ... and ending: Dicunt enim moveri. Yukio Iwakuma of the Prefectural University, Fukui, Japan, is editing this commentary and considers it to be of the twelfth century.

The *Helcep* is written in two hands: C^2, p. 113, line -11 (quantum est...) until the last word of the page, and p. 117, line -16 (quotiens hunc...) until the last word of the text; C^1, the rest of the text and all the tables. Two glosses have been added by C^2 (to sentences **30** and **32**); and a pen-trial on p. 111 (copying 'inter eos qui habitum' from the text) may be in the same hand.

P   Paris, Bibliothèque nationale, lat. 6626 is a quarto manuscript consisting of eleven quaternions of thick yellowing vellum with many folds and crinkles in it. The lines are ruled with a hard point, and there is writing on the top line. The entire MS has been written by one scribe, who appears to be English or at least Anglo-Norman[2] and writing ca. 1200.[3]

---

1 I am indebted to an unpublished typescript description of this manuscript made by Professor Marvin Colker of the University of Virginia. Colker dates the whole manuscript to the first half of the thirteenth century. See Plates XVI-XXIII.
2 The orthography (see below) is perhaps more Insular than French, but the form of Adelard's name – 'Aelardum' – and the scribe's initial mis-

258

1. Fols 1r–12v, Seneca, *De clementia* (incomplete).
2. Fols 13r–83v, Seneca, *De beneficiis.*
3. Fols 84r–87v, H. Ocreatus, *Helcep Sarracenicum.*

The scribe has added a 'nota' opposite the empty square at **104** (fol. 87r) and has added reminders for the rubricator which survive in the margins of fols 84v, 86r and 87r. Only fol. 84r has been rubricated. This means that the first letters of each paragraph on subsequent folios are missing. At least two of the numerical tables have been written before the surrounding text (fol. 86v bottom, fol. 87r top). Proverbs have been added in the upper margins of fols 55v ('Non tulit gratis qui cum rogasset accepit'), 82v ('Amasio dulcissimo clericorum fidissimum'), and 86v ('Nimia familiaritas parit contem⟨p⟩tum'). A (later?) scribe has jotted down various forms of what became the standard form of the Arabic zero (ϕ) in the right-hand margin of fol. 87r.

Neither manuscript is a copy of the other, but they both depend (at one or several removes) on a common archetype. This is indicated by the apparent intrusion of the same gloss into the text in **70**, and common errors: 'centies mille milia, decima unitas' omitted in **12**; 'scilicet' read as 'sunt' (**22**), 'sint' as 'sicut' (**84**), 'denominatio' as 'denominatione' (**101**), 'divisorum' as 'divisionum' (**103**), 'invicem' as 'in unum' (**107**) and '.v.' as '.cc.' (**128**); and a table missing after **107**. There is considerably variation in phraseology and style: *C* tends to omit connecting words (quidem **4**; autem **11**; vero **69**; ergo **99**), subjunctives are found in one manuscript where futures are in the other, etc. But the variants are not only due to scribal errors or stylistic variation. Further arithmetical glosses appear in one or other of the manuscripts (*C*: **30, 32, 62, 89**; *P* **28, 123, 124**) and some statements are expressed in different ways, both of which are acceptable arithmetically (see **14, 26**). These arithmetical variations would not appear to be due to the main scribes of *C* or *P* who both seem to be ignorant of the subject-matter: *C* does not recognize the word 'secundare' (**82**), and reads '.xvi.' as '.xiii.' (**118**) and the abbreviation for 'et' as '.i.' (= 'one' **123**); *P* reads the '.i.' (= 'one') as

take in describing him as a citizen of Bayeux ('Baiocensem') would suggest an Anglo-Norman context.

3　See Plates II, IX–XV. For a more detailed description see M.-T. Vernet, 'Notes de Dom André Wilmart sur quelques manuscrits latins anciens de la Bibliothèque nationale de Paris (fin)', *Bulletin d'information de l'Institut de recherche et d'histoire des textes*, 8 (1959), p. 14. Vernet considers that the hand that wrote *Helcep Sarracenicum* could perhaps be different from that of the rest of the MS and notes that, from fol. 84v onwards, spaces for initials have been left empty.

'idem' (**103**), abbreviations for 'unde' (**95**) and 'videlicet' (**122**) as 'unum', and 'minus' for 'numero' (**115**). The second scribe in C (C^2) does, however, seem to have more mathematical acumen. Not only does he add the two glosses in C, one, at the point where he takes over from C^1 (**30**), the other, within his own transcription; he also makes considerable more sense of the last passage in the text (**119–24**), where there is more variation between C and P than elsewhere. Significantly, the arithmetical terminology is the same in both manuscripts.

As for orthography, P regularly writes 'ti' before a vowel, even when 'ci' would be expected (infitiuntur **1**, fatiendum **76**; cf. 'tifre' for 'cifre' in **99**); C writes 'ci'. P writes 'm' before 'd/t' (eamdem **24**; quamtum **32, 42**; tamtum **46**), and shows some uncertainty about the use of 'p' or 'b' (optineant **56**, but 'scribsi' **68**, and the omission of 'b' in 'subduplum' **28**). C sometimes writes short vowels for long (Saracenicum **2**; apellari **7** etc.), but so does P (posint **78**). P sometimes writes syllables twice (de decuplicatione **15**, differerentiam **46**, ordininis **68**, denomiinationum **117**), as does C on one occasion (promomoveo **102**). C^2 writes 'cu' for 'qu' (relicum/relico **119**) and omits 'h' (detrai **119** and **121**). For the zero P generally writes 'cyfre', but occasionally 'cifre' (**58, 84, 97, 99, 100**); C writes 'cyfre' only once (**104**), but sometimes writes 'cifre' within a box, as if it were an abacus-counter (**97, 101, 104**; he once writes '.t.' in such a box: **103**). In the representation of Roman numerals C tends to use '.m.' where P uses '.ī.' (**12, 88**); both main scribes sometimes write macrons over .c. and .m. (erroneously). Curiously, in both manuscripts, 3 is always written as 'tres' until **52**, but thereafter it is written as 'tria'; the shift is signalled by C changing the last 'tres' into 'tria' in **52**.

Since neither manuscript can be regarded as authoritative, but each (a) give independent hints at what may have been in a more authoritative exemplar, (b) show how different scribes deal with mathematical texts in different ways, I have attempted to indicate accurately what can be found in both manuscripts. [ ] indicates editorial deletions; ⟨ ⟩ indicates editorial additions; \ / indicates additions made by the scribe. Where an emendation has been made in the text, the foot-note gives the manuscript reading. Wherever possible I have followed the orthography of the Paris manuscript, and have retained peculiarities in the syntax (e. g., supra + genitive **101**; cf. **83**). The punctuation is editorial and the sentences have been numbered for ease of reference.

I have tried to translate the text as edited, but I have tacitly altered the numbers in the tables in order to illustrate the text more clearly. Additions to make the meaning more obvious have been added in round brackets. The numeral 1 is always written out ('one') to avoid confusion. *Numerus* is usually translated 'digit' when it means a single digit used in calculation (when the Latin word *digitus* is used this is specifically mentioned), and as 'integer' when it means a whole number made up of one or more digits. The word 'number' is reserved for translating number in a general sense, and for *numerus* when this applies to the number of columns or number of digits in an integer. *Ordo* is translated as 'order' (indicating the decimal place), or 'row' (indicating the three tiers of products, multiplicands, and multipliers, or quotients, dividends and divisors), depending on the context. *Locus* is translated as 'column'. *Quotus* is translated 'what number of' (order, column, etc., i.e., expecting an answer such as 'the first order' or 'the second column'); *totus* is translated 'that number of', 'such a number of', or 'the (column, etc.) of that number'. A plural number may qualify a verb in the singular or in the plural; this inconsistency is reflected in the translation.[4]

---

4 Numbers conceived as 'collections of units' (Boethius's 'unitatum collectio') take a plural verb. Numbers conceived in themselves and appearing in the post-Classical formations with '-arius' ('binarius, ternarius, quaternarius,' etc.) govern a singular verb (the earliest examples of this are in Boethius and Calcidius). H. Ocreatus uses both forms.

H. OCREATI

HELCEP SARRACENICUM

262

EDITION

/ *C* 111, *P* 84r/ Prologus H. Ocreati in Helceph ad Aelardum Batensem[5] magistrum suum.[6]

1 Virtus amicitie inter eos qui eius habitu inficiuntur[7] hanc legem constituit, ut alterutro[8] precipiente, alter parere non pigritetur. **2** Iussus igitur ab amico – immo a domino et magistro – festino aggredi Helcep Sarracenicum,[9] tractare de multiplicatione scilicet[10] numerorum et divisione, necnon etiam de multiplicatione proportionum[11] que non nisi per numeros investigantur, licet non omnes in numeris reperiantur.[12] **3** Cuius quidem compendium ingenio vestro placiturum non ambigo, si Dominus dederit[13] proferre prout dedit intelligere. **4** Cuius quidem[14] invocato nomine vel auxilio[15] pie presumitur quod sine eo temere auderetur, in quo omnes[16] thesauri sapientie et scientie sunt absconditi,[17] qui sit[18] benedictus in secula. Amen.

Textus[19]

⟨I⟩

5 Ordines igitur numerorum sive limites a primis numeris, qui digiti vocantur et sunt novem, per decuplos in infinitum procedunt.[20] **6** Sunt autem in unoquoque limite numerorum novem termini, nec plures inveniri vel[21] excogitari possunt; unde, ut opinor, novenarium celestium spirituum ordinem Auctor omnium mutuatus est. **7** Omnes autem qui sunt in ceteris limitibus preter primum articuli solent appellari.

5 baiotensem *P*
6 Prologus...suum] *Rubric in P. C omits, but leaves sufficient space for a rubric.*
7 infitiuntur *P*
8 *P corrects from* alterutrum
9 Saracenic̄ *C, P corrects from* Saracenicum.
10 s; (= sed) *P*
11 *P corrects from* propositionum
12 inveniantur *C*
13 dedit *P*
14 *C omits*
15 vel auxilio] *C omits*
16 *C corrects from* omnis
17 sunt absconditi] ascunditi *P*
18 est *C*
19 *A rubric in P, C omits*
20 prodedunt *C*
21 nec *C*

The Prologue of H. Ocreatus to the *Helceph*: to Adelard of Bath[1], his Master.

1 The power of friendship has established this rule amongst those who are affected by its condition: that when one gives an order, the other is not slow to obey. **2** Ordered, therefore, by a friend – indeed by a lord and master – I hasten to attack the Saracen *Helcep* – i.e., to deal with the multiplication and division of integers, and also the multiplication of ratios (fractions) which are not investigated except through integers,[2] although not all of them are found in integers (?). **3** I have no doubt but that the summary of this (calculation) will please your intelligence, if the Lord grants the ability to explain as he has granted the ability to understand. **4** Having called on His name and help, that (task) is humbly assumed which would be dared rashly without Him, in Whom all treasures of wisdom and knowledge are hidden[3]; Who be blessed for ever, Amen.[4]

The Text.

⟨I⟩

5 The orders of numbers, then, or their boundaries proceed from the first numbers, which are called digits (*digiti*) and are nine, through multiples of ten to infinity. **6** In each order of numbers there are nine terms (*termini*) – nor can more be discovered or thought up; from this fact, I think, the Creator of all things borrowed the 'nine' as the order of celestial spirits.[5] **7** It is customary to call all (the numbers) which are in the orders other than the first (order) *articuli.*

---

1 The scribe of *P* originally described Adelard as a citizen of Bayeux ('Baiocensem').

2 For the common view that fractions were not 'numbers' see G. Molland (this volume, p. 1 ff.).

3 in whom ... are hidden] St Paul's Letter to the Colossians, 2.3: in quo sunt omnes thesauri sapientiae et scientae absconditi.

4 The language of the prologue recalls that of the dedication of Boethius's *Arithmetic*; cf. the use of the words 'ingenio', 'non ambigo' and 'pigram' on p. 5, lines 12–17 of G. Friedlein's edition (Leipzig, 1867).

5 The reference is to the common medieval view, ultimately deriving from *Pseudo-Denys the Areopagite's The Celestial Hierarchy*, that the angelic beings were arranged in the three groups of three.

264

8 Ut sit primus limes ab uno usque ad decem, secundus vero a decem per[22] decuplos primorum[23] digitorum usque ad centum, tertius vero[24] a centum usque ad mille per decuplos secundorum, et sic de ceteris. **9** Verbi gratia: prima unitas – videlicet primi[25] limitis principium – est .i., primus binarius .ii., primus ternarius .iii., primus quaternarius .iiii., primus quinarius .v., primus senarius .vi., primus septenarius .vii., primus octonarius .viii., primus novenarius .ix. **10** Ecce, assignavi primum ordinem numerorum! **11** Secunda autem[26] unitas, que est secundi limitis principium, est .x., et in eodem limite sunt ceteri numeri qui sunt primorum decupli – scilicet .xx., .xxx., .xl., .l., .lx., .lxx., .lxxx., .xc. **12** Tertia unitas .c., quarta unitas mille, quinta unitas .x.ī.,[27] sexta unitas .c.ī.,[28] septima unitas mille milia,[29] octava unitas decies mille milia, nona unitas ⟨centies mille milia, decima unitas⟩ milies mille milia. **13** Sed et reliquos numeros, quota fuerit[30] ipsa unitas, totis[31] assignabis,[32] ipsos precedentis quidem limitis decuplos. **14** Placuit igitur ad evidentiam ordines[33] predictos cum suis novenis terminis subnotare, /P 84v/ ut quotus sit unusquisque numerus locus designet, ut in secundo loco scriptus binarius pro .xx.[34] accipiatur, et sic de ceteris deorsum, **15** disposita decuplicatione[35] sinistrorsum [vero] naturali multiplicatione a prima specie multiplicitatis que est decupla usque ad octavam. /C 112/

22 perę P. *Signs after* perędecuplos *and before* primorum *indicate the correct order of the words in P.*
23 primorum] *bis P*
24 *C omits*
25 primi videlicet *C*
26 *C omits*
27 .xm. *C*
28 .cm. *C*
29 m̄m̄ *P*
30 fuerint *P*
31 totos *P*
32 assignabit *P*
33 ordinis *P*
34 binarius pro .xx.] secundus binarius, i.e. .xx. *P*
35 dispositi de decuplicatione *P*

8 So that the first order is from one to ten, but the second through the multiples-by-ten of the first numbers up to 100; but the third is from 100 to 1,000 through the multiples-by-ten of the second (numbers), and so on concerning the rest (of the orders). **9** For example: the first unity – namely the beginning of the first order – is one, the first binary is two, the first ternary is three, the first quaternary is four, the first quinary is five, the first senary is six, the first septenary is seven, the first octonary is eight, the first novenary is nine. **10** Look, I have assigned to their places the first order of integers! **11** But the second unity, which is the beginning of the second order, is 10, and in the same order are the rest of the numbers which are multiples-by-ten of the first (numbers), i.e., 20, 30, 40, 50, 60, 70, 80, 90. **12** The third unity is 100, the fourth unity is 1,000, the fifth unity is 10,000, the sixth unity is 100,000, the seventh unity is 1,000,000, the eighth unity is 10,000,000, the ninth unity (is 100,000,000, the tenth unity) is 1,000,000,000. **13** But of whatever number of (order) the unity itself is, you will assign all the other digits to the (orders) of that number – they being themselves ten times (the numbers) of the preceding order. **14** To make this clear we have decided to indicate below the aforesaid orders with their nine-fold terms, so that the position may show in what number of order each digit is, so that when a 2 is written in the second column, it should be understood as '20', and so on concerning the other digits, arranged from top to bottom; **15** while the multiplications-by-ten have been set out from right to left by natural multiplication from the first species of multiplicity which is (multiplying) by ten, up to the eighth (species).[6]

---

6 This description implies that the axes of the table should be swapped, so that the digits (1–9) are from top to bottom, and the decimal places (from 10 to 1,000,000,000; *ten* species not 'eight') from right to left. Note that the sequence from right to left is that of Arabic, not Latin, script. The table demonstrates that, e.g., the *binarius* ('.ii.') can be placed in any of the decimal columns, just as, in the abacus treatises we read that the *binarius* ('andras', ⟨*⟩) can be placed in any of the decimal columns; cf. Turchillus, ed. Narducci (n.91 above), p.136: Similiter, si andras in primo arcu statuatur, ibidem significat tantummodo duo, si in secundo, ibidem .xx., si in tercio, ibidem ducenta ... The same table, but with Arabic numerals, is entitled *Tabula abaci de opere practico numerorum* in Paris, Bibliothèque nationale, lat. 15461 (s.xiii; a MS of *Liber Alchorismi*), fol.50v, and reproduced as Fig. 9 in R. Lemay, 'The Hispanic Origin of Our Present Numeral Forms', *Viator*, 8 (1977), pp.435–62.

ix	viii	vii	vi	v	iiii	iii	ii	i	
nona-ginta	octo-ginta	septua-ginta	sexa-ginta	quinqua-ginta	quadra-ginta	tri-ginta[36]	vi-ginti[37]	decem	
ix	viii	vii	vi	v	iiii	iii	ii	i	decem
ix	viii	vii	vi	v	iiii	iii	ii	i	centum
ix	viii	vii	vi	v	iiii	iii	ii	i	mille
ix	viii	vii	vi	v	iiii	iii	ii	i	x. milia
ix	viii	vii	vi	v	iiii	iii	ii	i	c. milia
ix	viii	vii	vi	v	iiii	iii	ii	i	m̄. milia
ix	viii	vii	vi	v	iiii	iii	ii	i	x[es] m̄. milia
ix	viii	vii	vi	v	iiii	iii	ii	i	c[es] m̄. milia
ix	viii	vii	vi	v	iiii	iii	ii	i	m[es] m̄. milia

/C 113/ **16** ⟨N⟩unc dicendum est[38] quid[39] proveniat ex ductu[40] cuiuslibet terminorum primi ordinis ducti in semetipsum aut[41] ex quolibet[42] uno in quemlibet alium eiusdem limitis ducto. **17** Igitur cuiuslibet termini infra decem supra subduplum eius constituti quere differentiam quam habet ad decem et[43] eandem subtrahe ab eo quem ducis in se; ipse enim inter[44] reliquum et .x. medius erit ari⟨th⟩met-ica[45] medietate. **18** Itaque,[46] secundum regulam Nichomachi, quod ex duobus extremis in alterutrum et ex duabus differentiis invicem ductis provenit, hoc ex ipso medio ducto in se. **19** Verbi gratia: novies .ix. quot sint[47] interrogatus, respondeo octies .x.[48] et semel unum. **20** Sumo enim differentiam quam habet .ix. ad .x. – id est unum – et eam demo de .ix., et relinquuntur .viii. **21** Ecce, arithmet-ica[49] medietas .viii. .ix. .x.!

36 triaginta *C*, tringinta *P*
37 vinginti *P*
38 est dicendum *C*
39 qui *P*
40 *P corrects from* ductus
41 vel *C*
42 qualibet *P*
43 *P erases* quod *after* et
44 ipse enim inter] intra *P*
45 arismetica *C*, arimetica *P*
46 *P corrects from* iteque
47 sunt *C*
48 *C corrects from* octo, *in P* .x. *is preceded by an erasure*
49 arismetica *C*

Algorismi vel helcep decentior est diligentia   267

9	8	7	6	5	4	3	2	1	
ninety	eighty	seventy	sixty	fifty	forty	thirty	twenty	ten	
9	8	7	6	5	4	3	2	1	10
9	8	7	6	5	4	3	2	1	100
9	8	7	6	5	4	3	2	1	1,000
9	8	7	6	5	4	3	2	1	10,000
9	8	7	6	5	4	3	2	1	100,000
9	8	7	6	5	4	3	2	1	1,000,000
9	8	7	6	5	4	3	2	1	10,000,000
9	8	7	6	5	4	3	2	1	100,000,000
9	8	7	6	5	4	3	2	1	1,000,000,000

16 Now one must say what is produced by any of the terms of the first order multiplied by itself, or of any one multiplied by any other of the same order.[7] **17** Therefore for every term situated below ten, but above half of it (i.e., above 5) seek the difference between it and 10, and take that amount away from that (digit) which you are multiplying by itself; for that (digit) will be the mean between the resulting (digit) and 10, in accordance with the arithmetical mean. **18** Thus, according to the rule of Nicomachus, what results from the multiplication of the two extremes (of an arithmetical mean) by each other plus the multiplication of the two differences by each other is the (same as what results from) the middle term itself multiplied by itself. **19** For example: when asked how much are 9 times 9, I reply: '8 times 10 plus one times one'. **20** For I take the difference which 9 has to 10 – i.e., one –, and I take that from 9, and 8 are left. **21** Look, (we have) an arithmetical mean – 8, 9, 10!

7 In fact Ocreatus does not follow this plan, but discusses (1) terms of the first order multiplied by themselves (**16–24**), (2) terms of the second order multiplied by themselves (**27–42**), and (3) terms of any order multiplied by other terms of the same order (**43–53**).

22 Ergo duo extrema – id est[50] .viii. et .x. – tanto minus continentur quam[51] medium ex se[52] quantum due differentie – scilicet[53] unum et unum. **23** Simili ratione interroganti quantum est octies octo respondeo[54] sexies .x. cum bis binis. **24** At vero septies .vii. est quater .x. cum ter ternis, per eandem /P 85r/ regulam Nichomachi, quia[55] septem est medius inter .iiii. et decem ⟨et⟩ sunt differentie[56] tres et tres. **25** Eadem quoque ratione sexies sex sunt bis .x. cum quater quaternis. **26** Iam vero ex[57] solo usu scio quia quinqies .v. sunt .xxv.,[58] quater quatuor sunt .xvi., ter terni sunt[59] .ix., bis bini sunt[60] .iiii., semel unum unum est.[61]

27 Nunc descendamus ad secundum ordinem ubi per geometricam[62] medietatem perpenditur quod in primo limite per ari⟨th⟩meticam,[63] sed universalius. **28** Non enim solum per centum[64] ab ultimo usque ad subduplum[65] perpenditur quantum quisque ex se producat, sed etiam usque ad[66] .x. (vel etiam decies[67]), qui est secundi ordinis[68] principium, de quo ex usu scio quia decies .x. sunt semel .c., vel[69] ex arte quia .x.[70] proportionaliter est inter unum et centum; ergo semel centum tantumdem est quantum decies .x. **29** Similiter investigandum est quantum producat ex se quilibet terminus secundi ordinis. **30** Considero enim quomodo terminus de quo queritur se habeat ad centum, et quisnam ad eum similiter se habeat.[71]

50 *C corrects from* sunt, sunt *P*
51 quia *P*
52 *P adds* provocat
53 sunt *CP*
54 ℞ *C*
55 q; (= -que) *P*
56 differentie sunt *C*
57 *C omits*
58 scio….xxv.] sexies .v. sunt .xxx. *P*
59 *C omits*
60 *C omits*
61 est .i. *C*
62 geometriam *P*
63 arismeticam *C*, arimeticam *P*
64 Non…centum] Non solum enim *C*
65 suduplum *P*
66 *P omits*
67 vel etiam decies] *C omits*
68 *P corrects from* ordiniis
69 de quo…vel] *P omits*
70 *C omits*
71 *C²* adds between the lines: quod ex sui nominis simplicis numero ducto in se produci (?) possit (?).

22 Therefore, the multiplication of the two extremes – i.e., 8 and 10 – produces that much less than the mean (multiplied by) itself as (the multiplication of) the two differences – i.e., one and one. **23** In a similar way, to the person asking how much is 8 times 8, I reply: '6 times 10, together with twice 2'. **24** But 7 times 7 is 4 times 10 plus thrice 3, using the same rule of Nicomachus, because 7 is the mean between 4 and 10 (and) the differences are 3 and 3. **25** By the same reasoning also, 6 times 6 are twice 10 plus 4 times 4. **26** But (as for the rest of the digits multiplied by themselves) I know by rote alone that 5 times 5 are 25, 4 times 4 are 16, 3 times 3 are 9, 2 times 2 are 4, once one is one.

27 Now let us descend to the second order where one calculates through the geometrical mean what (one calculates) in the first order through the arithmetical (mean), but in a more universal way. **28** For not only through 100, from the highest integer to half (the hundred) does one calculate how much each (term multiplied) by itself produces, but also (from the highest integer) as far as 10 – or '10 times' – which is the beginning of the second order, from which (number) I know by rote that 10 times 10 are one times 100, or by the (rules of the) art that 10 is proportionally between one and 100; therefore, one times 100 is as much as 10 times 10. **29** In a similar way one must investigate how much each term of the second order produces from (being multiplied by) itself. **30** For I ask myself what is the ratio of the term in question to 100, and what (term) has a similar ratio to that (term).[8]

---

8   C^2 adds between the lines: 'which can be produced from the number of its simple name, multiplied by itself'; i.e., 20×20 can be worked out by multiplying 2×2. Compare **68**: 'sub nominibus digitorum'.

31 Qui autem sub eodem et centum continetur, contra[72] interrogationem respondeo. **32** Verbi gratia: interrogatus quantum[73] est vigies viginti, dico quadringenti[74] – qui numerus continetur sub quatuor et .c., inter quos .xx. proportionaliter continetur.[75] **33** Sed etiam[76] triginta inter .ix. et .c. **34** At vero quadraginta inter[77] .xvi. et .c. **35** Sed quinquaginta inter .xxv. et .c. **36** .Lx. vero cum sint tres quinte centenarii proportionaliter continetur inter suas ⟨.iii.⟩ partes quintas[78] que sunt .xxxvi. et .c. **37** Septuaginta vero[79] cum sint[80] .vii. decime partes centenarii, proportionaliter continetur inter suas .vii. decimas partes que sunt .xlix. et .c.[81] **38** At vero .lxxx. cum sit quatuor quinte[82] centenarii, proportionaliter continetur inter suas quatuor quintas, que sunt[83] quater .xvi. – id est .lxiiii. – et .c.[84] **39** Nonaginta vero cum sit .ix. decime in numero .c., proportionaliter continetur inter novies .ix. – id est .lxxxi. – et .c.[85] **40** Patet ergo[86] quantum ex se producat[87] quilibet terminus secundi limitis ex decimo[88] /C 114/ nono theoremate[89] septimi libri Euclidis. **41** Similis ratio est etiam in[90] ceteris ordinibus. **42** Cum[91] igitur ostensum sit[92] quantum producat quisque ductus in se, restat ostendere quantum producat quisque ductus in alium[93] sui ordinis.

72 eum ad *C*
73 *C*² *starts writing with this word.*
74 quadrugenta *C*
75 *C*² *adds between the lines:* cum sit secundus binarius, quatuor vero ex binario in se ducto producitur.
76 et *C*
77 *In P signs indicate that the order of* inter *and* quod *should be changed to this order.*
78 *P corrects from* quantas
79 *P omits*
80 sit *C*
81 .c̄. *P*
82 *P adds* que sunt
83 *P adds* centenarii...que sunt *in margin.*
84 .c̄. *P*
85 novies....c.] .ix.i. et .c. *C*
86 igitur *C*
87 producat ex se *C*
88 decimo] *the last word in the section transcribed by* C².
89 teoremate *C*, .t̄h. *P*
90 *C omits*
91 Unde *P*
92 est *P*
93 ductus in alium] in alterum ductus *P*

31 I reply to the question: 'The answer is the product of the (latter) term and 100'. **32** For example: when asked how much is 20 times 20, I say: '400' – which integer is the product of 4 and 100, between which 20 is proportionally positioned.[9] **33** But 30 is also (proportionally) between 9 and 100. **34** Yet 40 is between 16 and 100. **35** But 50 is between 25 and 100, **36** whereas 60, since they are three fifths of 100, has a proportional position between its own ⟨three⟩ fifth parts, which are 36, and 100. **37** But 70, since they are seven tenth parts of 100, has a proportional position between its own seven tenth parts, which are 49, and 100. **38** But 80, since it is four fifths of 100, has a proportional position between its own four fifths, which are 4 times 16 – i.e., 64 – and 100. **39** But 90, since it is nine tenths of the integer 100, has a proportional position between 9 times 9 – i.e., 81 – and 100. **40** How much each term of the second order produces (when multiplied) by itself is clear from the 19th Theorem of the 7th Book of Euclid. **41** The reasoning is similar, too, in the other orders. **42** Since, therefore, it has been shown how much each term multiplied by itself produces, it remains to show how much each produces when multiplied by another of its own order.

---

9   C^2 adds between the lines: 'since it is the second 2; 4 is produced by 2 multiplied by itself'.

43 Dico ergo quia omnis minor terminus cuiusque[94] limitis in maiorem eiusdem ordinis tantumdem producit quantum continetur sub ipso minore et principio sequentis ordinis, subtracto eo quod continetur sub eodem minore et differentia maioris et ipsius principii[95] sequentis ordinis. **44** Verbi gratia: septies .ix. est septies .x. septies .i. minus. **45** Similiter sexies .ix. est sexies .x. sexies uno minus. **46** Vel aliter: quisque in alium tantumdem[96] producit quantum minor[97] in se /P 85v/ et in ipsorum[98] differentiam.[99] **47** Verbi gratia: septies .ix. tantumdem est quantum[100] septies septem et septies duo.

48 Est etiam inveniri[101] numerum[102] qui ad minorem sic se habet[103] ut maior ad principem (sic enim vocamus principium sequentium limitum). **49** Erit igitur[104] ibi quod continetur sub extremis, hoc[105] et[106] sub ipsis contineri. **50** Sic est enim in omnibus .iiii. terminis[107] proportionalibus.[108] **51** Verbi gratia: quinquies sex sunt ter .x., quoniam tres ad quinque sic se habent ut sex ad decem. **52** Simili ratione decies .xxx. sunt .ccc.,[109] quoniam ut tres[110] ad .x. sic .xxx. ad .c. **53** Sed quadragies .lx. sunt vigies .c. et quater .c.[111] que sunt .īī. et .cccc., quoniam[112] sicut .xxiiii. se habet[113] ad .xl. sic .lx. ad .c. **54** Sed quia hec et his similia sunt quedam quasi anxiomata[114] et ad artem propositam eminus respicientia, accingamur[115] ad ipsam cominus[116] adgrediendam.

94 cuius *P*
95 principium *P*
96 tamtum *P*
97 *P omits*
98 et in ipsorum] *C bis*
99 differerentiam *P*
100 *P omits*
101 *P corrects from* inventuri
102 *P omits*
103 habent *P*
104 ergo *C*
105 *P corrects from* hec
106 *P omits*
107 terminis .iiii. *C*
108 Sic est…proportionalibus] *C¹ adds between lines.*
109 .ccc.^(ta) *C*
110 *C corrects* tres *to* tria.
111 vigies .c. et quater .c.] \vigies/ .cccc. *P*
112 quia *C*
113 *C omits*
114 anximata *P*
115 accingimur *P*
116 comminus *C*

43 I say then, that every smaller term of each order (when multiplied) by a larger term of the same order produces as much as the smaller term itself and the beginning of the next order, when that has been subtracted which is the product of the same smaller term and the difference between the larger term and that beginning of the next order.[10] **44** For example: 7 times 9 is 7 times 10 less 7 times one. **45** Similarly 6 times 9 is 6 times 10 less 6 times one. **46** Or by another method: each (multiplied) by another produces as much as the lesser term (multiplied) by itself plus (the term multiplied) by the difference between them. **47** For example: 7 times 9 is as much as 7 times 7 plus 7 times 2.

48 The (problem) is to find the term which has the same ratio to the smaller as the larger to the 'leader' – for in such a way we refer to the beginning(s) of the following (i. e., higher) orders. **49** Therefore it will (happen) in that situation that the product of the extremes is also the product of these terms. **50** For this is the case whenever four terms are in proportion. **51** For example: 5 times 6 are 3 times 10, since 3 is in the same ratio to 5 as 6 is to 10. **52** By a similar reasoning 10 times 30 are 300, since 3 to 10 is the same (ratio) as 30 to 100. **53** But 40 times 60 is (the same as) 200 times 100 and 4 times 100, which are 2,400, because the ratio of 24 to 40 is the same as 60 to 100. **54** But because all these and similar things are like certain axioms[11] and are looking at the art laid before us from a distance, let us gird ourselves up to attacking the (art) itself in hand-to-hand combat.

---

10 This is the 'first rule of multiplication' in Sacrobosco's *Algorismus vulgaris*, and is also found in MS British Library, Egerton 2261, fol. 226rb (see p. 305 below).

11 The phrase 'quasi axiomata' would have been known to Ocreatus from Boethius, *Music*, II.6, ed. Friedlein, p. 231: Nunc quaedam, quae quasi axiomata Graeci vocant, praemittere oportebit, quae tunc demum, quo spectare videantur, intellegemus, cum de uniuscuiusque rei demonstratione tractabimus.

⟨II⟩

55 ⟨C⟩um igitur voluerimus aliquot numeros, sive duos sive quotlibet[117] vel per se ipsos[118] vel per alios totidem aut etiam[119] per plures[120] sive per pauciores multiplicare,[121] [etiam] ipsos multiplicandos scribemus in locis diversis singulos in singulis sinistrorsum dispositis ut de quoto ordine[122] quisque fuerit, totum locum teneat. **56** Quod si non fuerint in ipsis multiplicandis numeri qui loca prima obtineant,[123] ponatur pro eis qui defuerint signum cifre,[124] ad locum vacuum designandum. **57** Ut si multiplicandi[125] fuerint .ī.cc.ti, cum eos oporteat quartum et tertium locum tenere, eo quod .ī. de quarto, .cc.ti vero de tertio ordine sint, idcirco secundum locum et primum obtinebunt due cifre,[126] hoc modo: .ī. .cc.ti[127] t.t. **58** Vel si .xxi. et mille[128] tertium locum habebit cifre, hoc modo: .ī.t.ii.i. **59** In hac igitur[129] dispositione quotum locum tenet[130] quisque, tota est unitas, vel totus binarius, et sic de ceteris.

60 Proponatur[131] igitur quod primum duo numeri multiplicandi sint[132] per semetipsos – scilicet[133].xxx. et .iii. – et scribantur in primo quidem loco primus ternarius, in secundo loco secundus ternarius[134] .iii. et .iii., deinde minimum multiplicantium sub maximo multiplicandorum, hoc modo:[135]

	iii.	iii.
iii.	iii.	

117   *C writes* alios *and then crosses the word out.*
118   *C omits*
119   *C omits*
120   plurimos *P*
121   sive per pauciores multiplicare] multiplicare vel etiam per totidem *C*
122   *C omits*
123   optineant *P*
124   cyfre *P*
125   multiplici *C*
126   cyfre *P*
127   c̄c̄ *P*
128   .xxii. mille *P*
129   ergo *C*
130   teneat *P*
131   Proponantur *CP*
132   sint multiplicandi *C*
133   *P omits*
134   in secundo...ternarius] *C omits*
135   hoc modo] sic *C*

⟨II⟩

55 When, therefore, we wish to multiply integers consisting of a
certain number of digits – whether two or as many as you like –
either by themselves or by other integers with the same number of
digits or with more or fewer digits, we will write those multiplicands
in their different columns, one per column, arranged from right to
left in such a way that from whatever number of order each digit
is, it occupies the column of the same number. **56** But if in the
multiplicands themselves there are no digits which fill the first
columns, one should put in place of those (digits) which are lacking
the sign of the zero, to show an empty column. **57** So that if 1,200
are to be multiplied, since they should occupy the fourth and third
columns, because 1,000 is from the fourth order, but 200 is from
the third order, therefore two zeros will fill the second and first
columns, in this way: 1 2 0 0. **58** Or if (the integer is) 1,021, the
zero will hold the third column, in this way: 1 0 2 1. **59** In this
arrangement, therefore, whatever number of column each digit
holds, of such a number is the unity, or of such a number is the
binary (digit), and so on in regard to the other (digits).

60 Let it be proposed then that first, two digits be multiplied by
themselves – 33 – and that the first 3 be written in the first column,
the second three in the second column – 3 and 3; then the smallest
of the multipliers (should be written) under the largest of the multiplicands, in this way:

	3	3
3	3	

61 Iuxta hanc regulam in omni multiplicatione minimus sub maximo ponendus est, sed et ceteri quotquot fuerint multiplicantes sinistrorsum disponuntur, in totis locis singuli quotorum ordinum fuerint – ita videlicet ut, quemadmodum supra dictum est, cifre[136] si opus fuerit vacuum locum designet. **62** Ducendus est ergo[137] sinistimus superioris ordinis /P 86r/ in sinistimum inferioris, et quique in quosque,[138] sed non nisi nominibus digitorum licet ipsi sint articuli, ut ter terni. **63** Si ergo digitus inde[139] excreverit, supra illum inferioris ordinis unde oritur in superiori ordine[140] ponetur. **64** Quod si articulus, ulterius, non supra illum, sed nec ibi remanebit, si articulus nascitur ex invento et apposito. **65** Iuxta hanc regulam, digitus de quo nascitur supra eum ponetur,[141] vel in quo[142] superiori nascitur ibidem remanet; si nascetur[143] articulus, scribetur[144] ulterius, ut hic.[145]

iii.	iii. iii.	iii.
	iii. iii.	
ix. ix. iii.		
iii. iii.		

/C 115/ **66** ⟨V⟩ides igitur quomodo disposui .xxxiii. que cum vellem ducere sinaphihi,[146] posui minorem sub[147] maiori – id est tria sub triginta – deinde a sinistris ipsius ternarii inferioris scripsi .xxx. **67** Et duxi .xxx. in[148] .xxx. sub nominibus digitorum faciendo[149] ter tria – id est novem.

---

136 cyfre *P*
137 igitur *C*
138 *C adds a table here:* iii iii. *over* iii. iii
139 *C omits*
140 *C omits*
141 ponitur *P*
142 qua *P*
143 nasceretur *P*
144 scriberetur *P*
145 ita *P*; hec *written above.*
146 sinaphi *P*
147 *C bis*
148 *C omits*
149 fatiendo *P*

61 According to this rule, in every multiplication, the smallest (digit) should be placed under the largest, but the other multipliers, however many they are, are placed to the left, each being in the columns of the same number as their orders – in such a way, that is, that, as has been said above, the zero, if needs be, designates an empty column. **62** Therefore the leftmost (digit) of the higher row (of digits) should be multiplied by the leftmost of the lower, and each (digit) by each, but only the names of the digits (*digiti*) are used even when the (integers) are *articuli* – e. g., (one always says) 'three threes'. **63** If, therefore, a digit (*digitus*) results from that (multiplication), it is placed in the higher row, above that (digit) of the lower row from which it originates. **64** But if an *articulus* (results, it will be placed) further over, not above it; but neither will the digit remain there if an *articulus* arises from the (digit) found (there) and what has been added to it. **65** According to this rule, the digit (*digitus*) will be placed above the (digit) from which it arises, or remains in the higher order in which it arises; if an *articulus* arises, it will be written further over, like this:

	3	3
3	3	
9	9	3
3	3	

66 You see, therefore, how I arranged 33: when I wished to multiply this by itself,[12] I placed the smaller digit under the larger – i. e., the 3 under the 30; then, to the left of that lower 3 I wrote 30. **67** And I multiplied 30 by 30, using the names of the digits (*digiti*), making 'three threes', i. e., 9.

---

12 For this transcription of an Arabic phrase 'fī nafsihi' ('by itself') or 'fī nafsihimā' (= 'by themselves [dual]') see p. 236 above.

68 Deinde[150] .ix. scripsi[151] super secundum ternarium inferioris ordinis, deinde secundum ternarium superioris ordinis[152] in primum ternarium inferioris sub nominibus digitorum, ut, secundum quod exigit ars helcep, scripsi .ix. rursus supra primum ternarium inferioris ordinis, ut patet in secunda formula. **69** In tertia vero[153] formula scripsi tria sub tribus – scilicet minimum sub maximo – id est quo non est maior ducendus cum ipse solus restet – scripsi, inquam, et a sinistris eius tria, ut tertia formula monstrat.[154]

iii.	iii. iii.	iii.	
ix. iii.	ix. iii.	iii.	
ix.	ix. iii.	iii. iii.	
ī.	t.	viii. iii.	ix. iii.

70 Cum ergo ter tria facerent novem [et .ix.] vel cum novem et .ix. facerent .xviii.,[155] reliqui digitum – id est .viii. – in sinu eius unde ortus fuerat[156] tulique cogitatione articulum – id est .x. – ulterius,[157] uti regula exigit, ad tertium locum tertie formule invenique ibi .ix. natusque[158] est articulus ex invento et addito. **71** Scripta ergo .ī.[159] unitate ulterius in[160] quarto loco, ibidem[161] in tertio loco scripsi[162] cifre.[163] **72** Rursus ter tria multiplicans[164] produxi .ix., quem[165] scripsi in primo loco quarte formule.

150 ducerem *P*
151 scribsi *P*
152 ordininis *P*
153 *C omits*; *P adds above the line*
154 *C adds two more rows above the first line of the following table; these appear to be a false start.*
155 .x. et .viii. *C*
156 ortus fuerat] est ortus *C*
157 articulum...ulterius] articulum, i.e. .x. ad novem *C*, .xx. articulum ad .ix. *P* (*perhaps a gloss* 'i.e. .x. ad .ix.' *has contaminated the texts of CP*).
158 natoque *P*
159 .s. (= scilicet) *C*
160 .i. (= id est) *C*
161 ibi idem *P*
162 in tertio loco scripsi] scripsi scilicet in loco tertio *C*
163 cyfre *P*
164 multiplicāns *P*
165 quod *P*

68 Then I wrote 9 above the second 3 of the lower row, then (I multiplied) the second 3 of the higher row by the first three of the lower, using the names of the digits (*digiti*), so that, according to the method which the art of the *Helcep* demands, I wrote 9 again above the first three of the lower row, as is shown in the second table. **69** But in the third table I wrote 3 under 3 – i.e., the smallest (digit) under the largest – i.e., that (digit) than which no larger (digit) is to be multiplied, since it alone remains; I say I wrote (this) and a 3 to the left of this, as the third table shows.

⟨First table⟩

	3	3
3	3	

⟨Second table⟩

9	9	3
3	3	

⟨Third table⟩

9	9	3
	3	3

⟨Fourth table⟩

1	0	8	9
		3	3

70 Since, therefore, three threes made 9, and since 9 plus 9 made 18, I left the digit (*digitus*) – that is 8 – in the lap of that (column) from which it had originated, and I took the *articulus* – i.e., 10 – in my head further over, as the rule demands, to the third column of the third table, and I found there a 9, and an *articulus* (i.e., 10) came into being out of the (digit) found (there) and the (digit) added (to it). **71** Having, therefore, written the thousand unity further over in the fourth column, in the same (row) in the third column I wrote a zero. **72** Again, multiplying 3 by 3 I produced 9, which I wrote in the first column of the fourth table.

73 Habes igitur quod provenit ex .xxxiii. ductis in se: .ī. t. viii. ix. **74** Quod ita esse divisione probetur. **75** Si enim cum divisero mille .lxxx. ix. per .xxxiii. exierunt[166] michi in denominationibus .xxxiii., recte multiplicatum cognoscam fuisse.[167] **76** Age ergo scribantur ut superius[168] .m̄. t. viii. ix. ponaturque, sicut[169] in omni divisione faciendum[170] est, maximus sub maximo, si inde[171] detrahi possit nominibus digitorum; alioquin ponatur dexterius sed et ceteri sub ceteris dextrorsum disponantur,[172] hoc modo:[173]

ī.	t.	viii.	ix.

/ P 86v/ **77** Nunc[174] ergo quia ternarius non poterat ab unitate prima detrahi, positus est sub cifre[175] ut a secunda unitate detrahatur. **78** Subtrahatur ergo ternarius a decem quotiens potest, ita tamen ut reliqui a reliquis totiens detrahi possint[176] – que determinatio si non hic, alibi tamen[177] erit necessaria. **79** Potest autem ter[178] detrahi et remanebit unum, quod unum quia diminutum est, non ibi remanebit. **80** Tunc enim nulla esset[179] determinatio, tantumdem remaneret quantum ibi erat. **81** Diminutum vero voco quecumque unitas relicta post detractionem; subdecupla est cuiusque erat[180] in eo loco unde detractio facta est.[181] **82** Itaque secundabis[182] reliquam unitatem in loco[183] cifre.[184] **83** Scribes autem cifre[185] in ipso primo loco iam vacuo, sed et denominationem ⟨supra⟩ cuiusque minimi divisoris affiges hoc modo:

166 exierit *P*
167 esse cognoscam *C*
168 ut superius] *P omits*
169 sic *P*
170 fatiendum *P*
171 tria *C*
172 deponantur *C*
173 *The following table is missing in P.*
174 *P expunges* in *after* Nunc
175 cyfre *P*
176 posint *P*
177 *P omits*
178 *P omits*
179 nulli esse *P*
180 erit *C*
181 facta erat detractio *C*
182 *C leaves a space for this word,* secunda .b. *P*
183 unitatem reliquam in locum *C*
184 cyfre *P*
185 cyfre *P*

73 You have, therefore, what results from 33 multiplied by itself: 1 0 8 9. **74** That it is like this may be proved by division. **75** For if, when I divide 1,089 by 33, there results for me in the quotients 33, I know that it has been multiplied correctly. **76** Well then, let 1,089 be written down, and, as has to be done in every division, let the largest (digit) be placed under the largest, if it can be substracted from it (using the) names of digits (*digiti*); otherwise it is placed to the right, but (in any case) the other (digits) are placed under the others to the right in this way:

1	0	8	9
	3	3	

77 Now, therefore, since 3 could not be subtracted from the first unity, it was placed under the zero, so that it should be subtracted from the second unity. **78** Therefore let 3 be subtracted from 10 as many times as it can be, in such a way, however, that the other (digits) can be subtracted from the other (digits) the same number of times; this result, if it is not necessary here (i. e., in this example), will nevertheless be necessary elsewhere. **79** It can be subtracted three times and one will remain. Since this one is incomplete, it will not remain there. **80** For then there would be no result, (for) the same amount would remain as was there (before). **81** But I call 'incomplete' (*diminutum*) whatever unity is left after subtraction; it is the tenth of whatever (digit) there was in that column from which the subtraction was made. **82** Thus you will move by one column the remaining unity, (and put it) in place of the zero. **83** But you will write the zero in the first column itself, being now empty; but you will also fix the quotient over the smallest divisor in this way:

282

		iii.	ix.
t.	i.	viii.	iii.

t.	ī.	iii.	ix.
	iii.	viii.	
		iii.	

84 Deinde reliquos a reliquis eadem denominatione[186] detrahes, ut hic ternarius[187] de .x. ⟨ter⟩ et reliquam unitatem quoniam diminuta est, secundabis[188] ad .viii., ut sint[189] .ix., et rursus scribes cifre, id est[190] post cifre, quia et ultimus et[191] penultimus vacui sunt, hoc modo:

		iii.	
t.	t.	viii.	ix.
		iii.	iii.

85 Deinde promovebis ipsos divisores quotquot sunt, et pones maximum sub[192] maximo et reliquos sub reliquis quemadmodum supradictum est, hoc modo:

		iii.	
t.	t.	viii.	ix.
		iii.	iii.

/C 116/ **86** Detrahes ergo secundum ternarium a secundo novenario: primum[193] a primo[194] eadem denominatione et pro illis novenis scribes .t.t. ad signanda[195] loca vacua, ponesque denominationem supra minimum divisorem, ut artis huius[196] postulat ratio, hoc modo:

t.	t.	t.	t.

---

186 denominationem *P*; *CP add* iii. iii. *above this word*
187 *P corrects from* termarius
188 *C adds* et
189 sicut *CP*
190 id est] scilicet *C*
191 *P omits*
192 *C omits*
193 primum] t. et (iii. *above* t.) *CP*
194 prima (iii. *above*) *CP*
195 significanda *C*
196 huius artis *C*

		3	
0	1	8	9
	3	3	

84 Then you will subtract the others from the others by the same quotient – i.e., here 3 is subtracted from 10 ⟨three times⟩, and the remaining unity, since it is incomplete, you will move by one column to 8, so that they become 9, and again you will write a zero, i.e., after the zero, because both the last and the penultimate (columns) are empty, in this way:

		3	
0	0	9	9
	3	3	

85 Then you will move forward the divisors themselves – however many they are – and you will place the largest under the largest and the others under the others as has been said before, in this way:

		3	
0	0	9	9
		3	3

86 Therefore you will subtract the second 3 from the second 9, and the first (3) from the first (9) by the same quotient (i.e., 3) and you will write in place of those 9s, 0 0, to indicate empty columns, and you will place the quotient above the smallest divisor, as the reasoning of this art demands, in this way:

		3	3
0	0	0	0

87 Rite ergo respondit[197] divisio multiplicationi[198] quoniam in denominationibus sunt .xxxiii. qui ducti fuerant in se ipsos ut[199] inde producentur .ī. t. viii. ix.[200]

88 Sint nobis propositi rursus .ī. cc.[201] per se multiplicandi. **89** Cum ergo duo cifre[202] primum ⟨et secundum⟩ locum obtineant, quecumque unitas tertium[203] locum tenet[204] centum est,[205] sed quaternum locum quarta tenet unitas scilicet[206] mille.[207] **90** Scribo ergo quasi minimum sub maximo -- id est primum cifre[208] sub .m̄. et[209] post hunc[210] sinistrorsum dispono .t. ii. i, hoc modo:

i.	ii.	t.	t.
i.	ii.	t.	t.[211]

91 Duco ergo primum – id est ultimum – superioris ordinis in primum inferioris ordinis[212] et secundum.[213] **92** Quoniam autem semel unum et semel duo digitos procreant, pono[214] eos qui procreati sunt[215] quasi in gremiis eorum ex quibus producti sunt, hoc modo:

i.	ii.	t.	t.	
i.	ii.	t.	t.	
ii.	t.	iii.	t.	t.
i.	ii.		t.	t.

197 respondedrit *C*

198 *P corrects from* multiplicatione

199 *P expunges* in *after* ut

200 .i.t.viii..ī.x. *C*

201 .m̄. cc. *C*

202 cyfre *P*

203 *Signs above* tertium *and* unitas *in P indicate that they should be in this order.*

204 tenens *P*

205 *C adds a table*: i.ii.tt *above* .ī.t.viii.ix.

206 *P corrects from* sed

207 unitas...mille] i.m̄. *C*

208 cyfre *P*

209 *P omits*

210 post hunc] post habet *P*

211 i.ii.t.t.ii.tt. *in place of the two lines of numerals, C*

212 *C omits*

213 secundi *P*

214 pone *P*

215 *C adds* īg

87 Therefore the division corresponds correctly to the multiplication, since in the quotients there are the 33 which had been multiplied by themselves to produce 1,089.

88 Again let 1,200 be proposed for us to multiply by itself. **89** Since then the two zeros occupy the first ⟨and second⟩ columns, whatever unity holds the third column is 100, but the fourth unity – i.e., 1,000 – holds the fourth column. **90** I write, therefore, as it were, the smallest under the largest – i.e., the first zero under the 1,000, and after this I arrange the 0, 2 and one to the left in this way:

			1	2	0	0
1	2	0	0			

91 I therefore multiply the first (digit) – that is the last – of the higher row by the first and the second of the lower row. **92** But since one times one and one times 2 gives birth to two digits (*digiti*), I place those which have been born as it were in the laps of those from which they have been produced, in this way:

1	2	0	0	2	0	0
1	2	0	0			

93 Sed quoniam in[216] inferiori[217] ordine inter duo et unum locus[218] erat vacuus, ideo ipse ego rite .t. cifre[219] po/*P* 87r/sui ad locum ⟨vacuum⟩ designandum,[220] pro quarta etiam unitate in eodem ordine posui[221] .t. i. ii. t. t.

94 Promoveo ergo terminos omnes inferioris ordinis ut in eos ducam tertium binarium et pono .t. sub ipso tertio binario et ceteros sinistrorsum dispono sic.[222]

i.	ii.	t. t.	ii.	t. t.
t.	i.	ii.	t.	t.
i.	ii.	t. t. t.	ii.	tt.
i.	ii.			tt.

95 Quoniam ergo[223] ex binario ducto in unitatem producitur binarius, pono illum[224] supra unum, quia digitus unde[225] ponitur. **96** Cum autem fuerint ibi prius[226] duo scribo alia duo. **97** Rursus duo duco in duo, et digitum – scilicet quatuor – qui nascitur pono supra eum ex quo nascitur, sublato cifre[227] quia[228] vacuus est locus, ut monstrat subiecta[229] descriptio.

i.	iiii.	iiii.	t.t.t.t.
i.	ii.	t.	t.
i.	iiii.	iiii.	t.t.t.t.
i.	ii.	t.	t.

98 Et hec multiplicatio divisione examinanda est.

216 *P omits*
217 inferiore *C*
218 *P corrects from* locum
219 cyfre *P*
220 ipse ego…designandum] rite .t. posui ad locum cifre designandum *C*
221 pono *C*
222 *C adds*: i.ii.t.t.ii.t.t
223 *C omits*
224 unum *P*
225 unum *P*
226 plus *P*
227 ⟨cifre⟩ *C*
228 qui *C*
229 subieta *C*, subiepta *P*

93 But since in the lower row, between (?) two and one there was an empty column, for that reason I have duly placed a 0 – zero – to indicate the ⟨empty⟩ column; for the fourth unity in the same row I wrote

0	1	2	0	0

94 Therefore, I move forward all the terms of the lower row in order to multiply the third 2^{13} by them, and I place 0 under the third 2 itself, and arrange the rest to the left in this way:

1	2	0	0	2	0	0
0	1	2	0	0		

95 Since, then, 2 is produced from 2 multiplied by one, I put it above one, because (that one) is the digit (*digitus*) from which it is taken. **96** But since there were 2 there before, I write another 2. **97** Again I multiply 2 by 2, and the digit (*digitus*) which arises – i.e., 4 – I place above that digit from which it arises, having taken away the zero (placed there) because the column was empty, as the description below shows:

1	4	4	0	0	0	0
	1	2	0	0		

98 And this multiplication should be tested by division.

13 This is how the author refers to a 2 in the third order (decimal place). Similarly, in **99**, he refers to the 'fourth one' (i.e., one in the fourth order) and the 'sixth 4' (i.e., 4 in the sixth order), etc.

99 Quoniam ergo[230] in omni divisione maximus supponendus est maximo si inde detrahi possit, et ego sic facio, et ceteros dextrorsum dispono – videlicet quartam unitatem sub septima unitate, et tertium binarium sub sexto quaternario, deinde duo cifre,[231] qui si deessent, nec tertium posuissem binarium, immo[232] primum, nec quartam unitatem, immo[233] secundam. **100** Sic ergo dispositis dividendis[234] et divisoribus,[235] detraho a maximo[236] maximum quotiens possum – videlicet[237] semel – et nichil remanet nisi quod ponitur ob signandum[238] cifre,[239] sed et reliquos a reliquis eadem denominatione – id est semel – detraho, videlicet duo de quatuor et relinquuntur .ii. **101** Deinde denominatio[240] – id est unum – supra minimi divisoris – videlicet[241] illius cifre[242] qui obtinet locum primum in ordine divisorum, hoc modo:

t.	ii.	iiii.	i.	tttt.,
i.	ii.			tt.[243]

102 Deinde omnes divisores uno gradu dextrorsum promoveo[244] et pono[245] hoc modo:

t.	ii.	iiii.	tttt.
i.	ii.		tt.

230 *C omits*
231 tifre *P*
232 uno *C*
233 uno *C*
234 videndis *C*
235 divisionibus *C*
236 maxima *C*
237 scilicet *C*
238 *P adds* ponitur
239 quod…cifre] quod ob signum cifre ponitur *C*
240 denominatione *CP*
241 divisoris – videlicet] divisoribus inde *P*
242 cifre *C*, cyfre *P*
243 tt.] tttt. *C*
244 promomoveo *C*
245 et pono] *C omits*

99 Since, then, in every division, the largest should be placed under the largest if it can be subtracted from it, I too do this, and I arrange the rest (of the digits) to the right – i.e., the fourth one under the seventh one, the third 2 under the sixth 4,[14] then two zeros; for, if they had been lacking, I could not have placed the 2 third, but rather (it would have been) first, nor (could I have placed) the one fourth, but rather (it would have been) second.

(1	4	4	0	0	0	0
1	2	0	0		)	

100 Having, therefore, arranged the dividends and the divisors in this way, I subtract the largest from the largest, as many times as I can – i.e., once, and nothing remains (in the first column) except what is placed (there) to indicate a zero, but also I subtract the others from the others by the same quotient – i.e., once – that is 2 from 4, and two are left. **101** Then (I place) the quotient – i.e., one – above the smallest of the divisors – that is that zero which occupies the first column in the row of the divisors, in this way:

			1			
0	2	4	0	0	0	0
1	2	0	0			

102 Then I move all the divisors one step to the right, and place them in this way:

			1			
0	2	4	0	0	0	0
	1	2	0	0		

14 See previous note.

103 Detrahatur ergo .i.[246] ab .ii. quotiens potest – scilicet bis – sed et duo de .iiii. similiter et nichil remanet[247] nisi .t.t. que pro locis designandis vacuis scribuntur,[248] sed etiam denominatio[249] – id est .ii. – supra minimum divisorem – scilicet supra[250] primum .t.[251] ordinis divisorum[252] – ponatur, videlicet iuxta predictam denominationem que erat .i. **104** Deinde compleantur .ii. t. in[253] ordine denominationum, hoc modo:[254]

i.	ii.	t.	t.		
t.	t.	t.	t.	t.	t.
i.	ii.	t.	t.		

/C 117/ **105** Patet ergo ex denominationibus recte multiplicatum[255] fuisse.

106 Breviter ergo colligende sunt regule huius artis: in omni multiplicatione minimus sub maximo et ceteri sinistrorsum disponendi sunt; digitus unde nascitur per multiplicationem supra eum ponitur, articulus ulterius. **107** Ut si ducantur[256] isti duo invicem,[257] fient .iiii. ut patet in formula supposita.[258] **108** Vel ubi nascitur per coacervationem, ibi remanet; si nascitur articulus transferatur ulterius; locum vacuum[259] designet .t.[260]

109 In omni divisione maximus sub maximo ponendus est ceterique[261] ponantur dextrorsum; maximus a[262] maximo[263] detrahatur quotiens poterit, ita /P 87v/ tamen ut reliqui a reliquis totiens detrahi possint.

246 idem *P*
247 remaneat *C*
248 .t.t....scribuntur] que scribentur pro locis vacuis designandis *C, P adds* notandis
249 denominato *P*
250 super *C*
251 |t.| *C*
252 divisionum *CP*
253 .ii. t. in] .ii. |cyfre| in *C*
254 *P gives an empty table.*
255 *C corrects from* multiplicationibus
256 ducant *C*
257 in unum *CP*
258 suppositi *followed by a lacuna of circa 10 letters in P.*
259 *C adds* .cifre.
260 *C omits*
261 ceteri *C*
262 *P corrects from* sub
263 minimo *C*

103 Therefore, one is subtracted from 2 as many times as it can be – i. e., twice – but 2 also is similarly (subtracted) from 4, and nothing remains except 0 0, which are written for indicating empty columns, but also the quotient – i. e., 2 – is placed above the smallest divisor – i. e., above the first 0 of the row of divisors, i. e., next to the aforesaid quotient which was one. **104** Then the two 0s should be filled in, in the row of quotients, in this way:

| 1 | 2 | 0 | 0 |

105 It is clear then from the quotients that it was multiplied correctly.

106 Briefly, therefore, the rules of this art should be brought together: in every multiplication the smallest (should be placed) under the largest and the other (digits) should be arranged to the left; the digit (digitus) is placed above the (digit) from which it arises through multiplication, the *articulus* is placed further over.[15] **107** So that if these 2s are multiplied by each other, they will become 4, as is clear from the table below.

| ⟨no table⟩ |

108 Or, when a digit arises through addition,[16] it remains there; if an *articulus* arises, it should be transferred further over; a 0 should indicate the empty column.

109 In every division the largest should be placed under the largest, and the others should be placed to the right; the largest should be subtracted from the largest as many times as it can be, in such a way, however, that the other (digits) can be subtracted from the others the same number of times.

15 I. e., one column to the left.
16 I. e., addition of the product and the digit already in the position immediately above the multiplier.

110 Si quid in dividendo diminutum fuerit,[264] secundetur et dextrorsum[265] in locum anteriorem transferatur. **111** Denominatio[266] supra minimum divisorem scribatur in tertio ordine. **112** Collige autem pro minimo divisore multotiens cifre[267] .t.[268] **113** Hoc igitur quod dictum est, ad multiplicandum et dividendum per integros patet posse sufficere.

⟨III⟩

114 ⟨N⟩unc de proportionandis minutiis dicendum est. **115** Si ergo completa[269] detractione ad modum supradictum[270] adhuc reliquum erit[271] aliquid de numero[272] dividendorum, fueritque illud reliquum minus toto numero divisorum,[273] proportionandum hoc illi est.[274] **116** Quoniam enim duobus numeris inequalibus ad invicem comparatis,[275] necesse est[276] ut minor maioris sit aut[277] pars aut partes,[278] si ad eum fuerit comparatus, videndum est reliquum illud quota pars aut quot[279] - id est quote - partes sit[280] totius numeri divisorum. **117** Dico itaque quia quantum fuerit hoc reliquum illius numeri, tota pars aut tot aut tote partes unius[281] contingit[282] singulos divisores de illo reliquo preter iam acceptam summam denominationum.[283]

264 fecerit *C*
265 dexterius *C*
266 denominato *P*
267 cyfre *P*
268 Collige...cifre .t.] *C omits, but leaves a space wide enough to accommodate this sentence.*
269 complecta *P*
270 *P adds* est
271 reliquum erit] relinquitur et fuerit *P*
272 de numero] denuo *P*
273 numero divisorum] minus divisore *P*
274 hoc illi est] est illi *C*
275 et paratis *P*
276 *C omits*
277 *C omits*
278 *C corrects from* pars
279 aut quot] *P omits*
280 si *P*
281 tota...unius] *P omits*
282 contigit *P*
283 denomiinationum *P*

110 But if there is anything incomplete in the division, it should be moved one column and transferred to the column in front to the right. **111** The quotient should be written above the smallest divisor in a third row. **112** On many occasions one should understand zero – 0 – as the smallest divisor. **113** It is clear, therefore, that what has been said can suffice for multiplying and dividing with integers.

⟨III⟩

114 Now one must speak of proportioning fractions. **115** If, therefore, when the subtracting has been completed in the aforesaid manner, there still remains some part of the number of the dividends, and that remainder is less than the whole number of the divisors, this should be made into a fraction of that (number). **116** For since, when two inequal integers are compared to each other, it is necessary that the smaller should be either one part or (several) parts of the larger, if it is compared to it, one must see what part or how many – i.e., what number of – parts of the whole number of divisors that remainder is. **117** I say then that however great (a part) of that number this remainder is, each divisor happens to receive that number of part or so many or such a number of parts of the remainder in addition to the sum of the quotients already received.

118 Verbi gratia: si .c.xl.viii. in .xvi.[284] dividere voluerimus,[285] scribemus supradicto modo[286] numerum[287] divisorum. **119** Si ergo, ut ars postulat dividendi, maximum sub maximo ponamus, ut quotiens[288] hunc[289] ab illo detraximus[290] – id est unum ab uno – totiens reliquum a reliquo – id est .vi. a quatuor;[291] quod quia non nisi semel facere poterimus, non poterimus procedere,[292] quoniam maius a minore non potest detrahi.[293] **120** Ideo ponemus[294] maiorem divisorem – id est .x. – sub quatuor, et minorem – id est .vi. – sub .viii.,[295]

iii.	xvi.	
i.	iiii.	viii.
i.	vi.	

121 Detra⟨hi⟩tur ergo unum a .x. quotiens potest[296] – ita tamen ut reliqui[297] a reliquis totiens detra⟨h⟩i possint – id est .vi. a .xl. et a reliquo primi.[298] **122** Detraho ergo[299] unum a decem novies – videlicet[300] quotiens poterit – et[301] relinquitur unum; quod quia[302] diminutum est, transferetur[303] ad[304] .iiii. ut sit in secundo loco .v., et[305] nichil in tertio nisi cifre.[306]

284 in .xvi.] in .xiii. *C, P omits*
285 voluerimus dividere *C*
286 *C omits*
287 numeri *P*
288 *C² begins writing with this word.*
289 *Marks have been placed over* hunc *and* ut quotiens *in P indicating that they should be in this order.*
290 *P corrects from* detracsimus
291 totiens...quatuor] *P omits,* ...relicum a relico...*C*
292 quia...procedere] quidem nisi poterimus semel facere non poteimus [*sic!*] producere *P, C puts* quod quia non...poterimus *before* totiens.
293 detrai *C, P corrects from* detraho
294 ponamus *C*
295 *P omits table.*
296 Detrahitur...potest] *P omits*
297 *P adds*: id est .vi.
298 a reliquis...primi] totiens detrahi a reliquis possent *P*
299 *C omits*
300 unum *P*
301 videlicet...et] *C omits*
302 quod quia] et quod *C*
303 transferatur *C*
304 a *CP*
305 sit....v., et] sint in loco secundo .vi. *C*
306 cyfre *P*

118 For example: if we wish to divide 148 by 16, we shall write the number of the divisors in the above-mentioned way. **119** Therefore if, as the art of dividing demands, we place the largest (of the divisors) under the largest (of the dividends), so that, however many times we can subtract this from that – i.e., one from one – so many times (we subtract) the other (digit) from the other – i.e., 6 from 4. Since we can only do this once, we cannot proceed, since a larger (amount) cannot be subtracted from a smaller (amount). **120** For that reason we shall place the larger divisor – i.e., the 10 – under the 4, and the smaller – i.e., the 6 – under the 8. **121** Therefore, one is subtracted from 10 as many times as it can be, in such a way, however, that the others can be subtracted from the others the same number of times – i.e., the 6 from the 40 and from the remainder of the first (number).

		9
1	4	8
	1	6

122 Therefore, I subtract one from 10 nine times – i.e., as many times as is possible – and one is left and, since it is incomplete it should be transferred to the 4, so that 5 is in the second column, and there is nothing in the third column except zero.

123 Pono igitur[307] denominationem – id est .ix.[308] – supra minimum divisorem – id est supra[309] .vi. – et secundum eandem denominationem aufero[310] sex de .lviii.[311] – id est novies.[312]

i.	iiii. .i.	viii. vi.
i.	iii⟨i⟩. i.	vi. .vi.
t.	v.	ix.

124 Relinquuntur[313] autem quatuor, qui numerus[314] – id[315] est quatuor – cum minor sit[316] numero divisorum, proportionandus est illi. **125** Invenitur autem esse quarta pars. **126** Dico ergo quod[317] unusquisque divisor accipiat[318] de numero dividendorum .ix. et quadrantem.

127 Omnis numerus tantumdem producit ex se quantum eius utraque pars et altera in alteram bis ducta. **128** Verbi gratia: si queratur quot sint[319] trigies quinquies .xxx. v.[320] respondeatur[321] trigies triginta et quinquies .v. et quinquies triginta bis vel trigies[322] quinque bis.

307 Pono igitur] Quo pono *P*
308 id est .ix.] *C omits*
309 *C omits*
310 aufer quinquies *P*
311 .lxviii. *P*
312 *P omits table.*
313 Relinquitur *C*, Reqlinquntur *P*
314 *P corrects from* mundus
315 *P omits*
316 minor sit] maior sit *C*, minor *P*
317 quia *C*
318 accipit *C*
319 sunt *P*
320 v.] cc. et *C*, cc. *P*
321 *P adds* per
322 trigies] trigesi *C*, trigies quinquies *P*

123 Therefore I place the quotient – i.e., 9 – above the smallest divisor – i.e., above 6 – and, using the same quotient, I take 6 from 58 – i.e., 9 times.

		9
0	5	8
	1	6

124 Four are left, which digit – i.e., 4 – since it is less than the number of the divisors, must be made a fraction of that (integer). **125** But it is discovered to be a fourth part. **126** I say, therefore, that each divisor should receive from the number of the dividends nine and a quarter.

127 Every integer (multiplied) by itself produces as much from itself as each part and one of the two parts multiplied by the other twice. **128** For example: if the question is 'How many are 35 times 35?', the reply should be: '30 times 30 and 5 times 5 and 5 times 30 twice or 30 times 5 twice'.

Appendix II.
The *Anxiomata Artis Aritmetice.*

THE MANUSCRIPT.

Cambridge, Trinity College, MS R.15.16 has been described in some detail by M. R. James, *The Western Manuscripts in the Library of Trinity College, Cambridge* (Cambridge, 1901), II, pp. 354–5 (no. 940).[1] The following notes supplement and bring up to date James's information.

Except for some later marks of ownership and scribbles, the MS is written entirely in two hands: hand 'A' appears to be English of the mid-twelfth century, hand 'B' may be French, of the same period. The manuscript is a quarto volume consisting of seven quaternions to which a bifolium has been added at the beginning and at the end.

1. Fol. Ar. Blank.

2. Fol. Av.[2] Two lines of hexameters giving the names of the abacus numerals, with the equivalent Arabic and Roman numerals written below the names:

> Primus. Igin. Andras. Ormis. quarto subit Arbas
> Post Quimas. Iermas. Zenis. Zenoma. Zelentis.

To the left of these verses are (1) different names for zero, and variant forms for writing zero (see p. 251 above), and (2) brief arithmetical instructions: locus limitem ponit. putetur. ñ. unitas imparis; sullevetur denarius (complete; meaning unclear).

To the right of the verses are the only instructions for adding and subtracting in the MS: In addendo, a prima (*supra*: ab ultima), id est dextra, cetera ut in multiplicatione, de sedibus scilicet natorum. In diminutione, quod non potest a prima, fiat vel ab ultim(a) sed retracto (complete). The whole of fol. Av is in hand 'A'.

3. Fol. 1r. A chess board, eight squares by eight, chequered with squares marked with red circles.

---

1 This MS was used for its numeral forms by G. F. Hill, *The Development of Arabic numerals in Europe* (Oxford, 1915), pp. 30–31 (Table III, example 1).

2 See Plate VIII.

4. Fols 1v–3r. The 'Coventry Introduction to Arithmetic': Arcium liberalium doctrina ... ut Alardus, Iohannes, Willelmus. Musicam (text ends abruptly).

Edited Burnett, 'Innovations in the Classification of the Sciences in the Twelfth Century', in *Knowledge and the Sciences in Medieval Philosophy. The Proceedings of the Eighth International Congress of Medieval Philosophy (S.I.E.P.M.)*, ed. S. Knuuttila et al. (Helsinki, 1990), II, pp. 25–42. Hand 'A'.

5. Fol. 3v. Marks of ownership, which may be listed from the most ancient to the most recent:

i) cas^s fratris radulfi de lychfeld et ysay. ⟨h⟩osato de cabi ... (partially erased and barely readable under ultraviolet light). ii) another sign of ownership originally written over (i) and now entirely erased. iii) Arismetica boecii de communitate fratrum minorum coventr^s et registratus sicut Boecius .a.. The degrees of numbers in English in a s.xvi hand.

6. Fols. 4r–59v. Boethius, *Arithmetica*. Hand 'B' with notes in hand 'A', and possibly one other hand.

7. Fol. 59v. Secunda pars artis que est practica eiusdem secundum Grecos, Arabes et Indos: Artis numerandi due sunt partes ... See pp. 249–50 above. Hand 'A'.

8. Fol. 60r. Rhythmomachy board.

9. Fol. 60v–61r. *Anxiomata Artis Aritmetice*: Omnis numerus aut par aut impar ... fiunt xii.ii.iiii.vi. This text has been written in the wrong order, with the last section ('Nota v. species inequalitatis ... fiunt xii.ii.iiii.vi.') coming before the beginning.[3] Edited below. Hand 'A'.

10. Fol. 61v–62r. Rhythmomachy: Fit tabula ad longitudinem et latitudinem distincta ... armonici .cc.lxxxix. This is an elaborated version of the rhythmomachy of Odo of Tournai edited by A. Borst in *Das mittelalterliche Zahlenkampfspiel* (Heidelberg, 1986), pp. 344–55.[4]

This is followed on fol. 62r by a paragraph on means, misleadingly entitled 'De dimidiatione et decimatione': Omnis numerus circum se naturaliter locatorum ... Per hec enim scito huiusmodi invenire duo media.

Instructions for rhythmomachy resume after a line drawn across the page: Ratio vero positionis calculorum secundum .x. inequalitates arismetice et tres medietates in eodem explicitas, que maximam

---

3 A custos at the bottom of fol. 60v ('per x. per dimidium') appears to relate to the heading at the top of fol. 62v (q.v.).

4 A custos at the bottom of fol. 61v ('Raro capiuntur⸴ per') does not correspond to the beginning of fol. 62r.

III

300

constituunt armoniam, sic se habet. Quatuor ad duo duplus est ... et .ccclxi. ad .cxc. supernoniparciens. This is a version of the text edited in Borst, *ibid.*, pp. 356–7.

This is followed by a description of the arithmetical, geometrical and harmonic means and their application to the playing of rhythmomachy: Tres autem medietates sive proportionalitates sic perpendentur ... et sic habet noticia de armonia in campis adversarii statuenda. Not edited.

Brief rules for rhythmomachy: Si quilibet numerus adverse partis sic compositus fuerit ut omnis ei rectus tractus denegetur, tollatur. Quod si unum sue partis in suo tractu stantem sit, liber dimittatur (complete). Edited in Borst, *ibid.*, p. 371. The whole of item 10 is in hand 'A'.

11. Fol. 62v. Arithmetical games: i) Cogita, triplica, divide, iteraque. R(esponsio) ad eius practicam. Si partes fuerint equales ... sed si per divisionem primum advertisti imparem, unum adde binariis. ii) Sumptis tribus ternis qualibet eadem tamen differentia distantibus, medius est summa ter ductus appositis ... impares suspecti. Not edited. The whole of fol. 62v is in hand 'A'.

A s.xvi hand has added lower down the page: good mr comberforde parson of ȝelvertofte taught me the arte of numeration. The condition of this obligacion is suche (complete).

12. Fol. 63r. Calculations involving marks, *denarii* and *libre*: Demptis .c.6.o. sunt .xv. m(arche). Iterum .mm.c.6.o. quad(rantes) .iiii. mar(che) den(arii) .xvi. m(arche) ... An isolated remark, cut off by edge of page: nescio quia nichil ⟨est⟩. Not edited. The whole of fol. 63r is in hand 'A'.

EDITION

Incipiunt anxiomata artis aritmetice[5]

Omnis numerus aut par[6] aut impar,[7] ut .viii. et .iii[i]. Omnis etiam numerus aut simplex[8] vel incompositus, ut .iii. et .xi.xii., aut mixtus vel compositus, ut .ciii. et .xxxiii. Item: omnis numerus aut est primi

---

5 The editorial principles followed here are the same as those used in Appendix I. In the MS the headings are incorporated into the text. Bold numbers indicate the numbers of those sentences of H. Ocreatus's *Helcep Sarracenicum* which correspond to sentences of this text.
6 *above*: perfectus
7 *above*: habundans
8 *above*: ut .i.x.c.

limitis vel ordinis, vel secundi, vel deinceps. Est autem primus limes[9] ab uno usque .x., qui digiti dicuntur. Secundus limes[10] a .x. usque .c. et sic[11] deinceps decuplando,[12] qui omnes dicuntur articuli. Singulorum autem limitum est proprialis prima unitas quam principem vel principium dicimus, ut in primo limite .i. et in secundo .x., in tercio .c. et sic deinceps. Item: omnis numerus ei quem non numerat est primus, ut .v. ad .ix. Omnis etiam numerus aut primus aut a primo.[13] Item: numerorum parium vel imparium pariter vel impariter quidam sunt superficiales, qui duobus numeris continentur, quidam trigoni – id est trianguli – qui equali et \inequali/ flectitur, ut ter quater, vel quater ter, quidam tetragoni – id est quadrati – qui equalibus numeris in seipsis producitur, ut quater quater.[14] Quidam cubici – id est solidi – qui ex ductis numeris bis in seipsis ducitur et tribus continetur numeris, ut bis bini bis, vel ter terni ter, et omnes hi habent latera proporci(o)nalia.[15] Ex cubo autem vel ex equali nonnisi cubus vel equalis producitur.

Numerus in alium ductus est is qui multiplicatur; in quem ducitur est multiplicans, scilicet is qui multiplicat. Multiplicare est duci numerum in alium, quod nichil aliud est quam eum qui ducitur – id est qui multiplicatur – tociens numerare illum quem procreat, quociens est unitas in ipso in quem ducitur – id est qui multiplicat.[16] Dividere est partem vel partes convenienter distributas[17] cuilibet

9   *above*: singulares
10 *above*: deceni
11 *above*: centeni
12 *above*: milleñ
13 Compare Euclid, *Elements*, VII, def.viii, Version I: numerus primus est ille qui unitate sola numeratur; Version II, MS *B*, fol.19r: numerus primus dicitur qui sola unitate numeratur.
14 Compare *ibid.* VII, def.xiii, Version I: Numerus quadratus est qui ex ductu numeri in se ipsum producitur quem duo numeri equales continent; Version II, MS *B*, fol.19v: Numerus quadratus dicitur qui ex ductu numeri in se ipsum producitur eumque duo numeri equales continent.
15 Compare *ibid.* VII, def.xiv, Version I: Numerus cubicus est qui ex ductu numeri bis in se ipsum producitur a tribus numeris equalibus contentus; Version II, MS *B*, fol.19v: Numerus cubicus dicitur qui ex ductu ... (as in Version I).
16 Compare *ibid.* VII def.xii, Version I: Numerus ductus in alium numerum est qui totiens ducitur quotiens in ducente unitas; Version II: Numerus ductus in alium dicitur quociens eum multiplicat quociens in se est unitas, id est multiplicante qui ductus est.
17 partem ... distributas] *above*: ut in ea que per caracteres – id est figuras – est scientia.

302

divisori impertiri. Divisor dicitur numerus per quem dividimus – scilicet qui dividit; dividendus qui dividitur. Denominatio dicitur ille numerus qui exit in divisione. Termini, id est summe. Differentias vocamus quantitates numerorum, aliquando autem locum vel figuram eorum. Denominatus numerus est qui ab alio nomen sumpsit, ut \a/ .iii. xxx. Sunt denominationes que adverbialiter dicuntur, ut bis, ter a duobus et tribus. Maior et minor extremitas – id est fines numerorum.

In imparibus resolutoriis.

Imparium numerorum quidam superficiales ut supra, quidam lineares, qui unitate naturaliter crescunt, quidam trianguli, qui tercio superveniente crescunt, quidam primi et incompositi, et quantum ad alios quos non numerant et quantum ad se qui nullo numerantur, ut .iii.[18] ad .xi., quidam secundi et compositi, ut .x.xv., quidam compositi sed contra se primi, quidam longilateres, cum unum latus aliud unitate vincit. Omnis numerus sit[19] perfectus, qui constat partibus suis conglomeratis, ut .vi. et .xxviii., quidam his superflui, quidam diminuti, quidam communes invicem, quidam proporcionales.[20]

/fol. 61r/(**17**) Si vis multiplicare aliquem numerum infra .x. in se, subtrahe differentiam quam habet ad .x. et quod remanet duc in .x. et differentiam in se, (**23**) ut octies .viii. ablata[21] differentia ad .x. – scilicet duo – et ductis extremis in alterutrum, sunt sexies .x. et bis duo. [Aliter: (cf. **43–45**) Ab articulo secundi limitis quem multiplicator denominat auferatur id quod ex ductu differentie multiplicati in ipsum provenit multiplicatorem et quod relinquitur, id est quod ex multiplicatione digitorum procreatur, ut octies .ix. ab .lxxx.ª tolle semel octo et patet. Omnis etiam numerus per se multiplicatus resolvitur in illos a se toto distantes quoto ipsi distant a singularibus – id est digitis – ut .x. in .c.][22] (**43**) Si autem minor in maiorem eiusdem ordinis ducitur, tantum producit quantum ductus in prin-

---

18 *corrected from* .ix.
19 alit (?)
20 Custos in lower margin: per .x. per dimidium. This must refer to the heading in the upper margin of fol. 62r: de dimidiatione et decimatione. 'Division by two' is a chapter in all the early algorisms; 'division by ten' does not seem to occur elsewhere. The text which follows this heading, however, is on neither of these subjects. The arithmetical notes have been written in a random order.
21 abbata
22 Ab articulo ... in .c.] *an addition at the bottom of fol. 61r signalled by the words in the margin after* bis duo: Aliter 3

cipem proximum, id est principium sequentis limitis, sublato minore per differentiam quam habet maior ad predictum principem. (44) Ut septies .ix.[23] sunt septies .x. semel septem[24] minus, et septies .viii.° sunt septies .x. bis .vii. minus. Hec etiam R(esponsio) ceteris convenit ordinibus, et apercior est in secundo et tercio, propter decenarium per quem se metiuntur, ut octogesies nonageni, octogesies .c. – id est .$\overline{viii}$. – semel octogies minus.

(46) Alia R(esponsio) omni conveniens limiti: omnis numerus ductus in aliquem sui ordinis tantum efficit quantum ductus in se et in eorum inter se differentiam, (47) ut septies .viiii. sunt .vii.^{es} .vii. et bis .vii. Duobus enim differt .vii. a .ix. Sed hec R(esponsio) insufficiens est; alia enim indiget. (cf. 17) Secundum aritmeticam medietatem que numerum de quo queritur medium facit hec diximus. (cf. 27) Dicamus secundum geometricam medietatem que numerum ad numerum sui ordinis proporcionatur, vel limitem ad limitem, et sit medius[25] cui contigerit semel et semper propositum geometricale.

(48) Est inveniri quis numerus ad minorem sic se habet ut maior ad principem (is est principium sequentis limitis, et medium limitum). Disponantur proportionaliter – id est minor cum minore, maior cum principe[m]. (49) Erit ergo ibi quod continetur sub extremis, hoc et sub ipsis mediis continetur. (50) Sic enim est in omnibus .iiii. terminis proportion(al)ibus. (51) Verbi gratia: quinquies .vi. sunt ter .x., quoniam .iii. ad .v. ut sex ad .x. Probatio: proposuisti quinquies .vi. et ego apposui principem ulteriorem, scilicet .x. Considerata ergo proportione maioris numeri ad principem, eandem propone inter maiorem[26] et minimum. Est igitur inveniri qui minor ad minorem sic se habeat ut maior ad principem. Porisma .iii. v. vi. x. Eadem est proporcio inter .vi. et .x. qui eum totum continet et eius duas partes, que est inter .iii. et .v. Similiter .v.^{es} viii. sunt quater .x. et .viii. .viii. sunt .vi^{es} .x. et .iiii. quia .x. est ad .viii. sesquiquarta, et non est qui ita se habeat ad .viii. sub eo. Ideo quatuor ausisti.[27] In maioribus autem melius patet; (52) ut .x.^{es} .xxx.^{a} sunt trecenta, quoniam .iii. ad .x. ut .xxx. ad .c. Sic .xxx. .xl. sunt .$\overline{m}$. et .cc. quia sicut .c. continet .xl. bis et eius mediam, sic .xxx.^{es28} numerat .xii.[29] bis et eius mediam, scilicet .vi. (53) Sic .xl. .lx. xxiiii.c. – id est .$\overline{ii}$.

---

23 *corrected from* .vii.
24 *corrected from* septies
25 *read*: melius ?
26 *read*: minorem ?
27 *i. e.*, auxisti
28 *i. e.*, triginta
29 .xxi.

et .cccc. – quia .c. semel habet .lx. et eius duas partes, et .xl. semel habet .xx^{es}iiii.^{ter30} et eius duas partes, scilicet .xvi.

Adhuc de articulis. **(28)** Si in se multiplicatur articulus, per geometricam medietatem sic habebis: .x.^{es} decem sunt semel .c., quia .x. est inter .i. et .c. proportionaliter – id est eandem proportionem habet unitas[31] ad denarium quam denarius ad centenarium, quia decimam eius partem. **(30)** Considerandum ergo quomodo terminus de quo queritur se habeat ad principem ulteriorem, et quisnam ad terminum similiter se habeat et tunc conduc extremitates. **(32)** Ut vigies .xx. sunt quater .c. – id est quadringenta, quia inter eos continetur proporcionaliter; quinta enim pars centenarii .xx. ut .iiii. vigenarii. **(36)** Et sexagies .lx. sunt .xxx.^{es}vi.^{es} c. – id est .$\overline{\text{iii}}$. et .dc.; continent enim se invicem et duas sui partes.

Idem disciplinalius.

Quoto distat differentia in primo limite[32] a principe ulteriori, toto[33] in quolibet limite eadem differentia ab ulteriori principe, ut trigesies .xxx. sunt .dcccc.; sicut enim tres in .x. sic .xxx. in .c. Et quinquagies .l. sunt quinquies quingenta. Eadem enim est differentia .v. in primo limite que et .l. in secundo. Sic centies .c. sunt .$\overline{\text{x}}$. et ducenties ducenta .$\overline{\text{xl}}$.[34] et deinceps.

Item de articulis.

Omnis articulus tantum efficit in se quantum eius simplex in se et in principem ultimum[35] ut .xxx.xxx. cuius simplex ternarius faciet ter ternos – id est .ix. – et hunc iterum duc in principem, et habes novies .c. Hec vero de predicta est extracta.

De mixto et simplici.

Si mixtus sit multiplicandus per simplicem, multiplica mixti utramque partem per simplicem, ut ter duodecim sunt ter .x. et ter duo – id est .xxxvi. – et ter .xxv. sunt ter .xx. et ter .v. – id est .lxxv.

De mixtis reciprocis.

Secundum primam regulam digitorum, pone principem omnem denarium, et fac aritmeticam medietatem, et subtrahe differentiam et reliquum duc in principem. Verbi gratia: .xvii. xvii. sunt .xiiii.xx.

---

30 *i.e.*, vigies quater
31 *corrected from* p ...
32 *corrected from* limita
33 *corrected from* tanto
34 .$\overline{\text{xx}}$.
35 *corrected from* medium

et ter terni – fuit hec enim differentia ablata – cuius summa est
.cc.lxxxix. Vel aliter ad secundum regulam:[36] duc in principem ulteriorem et subtrahe ductum[37] per differentiam. (**127**) Item: Omnis
mixtus numerus ductus in se tantum efficit quantum utraque pars
sui in se et altera in alteram bis. (cf. **128**) Verbi gratia: vigies quinquies .xxv. sunt vigies .xx. et quinquies .v. et quinquies bis .xx. Sed
tritata est hec maxima et usus discipula.[38] Similiter si mixtus numerus
per alium mixtum ducis, duc utrumque in utrumque, ut duodecies
.xiii. sunt decies .x. et decies .iii. et bis .x. et bis .iii., ut supra.

Summa multiplicationis.[39]

Si digitus digitum multiplicat, aufer a multiplicato differentiam
multiplicantis tocies quoto loco multiplicans distat a denario, et duc
in ipsum.[40] Si quilibet digitus multiplicat quemlibet articulum,
[adde] multiplica digitum articuli et[41] multiplicantem digitum, et fit
tota denominatio quotus est ordo articuli. Ut quinquies .dc. dic quinquies .vi.; fiunt .xxx. Pones[42] ergo iii̅. Si articulus primi ordinis multiplicet articulum quemcunque, multiplica digitos ut supra, et erit

36 i.e. **43–45**.
37 duptum
38 d̅cipl̅a
39 Much of the rest of this text corresponds to the section (treatise?) on
    multiplication and division in the collection of works on arithmetic and
    the algorism in the English MS London, British Library, Egerton 2261
    (s.xiii). This untitled section covers fols 226rb–227rb, and begins: 'Si
    digitus digitum multiplicat ... ' The text which immediately precedes
    this section has been edited in L.C. Karpinski, 'Two Twelfth Century
    Algorisms', *Isis,* 3 (1921), pp.396–413; Karpinski summarizes the con-
    tents of the section on multiplication and division on pp.410-1. In the
    following notes I refer to the sentence-numbers of this text *(Eg.)*. The
    rules for multiplication in the Egerton MS recur, in a different order,
    as 'six rules for multiplication' in Sacrobosco's *Algorismus vulgaris,*
    which also includes the rules for progression; see MS British Library,
    Harley 4350 (s.xiii), fols 15v–25v, and the English translation in
    E.Grant, *A Source Book in Medieval Science* (Cambridge, Mass., 1974),
    pp.94–101 (98).
40 Cf. *Eg.*1 : Si digitus digitum multiplicat, vide quota sit differentia mul-
    tiplicandi ad denarium et totiens aufer multiplicantem a summa quam
    reddiderit denarius multiplicatus per eundem. This is Sacrobosco's first
    rule, Harley MS, fol.19v: Quando enim digitus multiplicat digitum,
    subtrahendus est minor digitus ab articulo sue denominationis per
    differentiam (-tias MS) maioris digiti ad denarium, denario simul com-
    putato.
41 multiplica ... et] digitum articuli et multiplica
42 primum

denominatio que excedit numerum ordinis multiplicati articuli unitate, et nota si excreverit digitus ex utroque, faciendum est idem. Si articulus secundi ordinis multiplicat articulum, multiplica digitos ut supra, quos denominatio excedit binario. Si autem tercii ordinis sit articulus, ternario, et sic deinceps.

Aliter (cf. **43–45**): si digitus multiplicet digitum, ducatur differentia multiplicati in multiplicantem, et productus auferatur ab articulo denominato a multiplicante; residuum est summa. Si digitus multiplicet articulum secundi limitis, ducatur multiplicans in digitum a quo multiplicatus denominatur, ut supra; quot unitates, tot denarii.[43] Si articulus secundi limitis multiplicet articulum eiusdem limitis, ducatur a quo denominatur multiplicans in eum a quo denominatur multiplicatus; quot unitates, tot centenarii. Si articulus secundi ducatur in tercii, ducatur denominans in alterum; unitates significa⟨n⟩t millenos.[44] Si quis articulus ducatur in numerum compositum, ducetur articulus in articulum, deinceps in digitum, ut supra.[45] Si compositus in compositum utrumque multiplicantium ducatur in utrumque multiplicandorum et patet summa.[46]

De divisione.

(**117**) De proportionando vero dico quod quantum fuerit reliquum numeri dividendi ad integrum, tota pars vel tote partes continget singulos divisores, preter iam acceptam summam denominationum – sed eorum que ad aliquid non simpliciter sed relative dicuntur.

/fol. 60v/ Nota[47] .v. species inequalitatis esse quando numerus maior fuerit respectu minoris. Aut enim est multiplex, aut superparticularis aut superparciens, aut multiplex superparticularis, aut multiplex superparciens. Multiplex dicitur maior ad minorem quociens maior minorem continet bis vel ter vel deinceps, et nichil supra, ut

43 Cf. *Eg.*3: Si digitus articulum multiplicat, eorum digitos multiplica, et ⟨quot⟩ hic erunt unitates, tot illic erunt denarii. Cf. Sacrobosco's second rule, Harley MS, fol. 19v: Quando digitus multiplicat articulum, ducendus est digitus in digitum a quo denominatur ille articulus, et quelibet unitas valet decem et quilibet articulus valet centum.

44 Cf. Sacrobosco's fourth rule, Harley MS, fol. 19v: Quando articulus ⟨in⟩ articulum ducendus, digitus a quo denominatur unus illorum in digitum a quo denominatur reliquus, et quelibet unitas valet centum et quilibet denarius mille.

45 Cf. Sacrobosco's fifth rule, Harley MS, fols 19v–20r.

46 Cf. Sacrobosco's sixth rule, Harley MS, fol. 20r.

47 That this is the continuation is indicated by two crosses (+), one at the end of the previous section, one before the word 'Nota' here. In the top margin of fol. 60v the following heading has been written: Aliter (?) de (?) divisione, inequalitate, et minutiis, et medietate.

senarius ad binarium.[48] Superparticularis maior ad minorem dicitur quociens maior minorem continet totum et insuper minoris aliquam partem, ut idem senarius ad quaternarium.[49] Superpartiens est numerus quociens maior minorem continet totum et insuper et eius aliquas partes, ut septenarius ad quinarium.[50] Multiplex superparticularis dicitur quando maior minorem aliquociens continet et insuper eius aliquam partem, ut denarius ad quaternarium. Multiplex superparciens dicitur quociens maior minorem continet bis vel ter vel deinceps et eius aliquas partes, ut octonarius ad ternarium. Iuxta has .v. species inequalitatis omnem maiorem numerum [numerum] minori proporcionaliter dividere poterimus.

De multiplici.

Quocienscunque in multiplici genere aliquis numerus dividendus fuerit, considerandum quociens minor contineatur in maiori, et tot integra contingent cuilibet divisori.[51] Verbi gratia: Senarius ter continet binarium; ergo si dividendus fuerit senarius binario cuilibet divisori, tria contingunt proportionali rationali.

De superparticulari.

Quocienscunque aliquis numerus in superparticularitatis proporcione fuerit dividendus, considerandum quota pars minoris a maiore preter quantitatem minoris supra contineatur, et tota pars cum uno integro cuilibet divisori continget.[52] Verbi gratia: Si senarius quaternario fuerit dividendus, quoniam senarius minorem totum habet et eius medietatem, quilibet minorum habebit unum integrum, et medietatem unius integri.

De superparciente.

Quocienscunque in superparcienti genere aliquid fuerit dividendum cuilibet divisori unum integrum continget habendum et tote partes unius integri, quote fuerint minoris in maiore, que supra continentur.[53] Ut si septenarius [ut si septenarius] dividendus fuerit qui-

---

48 Cf. *Eg.* 20.
49 Cf. *Eg.* 21.
50 Cf. *Eg.* 22.
51 Cf. *Eg.* 8: Quociens maior numerus fuerit dividendus, qui ad minorem fuerit multiplex, consideretur quociens minor contineatur a maiori, et tot integra contingent cuilibet divisori.
52 Cf. *Eg.* 9: Quociens maior numerus fuerit dividendus minori qui superparticularis sit ad minorem, consideretur quota pars supra contineatur: tota pars continget cuilibet divisori preter unum integrum.
53 Cf. *Eg.* 10: Quociens maior numerus minori fuerit dividendus qui ad

nario, cuilibet divisori continget unum integrum et due quinte partes unius integri.

De multiplici superparticulari.[54]

Quocienscunque aliquis numerus in multiplici superparticulari genere fuerit dividendus, considerandum quociens minor in maiore contineatur et quota pars supra, et utere predictis regulis, ut tociens cuilibet divisori unum contingat integrum, et talis pars cuiusmodi supra in dividendo continetur.[55] Verbi gratia: Dividatur denarius quaternario. Quoniam ergo quaternarius bis continetur in maiore et insuper eius medietas, cuilibet divisori duo contingunt integra et medietas integri.

De multiplici superpartiente.

Quocienscunque in multiplici superparciente aliquis numerus minori fuerit dividendus, consideretur quociens minor in maiori contineatur, et quote partes supra. Deinde ut supra agendum.[56] Verbi gratia: Si octonarius dividendus fuerit ternario, quoniam ternarius bis continetur in eo et insuper eius due tercie, cuilibet divisori contingunt duo integra et due tercie unius integri.

De impari.

Ad inveniendam summam cuiuslibet numeri. Numero naturali usque ad quemlibet imparem aggregato, ultimum ordinem per sui maiorem medietatem multiplica et invenies summam.[57] Verbi gratia: .i.ii.iii.iiii.v. dic ter quinque et fiet summa, scilicet .xv.

De pari.

Numero naturali usque ad quemlibet parem coacervato, per medietatem ultimi sequentem non positum inparem multiplicabis, vel per sui minorem medietatem eundem imparem et erit summa omnium,[58] ut .i.ii.iii.iiii. dic bis .v., fiunt .x.

minorem fuerit superparciens, consideretur quote partes supra contineantur, et tote partes contingent cuilibet divisori.

54 *corrected from* superpartiente
55 Cf. *Eg.* 11.
56 Cf. *Eg.* 12.
57 Cf. *Eg.* 13: Si quorumlibet numerorum hoc modo aggregatorum – scilicet ab unitate procedendo usque ad quemvis numerum – summam scire volueris, vide si impar fuerit extremus numerus. Quod si fuerit, duc medium in eundem et erit summa tocius. Compare Sacrobosco, Harley MS, fol. 22r–v.
58 Cf. *Eg.* 13 (cont.): Si vero par, duc medium eiusdem in imparem qui proximo sequitur ultimum parem, et erit summa tocius. Compare

De imparibus.

Numeris ab unitate omnibus (imparibus) aggregatis, maiorem medietatem ultimi per seipsum more tetragoni multiplicabis, ut .i.iii.v.
dic ter tria; .ix. habes.[59]

De paribus.

A binario omnibus paribus congregatis, per medietatem ultimi
maiorem medietatem sequentis non positi imparis multiplica, et
habebis summam.[60] Vel apercius: a binario omnibus imparibus in
unum ductis, sequentis imparis medietatem per eius alteram partem
multiplica, ut si .vi. est ultimus par, divide .vii. et eius quamlibet
alteram per alteram partem multiplica, et fiunt .xii.ii.iiii.vi. + [61]

Sacrobosco, Harley MS, fol. 22r: Quando enim progressio naturalis ter-
minatur in numerum parem, per medietatem ipsius multiplica numerum
proximum totali superiorem. Verbi gratia 1.2.3.4., multiplica quinarium
per binarium: sic bis quinque, et exibunt 10, summa totius progressionis.

59 ħs. Cf. *Eg.* 15: Si impares tantum, maiorem medietatem ultimi in se
    duces. Compare Sacrobosco, Harley MS, fol. 22v.
60 Cf. *Eg.* 14: Si tantum per pares (MS partes) sibi continuos fiat processus,
    ultimum divide, cuius medietatem multiplica per numerum proximo
    sequentem eandem medietatem, unitate addita. Compare Sacrobosco,
    Harley MS, fol. 22v: Quando enim progressio intercisa terminatur in
    numerum parem, per medietatem illius [illius] multiplica numerum pro-
    ximum medietati superiorem ... .
61 A reference sign at the end of the text would lead one to expect a
    continuation elsewhere in the MS, signalled by the same reference sign.
    This cannot be found.

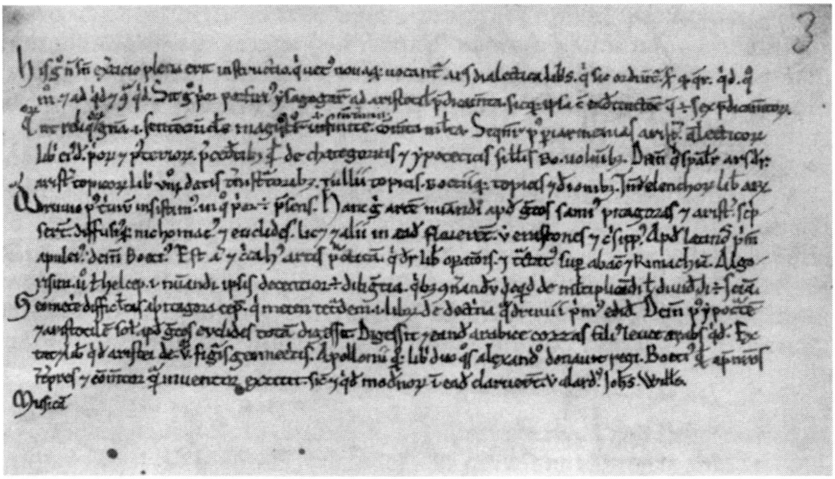

Plate I. 'Algorism, 'helcep' and the geometers 'Alardus', 'Johannes' and 'Willelmus' in MS Cambridge, Trinity College, R.15. 16, fol. 3r (reproduced by permission of the Master and Fellows of Trinity College).

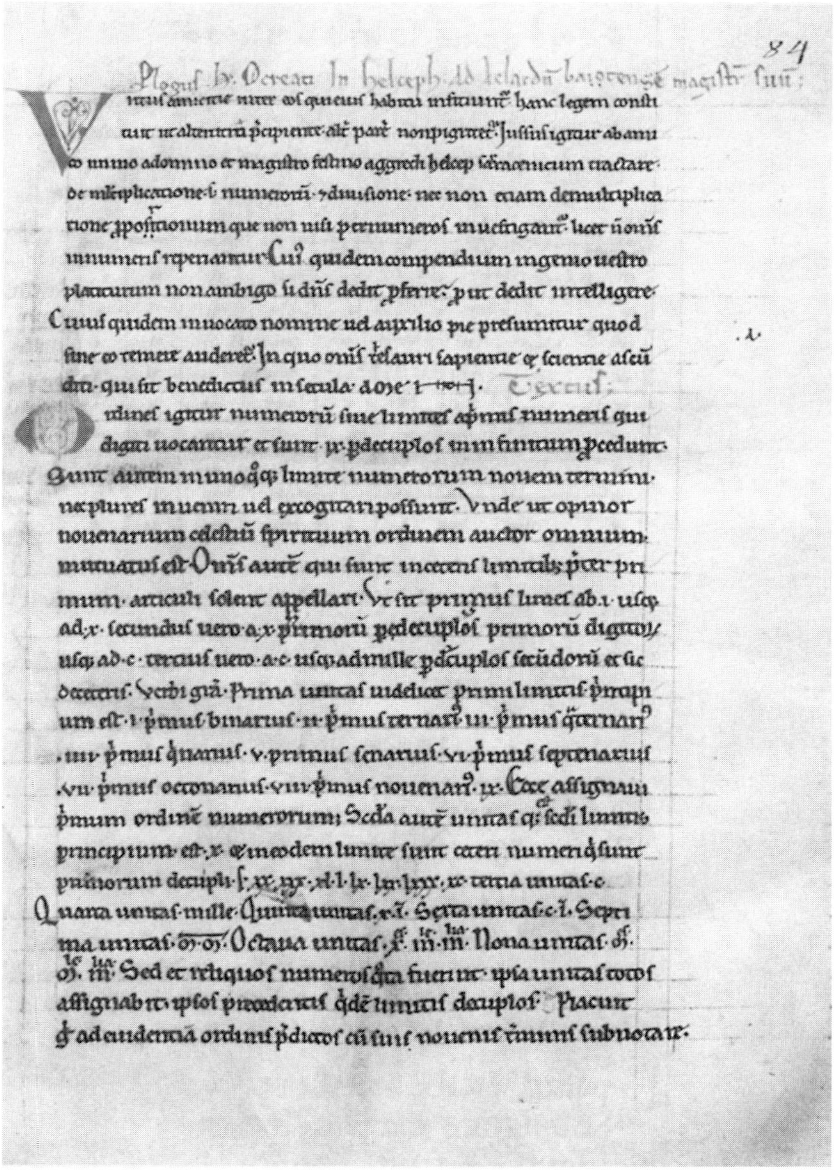

Plate II. The first page of *Helcep Sarracenicum*, Paris, Bibliothèque
nationale, lat. 6626, fol. 84r (reproduced by permission).

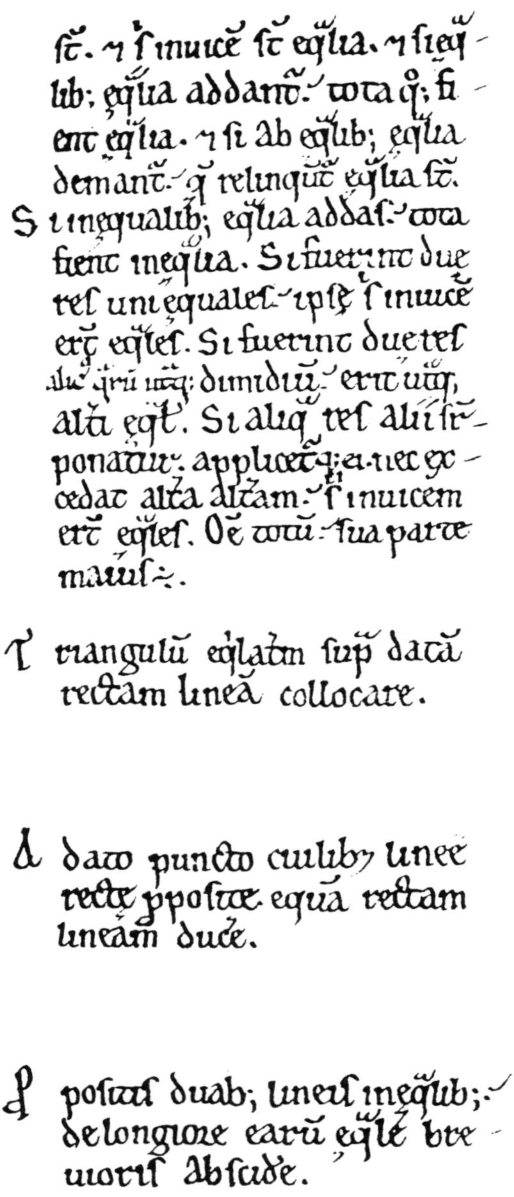

fc̅. ꝫ ſ inuice ſc̅ eq̅lia. ꝫ ſieꝗ-
lib; eq̅lia addant͛·ꝛ tota q̛; ſi-
erit eq̅lia. ꝫ ſi ab eq̅lib; eq̅lia
demant͛·q̛ relinqut̅ eq̅lia ſc̅.
S i inequalib; eq̅lia addaſ·ꝛ tota
fient ineq̅lia. Si fuerint due
res uniequaleſ·ipſꝯ ſ inuice
erꝷ eq̅leſ. Si fuerint dueres
alie q̛rū ūꝗ: dimidiū·ꝛ erit ūꝗ,
alia eqt. Si aliq̛ res alii ſr-
ponatur· applicat̛q̛a·nec ex-
cedat alia alteram·ſ inuicem
erꝷ eq̅leſ. Oe totū·ſua parte
maiuſꝫ.

T triangulū eq̅latm ſuꝑ datā
rectam lineā collocare.

A dato puncto cuilibꝫ linee
rectꝯ ꝓpoſitꝯ equā rectam
lineam ducere.

S poſitis duab; lineiſ inequlib;·
de longiore earū eq̅lē bre-
uioriſ abſcidere.

Plate III. A page from the version of Euclid's *Elements* once attributed to 'Joan. Ocreatus'. MS London, British Library, Royal 15 A.XXVII, fol. 2r (reproduced by permission).

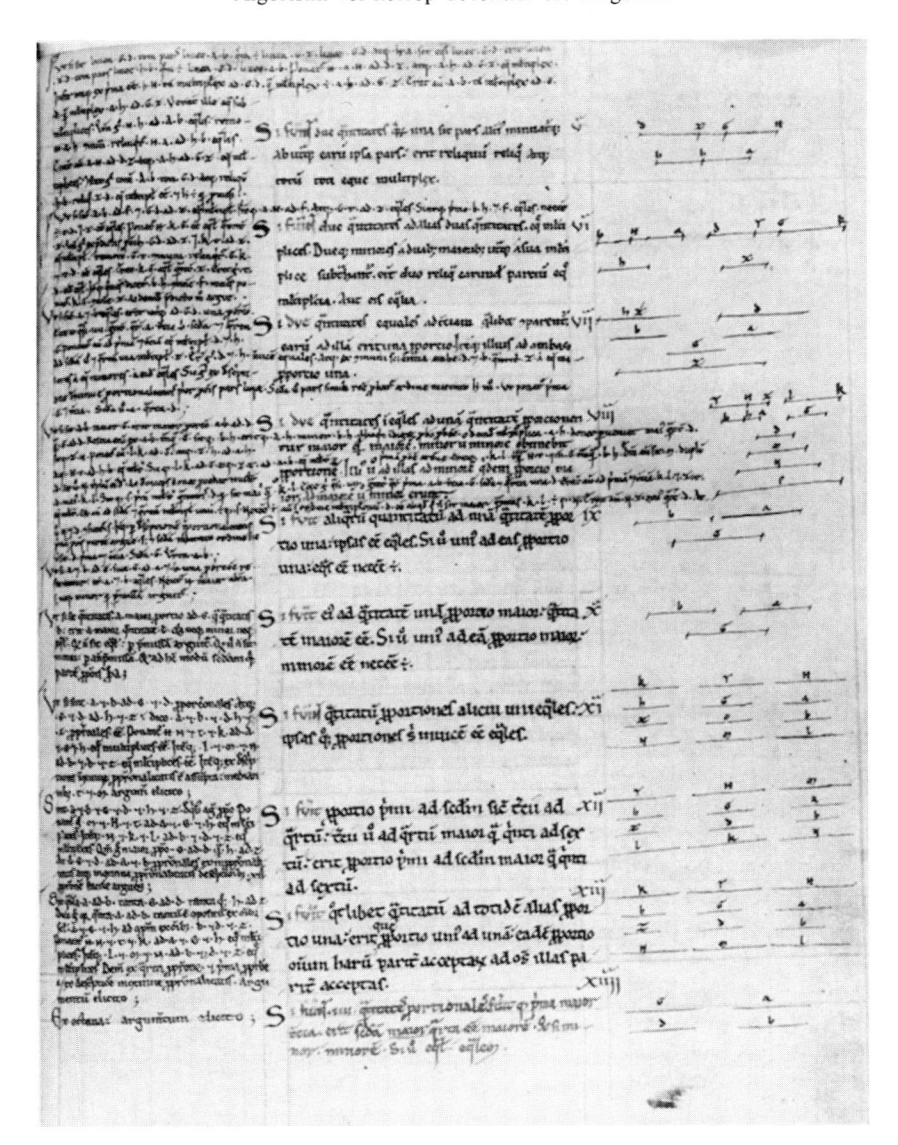

Plate IV. An example of the *mise-en-page* of Euclid, *Elements*, Version II in MS Oxford, Trinity College, 47, fol.113r (*Elements*, V.5–14) (reproduced by permission).

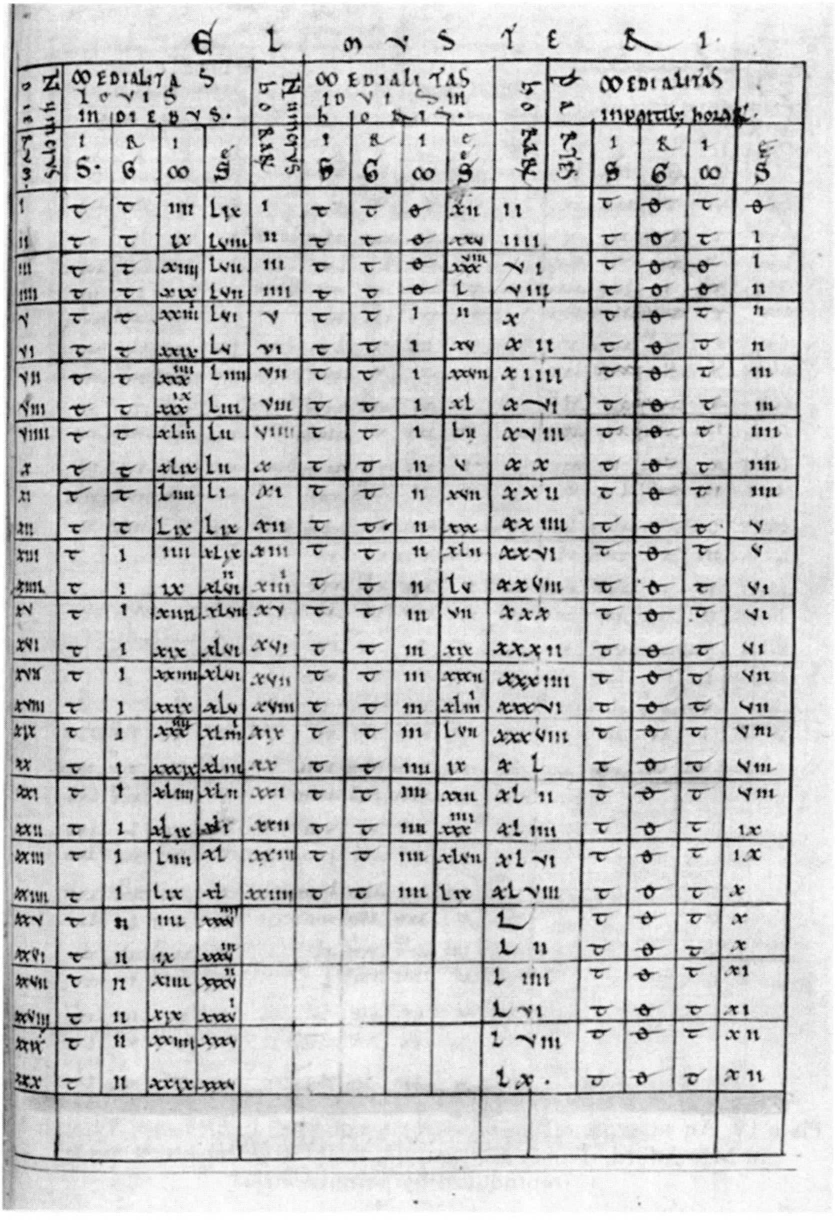

Plate V. The different symbols for zero in MS Oxford, Auct. F.1.9, fol. 106r (reproduced by permission).

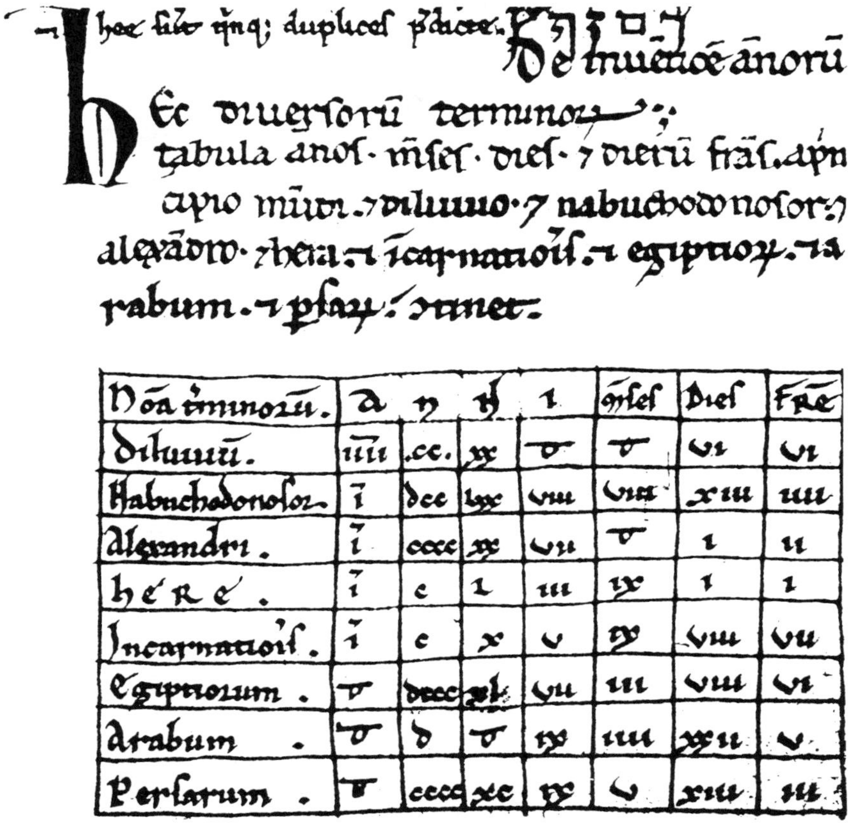

Nomina terminorum.	a	n	H	1	Menses	Dies	Fre
Diluuii.	iiii	.cc.	xx	v	v	vi	vi
Nabuchodonosor.	i	dcc	lxx	vii	viii	xiii	iiii
Alexandri.	i	ccc	xx	vii	v	i	ii
Here.	i	c	i	iii	ix	i	i
Incarnationis.	i	c	x	v	ix	viii	vii
Egiptiorum.	v	dccc	xl	vii	iii	viii	vi
Arabum.	v	d	v	ix	iiii	xxii	v
Persarum.	v	cccc	xc	ix	v	xiii	iii

Plate VI. The table of eras in the *Liber ysagogarum Alchorismi* in MS Paris, Bibliothèque nationale, lat. 16208, fol. 70v (reproduced by permission).

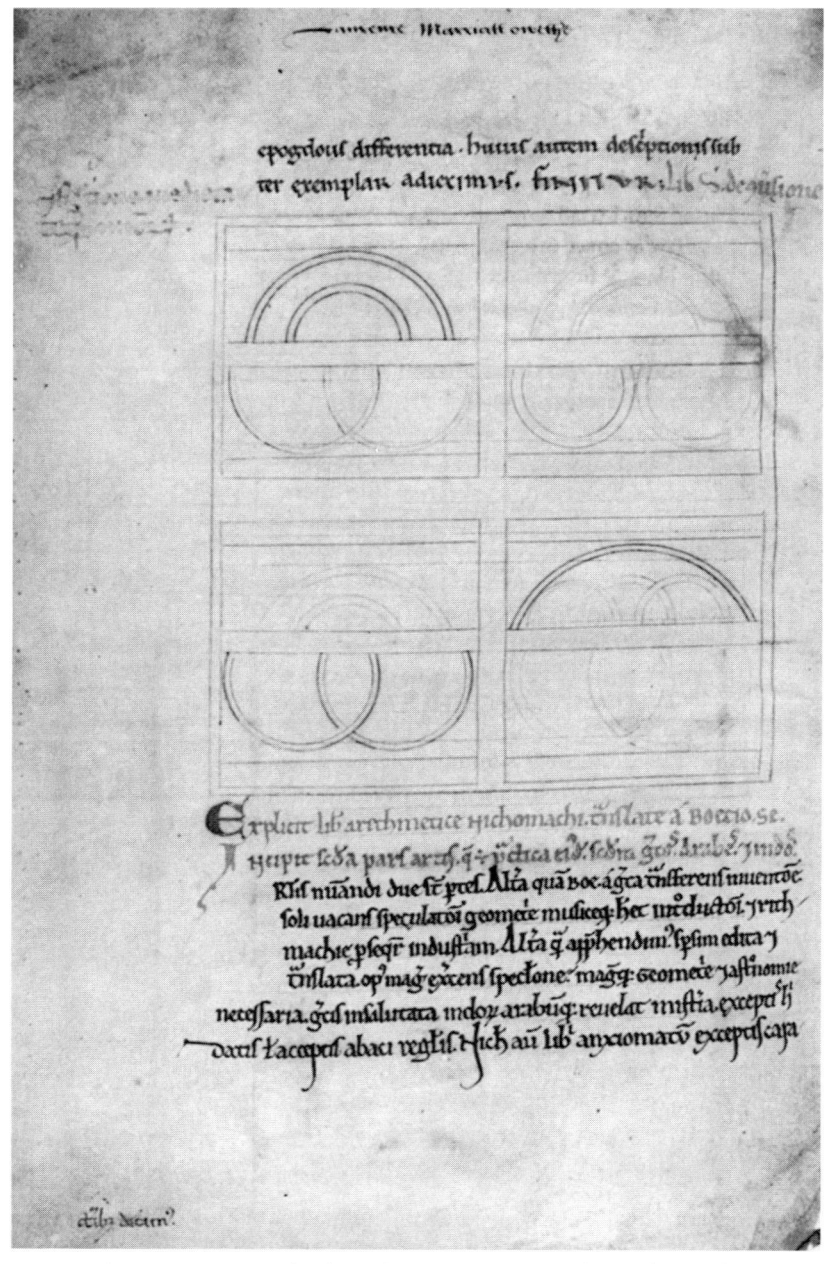

Plate VII. The beginning of the text on 'practical arithmetic'
in MS Cambridge, Trinity College, R.15.16, fol. 59v
(reproduced by permission of the Master and Fellows of Trinity College).

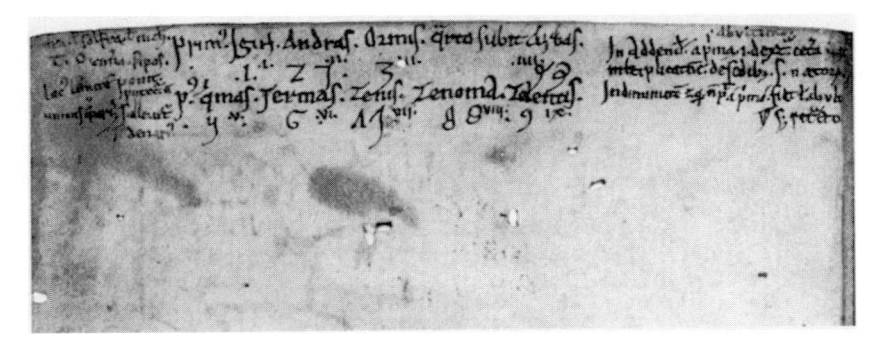

Plate VIII. The names and symbols for the numbers in MS Cambridge, Trinity College, R.15.16, fol. Av (reproduced by permission of the Master and Fellows of Trinity College).

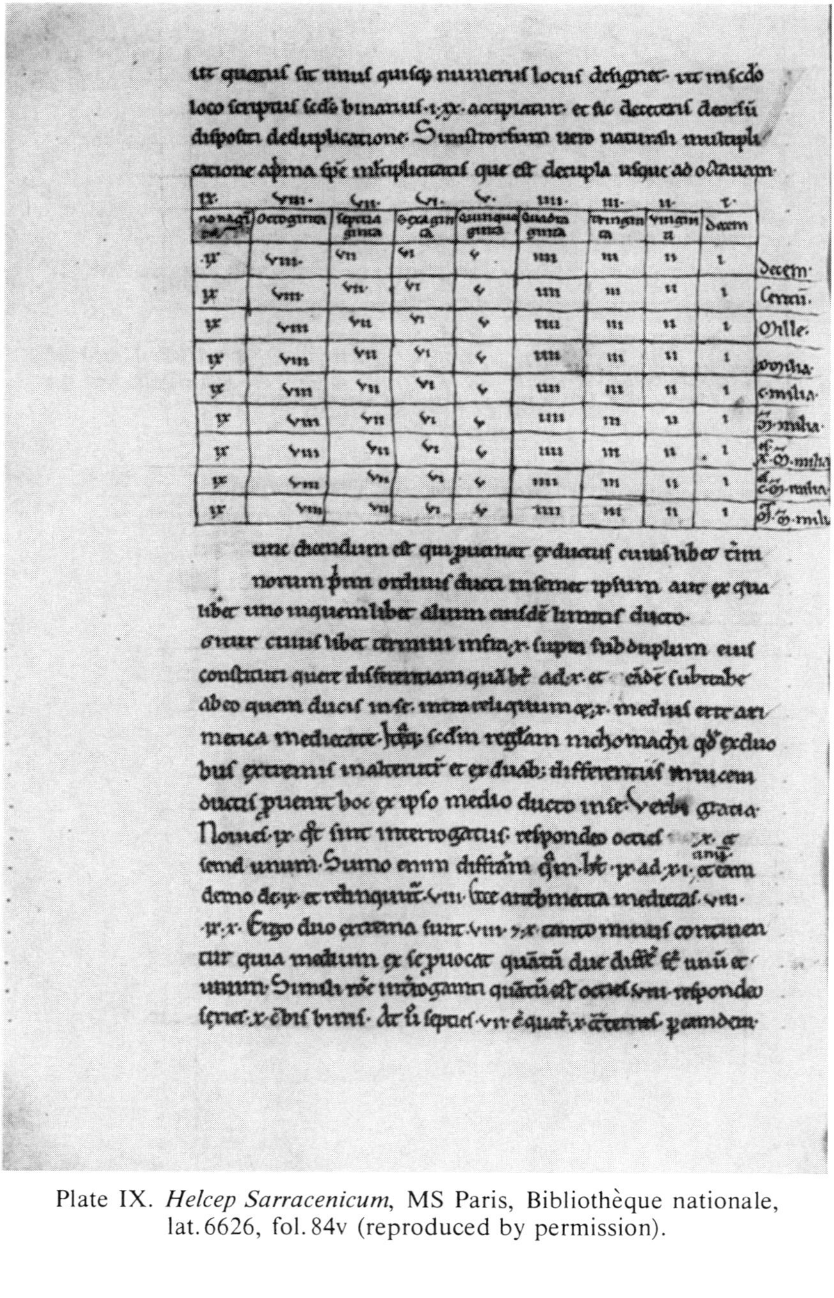

[Facsimile of a medieval Latin manuscript page — Helcep Sarracenicum]

utr quatuns sir unus quisq numerus locus designet· uir in scdo
loco scriptus scdo binarius·1·3x· accipiatur· er sic decccxui deorsu
disposita deduplicatione· Similis rorsum uero naturali multipli
catione apma tipe multiplicamus que est decupla usque ad octauam·

X	VIII	VII	VI	V	IIII	III	II	I	
nonagin ta	Octoginta	septua ginta	sexagin ta	quinqua ginta	Quadra ginta	tringin ta	uingin ti	decem	
X	VIII	VII	VI	V	IIII	III	II	I	Decem·
X	VIII	VII	VI	V	IIII	III	II	I	Centu·
X	VIII	VII	VI	V	IIII	III	II	I	Ơille·
X	VIII	VII	VI	V	IIII	III	II	I	𝔡 milia
X	VIII	VII	VI	V	IIII	III	II	I	c milia
X	VIII	VII	VI	V	IIII	III	II	I	Ɔ· milia
X	VIII	VII	VI	V	IIII	III	II	I	X· Ɔ· milia
X	VIII	VII	VI	V	IIII	III	II	I	c Ɔ· milia
X	VIII	VII	VI	V	IIII	III	II	I	Ɔ·Ɔ· milia

unc dicendum est qui puentur eductus cuius liber tim
norum prin ordinis ducta in semer ipsum aut ex qua
liber uno inguemliber alium einsde limnis ducto
orcur cuius liber carunin infra·x· supra subduplum eius
constitui quere differentiam qua bt ad·x· et eade subtrabe
ab eo quem ducis in se inra reliquum ex·x· medius ea ari
merica mediecate hiqis scdm regulam michomachi qd ex duo
bus extremis nascuntur et ex duab; differentiis minuem
ducas puentr hoc ex ipso medio ducto inse Verbi gracia
Nouem·ix· est sint interrogatus respondeo octue·x· e
semel unum Sumo enim differtin du·i·bt·ix·ad·xi· et eam
demo de·ix· et relinquunt·viii· sunt anctmenta mediecat·viii·
·ix·x· Ergo duo extrema sunt·viii·ix· canto minus continen
tur quia medium ex se puocat quantum dur diffe ft unu et
unum· Simili rde introgamus quantum est octuel scm respondeo
scriet·x· ebis binni· di si scriet·vii· equat·x· externus· penmdem·

Plate IX. *Helcep Sarracenicum*, MS Paris, Bibliothèque nationale,
lat. 6626, fol. 84v (reproduced by permission).

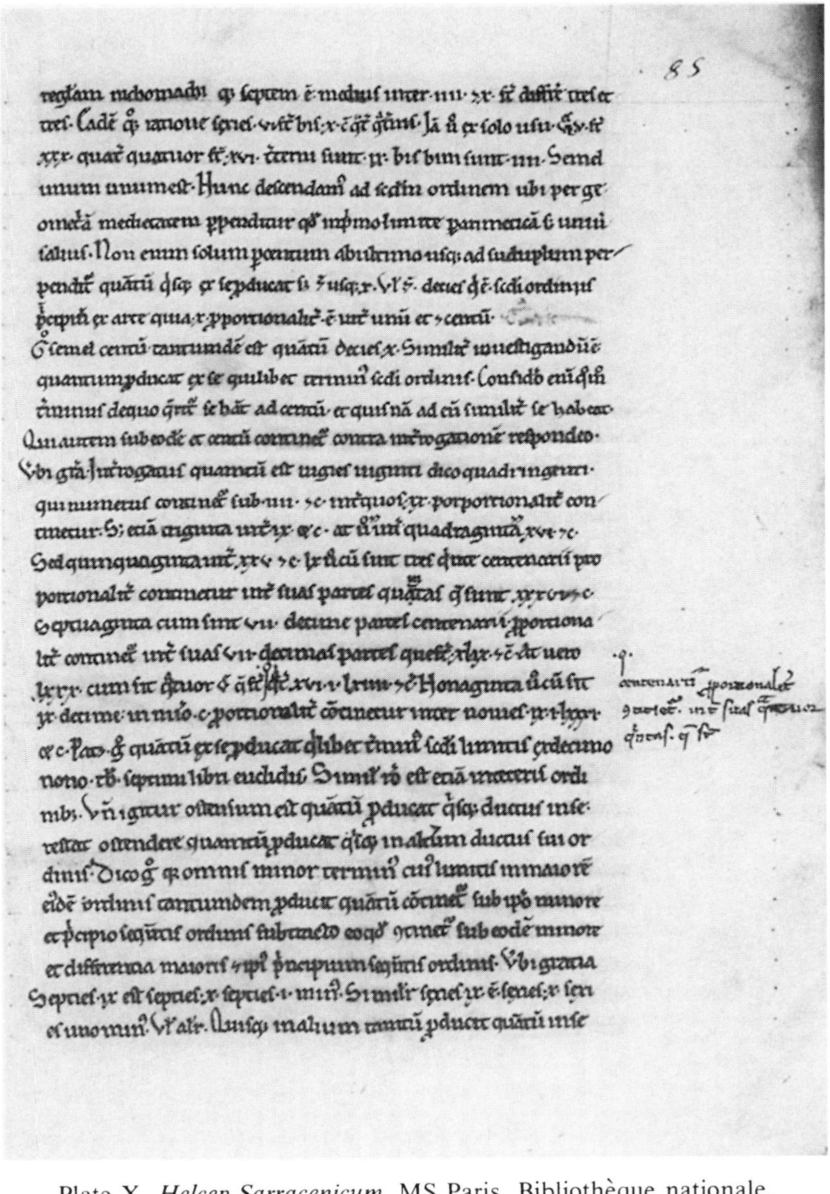

III

Plate X. *Helcep Sarracenicum*, MS Paris, Bibliothèque nationale, lat. 6626, fol. 85r (reproduced by permission).

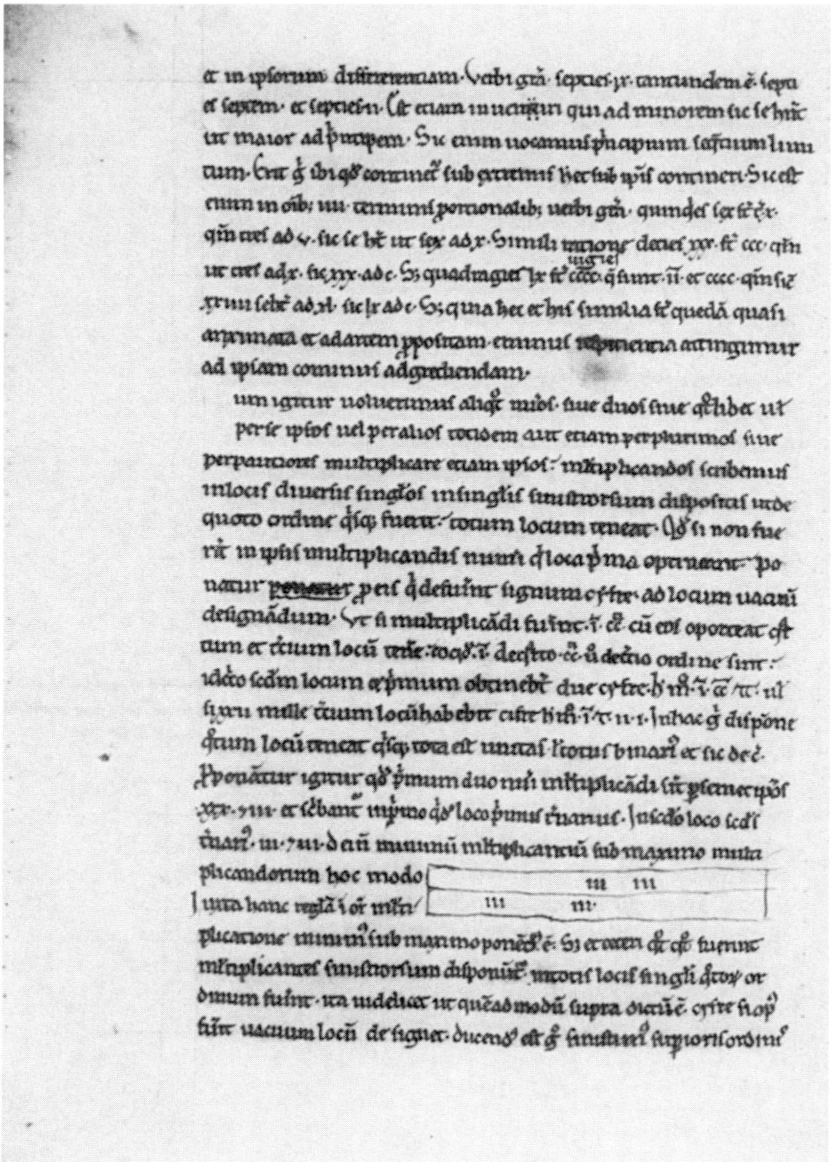

Plate XI. *Helcep Sarracenicum*, MS Paris, Bibliothèque nationale, lat. 6626, fol. 85v (reproduced by permission).

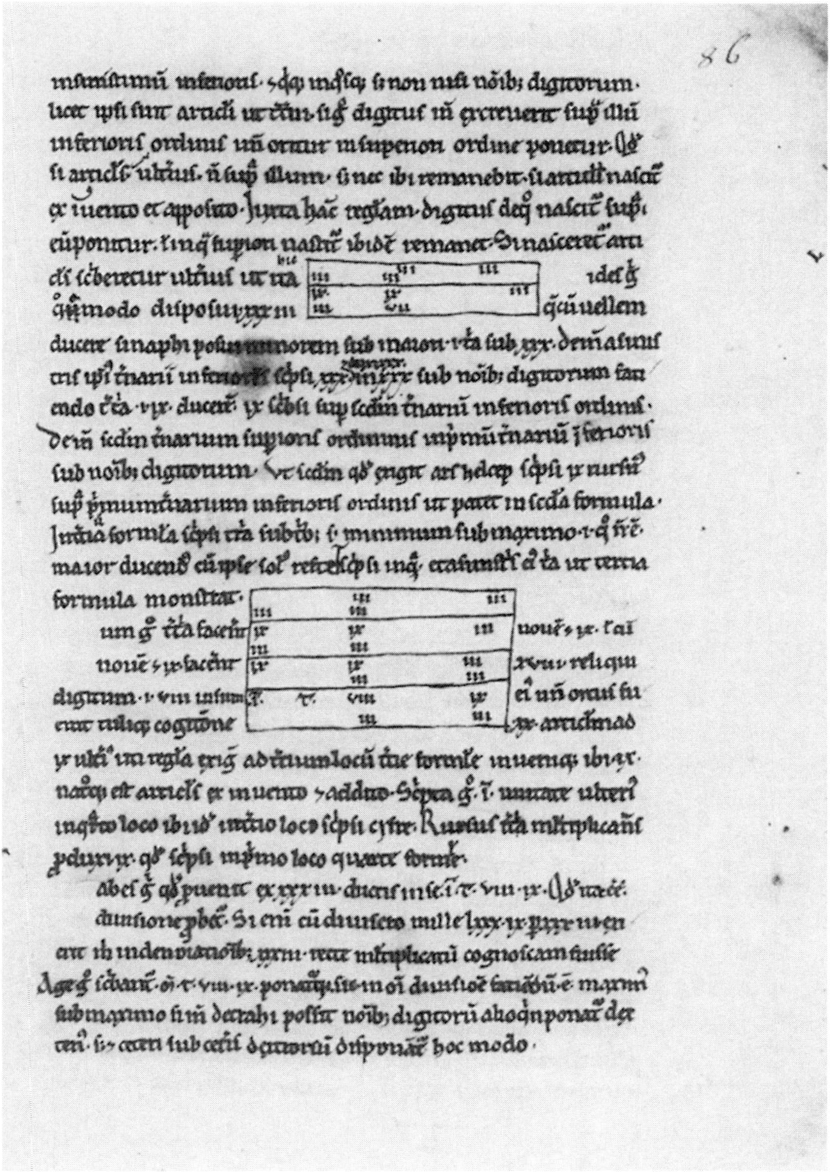

Plate XII. *Helcep Sarracenicum*, MS Paris, Bibliothèque nationale, lat. 6626, fol. 86r (reproduced by permission).

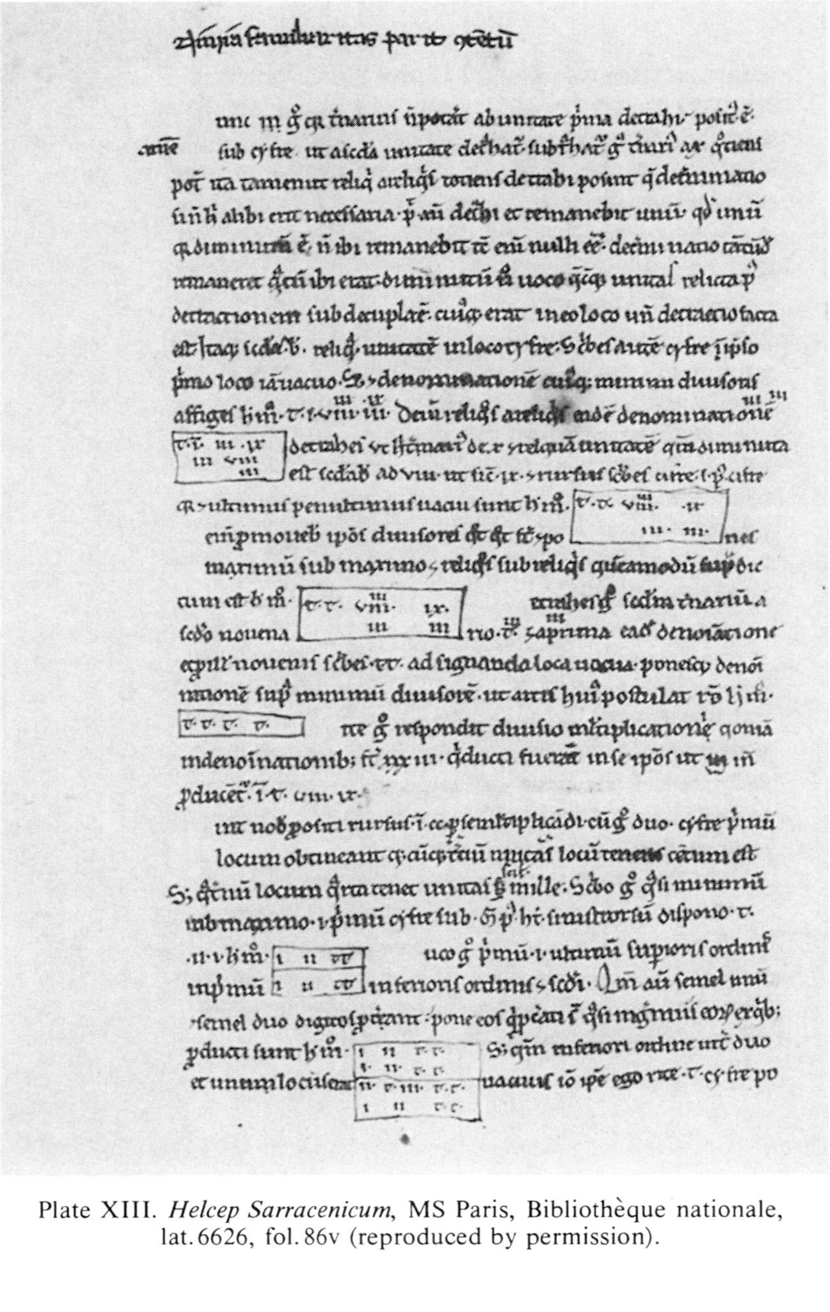

Plate XIII. *Helcep Sarracenicum*, MS Paris, Bibliothèque nationale, lat. 6626, fol. 86v (reproduced by permission).

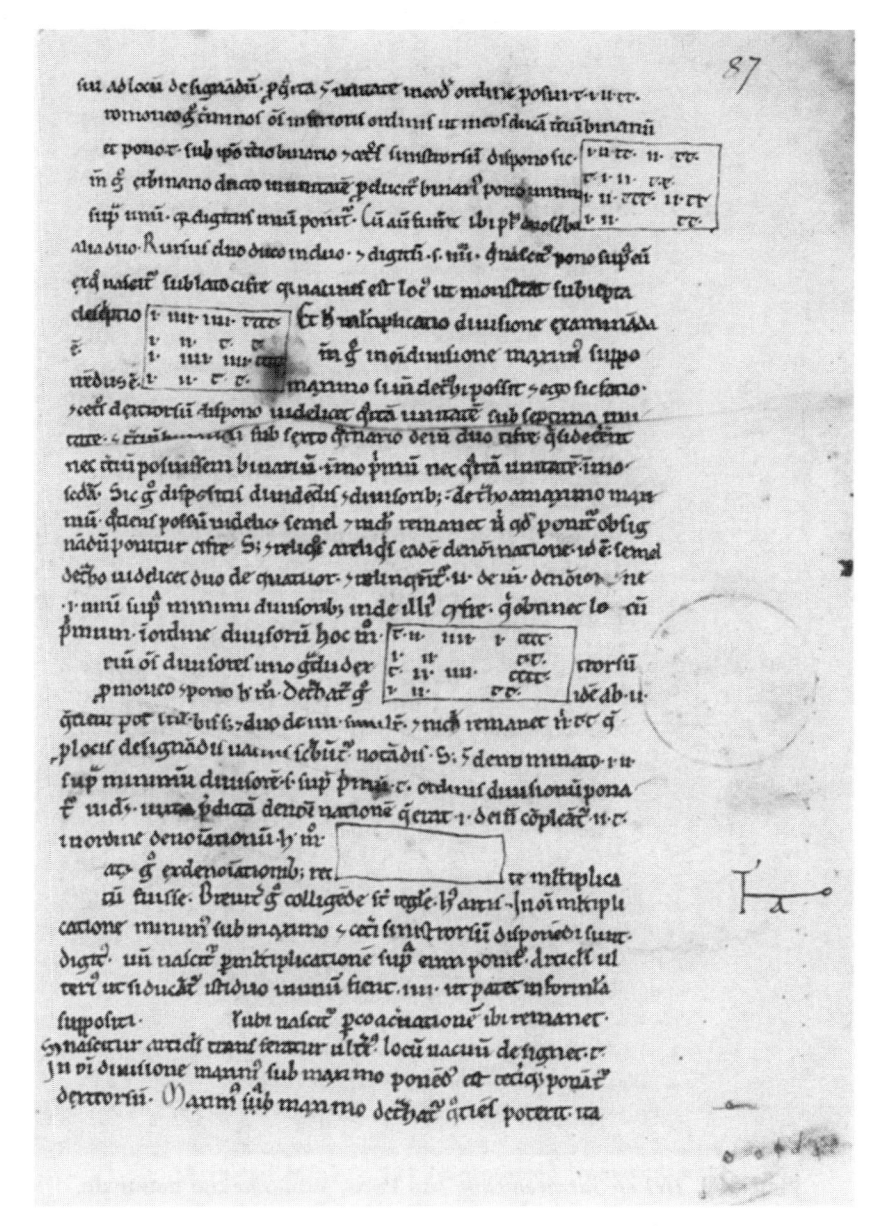

Plate XIV. *Helcep Sarracenicum*, MS Paris, Bibliothèque nationale, lat. 6626, fol. 87r (reproduced by permission).

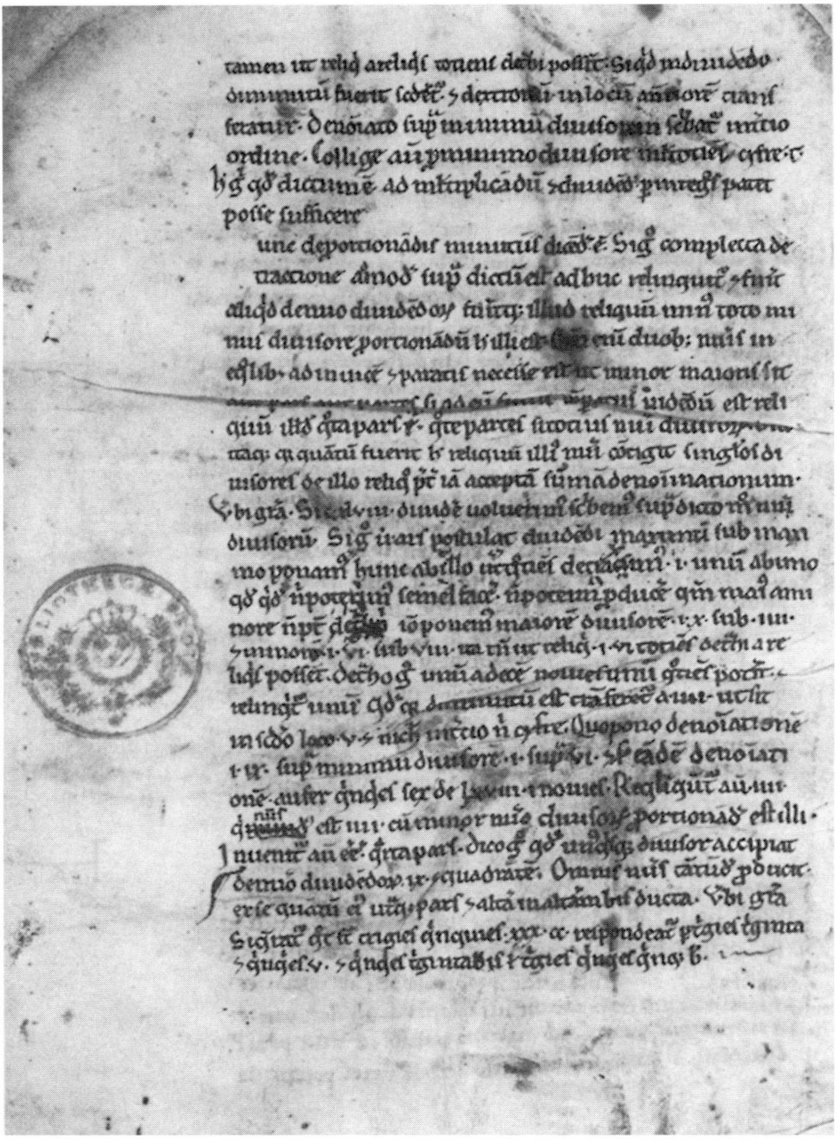

Plate XV. *Helcep Sarracenicum*, MS Paris, Bibliothèque nationale, lat. 6626, fol. 87v (reproduced by permission).

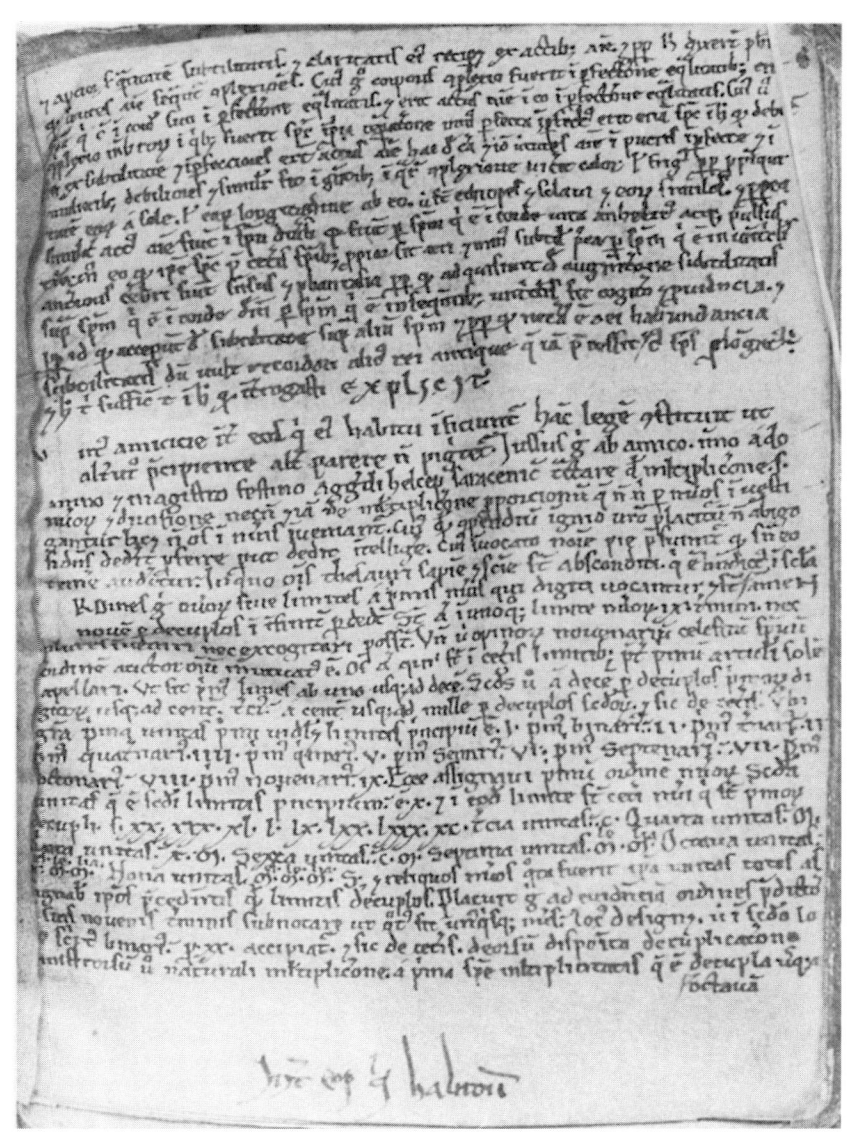

Plate XVI. *Helcep Sarracenicum*, MS Cashel, GPA Bolton Library,
Medieval MS 1, p.111 (reproduced by permission).

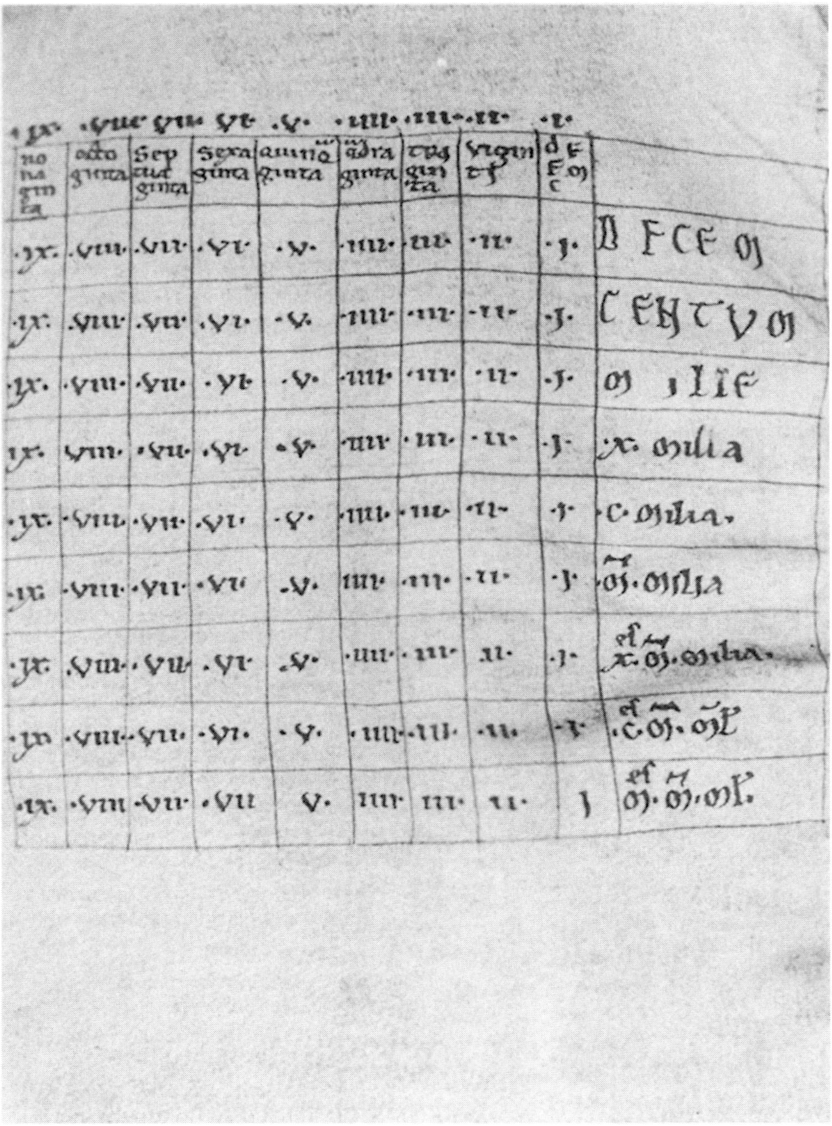

.ix.	.viii.	.vii.	.vi.	.v.	.iiii.	.iii.	.ii.	.i.	
nonaginta	octoginta	septuaginta	sexaginta	quinquaginta	quadraginta	triginta	viginti	decem	
.ix.	.viii.	.vii.	.vi.	.v.	.iiii.	.iii.	.ii.	.i.	DECEM
.ix.	.viii.	.vii.	.vi.	.v.	.iiii.	.iii.	.ii.	.i.	CENTUM
.ix.	.viii.	.vii.	.vi.	.v.	.iiii.	.iii.	.ii.	.i.	M ILLE
.ix.	.viii.	.vii.	.vi.	.v.	.iiii.	.iii.	.ii.	.i.	.x. milia
.ix.	.viii.	.vii.	.vi.	.v.	.iiii.	.iii.	.ii.	.i.	.c. milia
.ix.	.viii.	.vii.	.vi.	.v.	.iiii.	.iii.	.ii.	.i.	m. milia
.ix.	.viii.	.vii.	.vi.	.v.	.iiii.	.iii.	.ii.	.i.	x. m. milia
.ix.	.viii.	.vii.	.vi.	.v.	.iiii.	.iii.	.ii.	.i.	.c. m. mil
.ix.	.viii.	.vii.	.vii	v.	iiii.	iii.	ii.	i	m. m. mil.

Plate XVII. *Helcep Sarracenicum*, MS Cashel, GPA Bolton Library, Medieval MS 1, p. 112 (reproduced by permission).

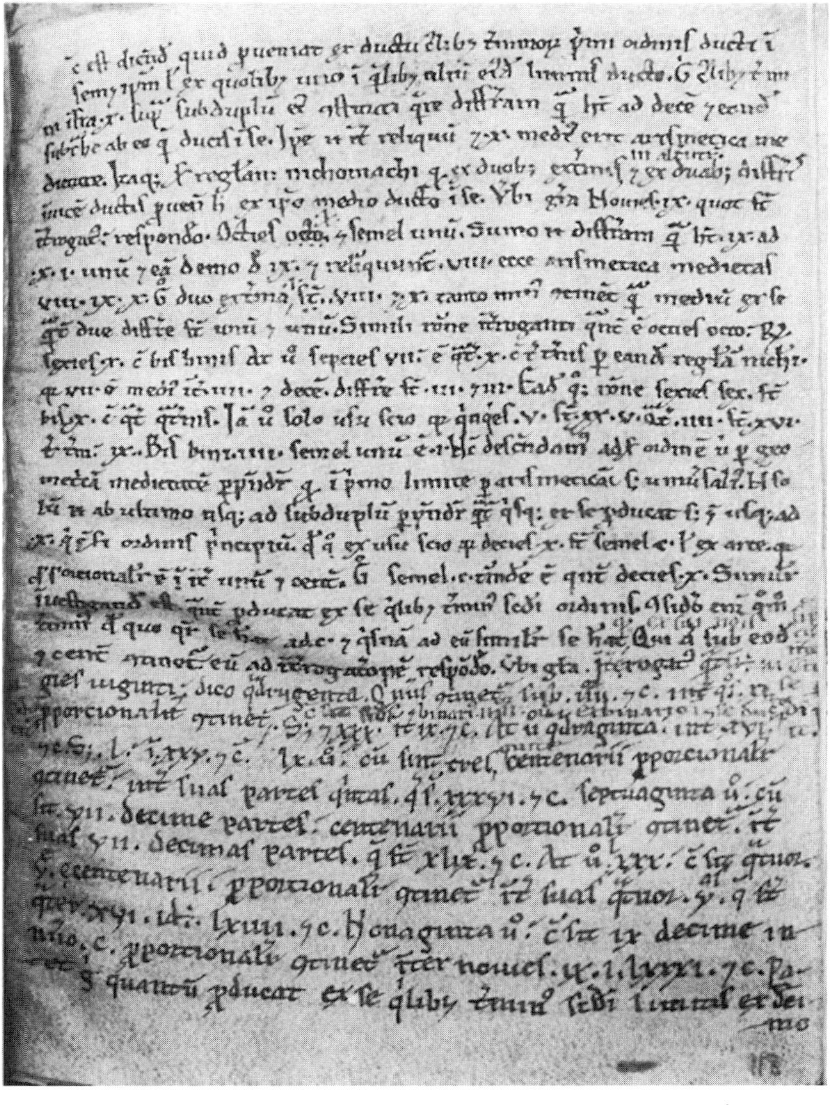

Plate XVIII. *Helcep Sarracenicum*, MS Cashel, GPA Bolton Library, Medieval MS 1, p.113 (reproduced by permission).

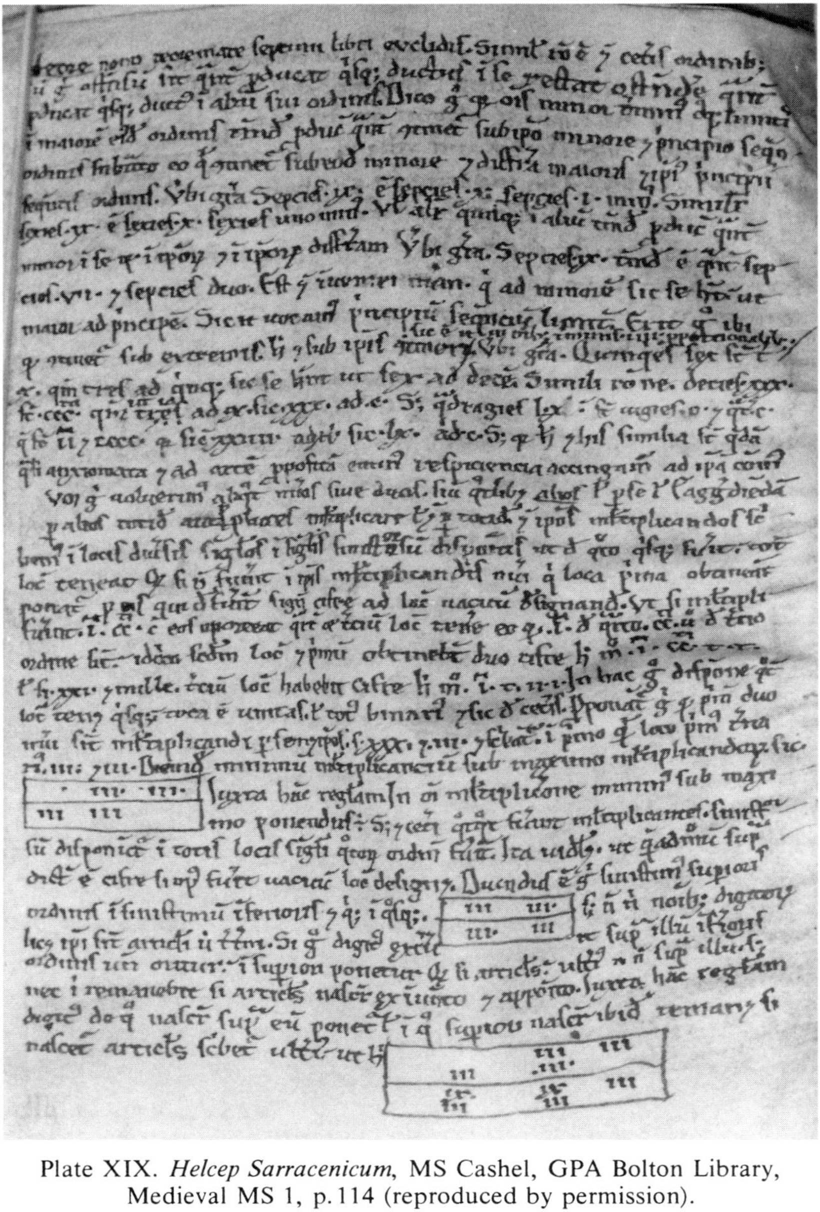

Plate XIX. *Helcep Sarracenicum*, MS Cashel, GPA Bolton Library, Medieval MS 1, p. 114 (reproduced by permission).

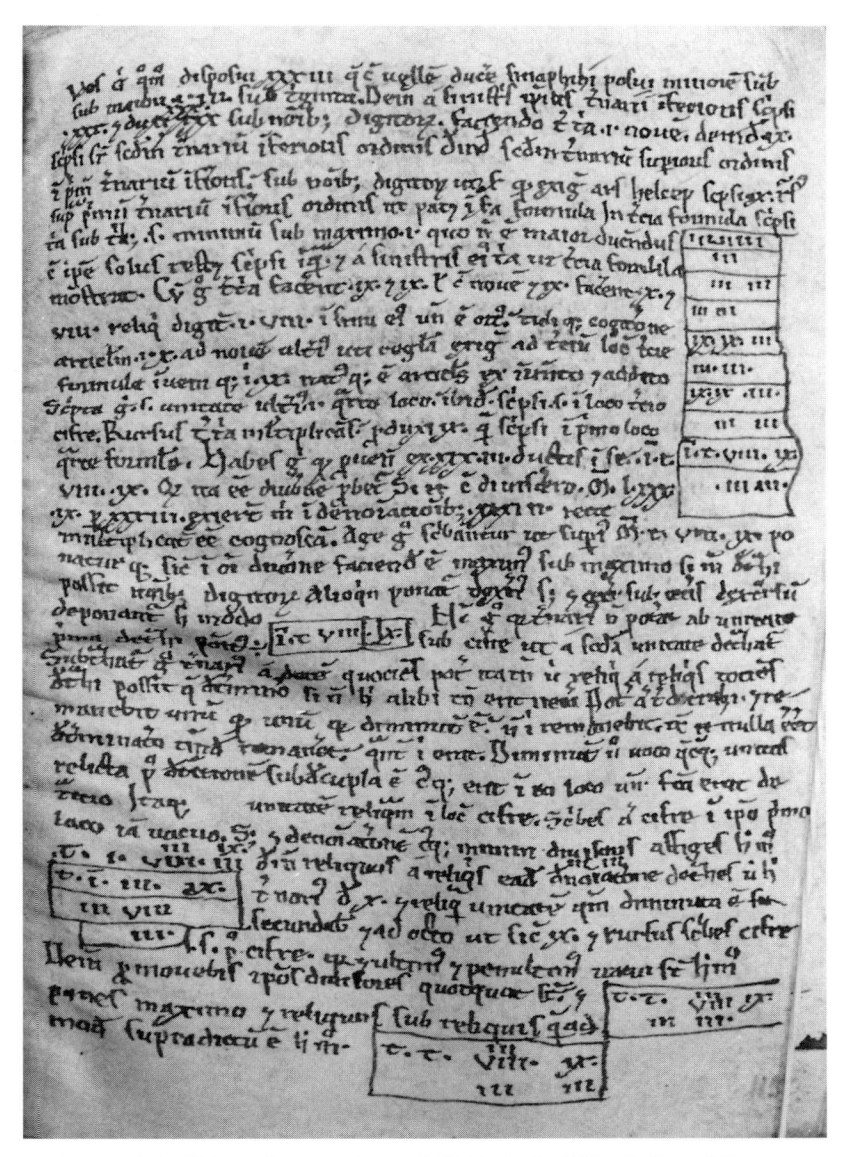

Plate XX. *Helcep Sarracenicum*, MS Cashel, GPA Bolton Library, Medieval MS 1, p.115 (reproduced by permission).

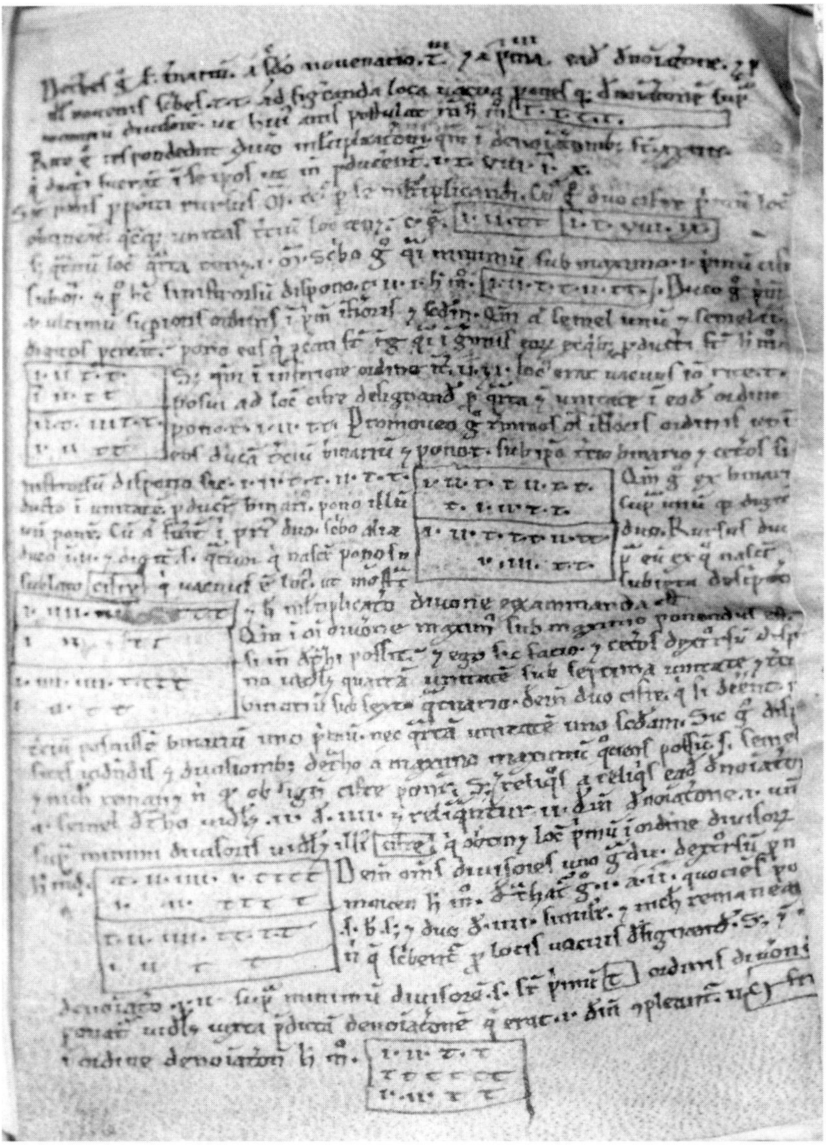

Plate XXI. *Helcep Sarracenicum*, MS Cashel, GPA Bolton Library, Medieval MS 1, p.116 (reproduced by permission).

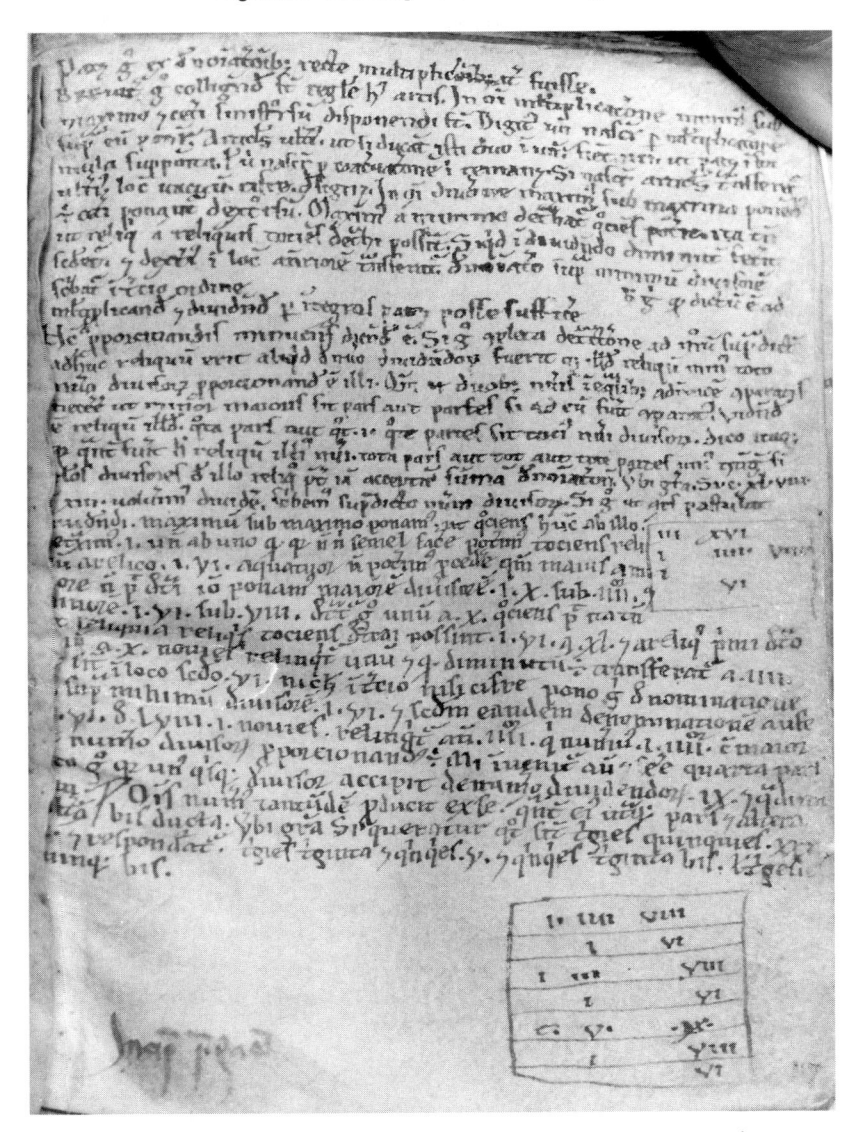

Plate XXII. *Helcep Sarracenicum*, MS Cashel, GPA Bolton Library, Medieval MS 1, p.117 (reproduced by permission).

IV

Ten or Forty?
A Confusing Numerical Symbol in the Middle Ages[*]

In the twelfth century several different systems of numeration were tried out in the West, especially by mathematicians and astronomers who needed to write high numbers and make complicated calculations. Some scholars used Arabic (more strictly, Hindu-Arabic) numerals, which manifest an 'Eastern' and a 'Western' form, and variations in between.[1] Others imitated in the Latin alphabet the alphanumerical systems used in Greek, Arabic and Hebrew.[2] A cipher system was invented in which each number, however complex, was represented by a single symbol.[3] The trouble with these systems was that the forms of the numerals were unfamiliar, and could result in dangerous errors for those who were not initiated. Another strategy was to modify the well-known Roman numerals so that they could be written more quickly and take up less space. Thus initial letters were used for the longer numbers: q(uatuor) for .iiii., o(cto) for .viii., and n(ovem) for .viiii.[4] Subtractive systems, used since Classical times – e. g. .iv. = 5 – 1, .ix. = 10 – 1 – were more economical

[*] I am very grateful to Menso FOLKERTS for his support and advice over many years, and for making it possible to consult a large number of microfilms of medieval mathematical manuscripts, including several of those mentioned in this article. The article has benefited from comments by Paul KUNITZSCH, Peter LINEHAN, Simon LOSEBY, Fritz Saaby PEDERSEN and Hanna VORHOLT, to whom I express my thanks.

[1] BURNETT, Charles: Indian Numerals in the Mediterranean Basin in the Twelfth Century, with Special Reference to the "Eastern Forms". In: From China to Paris: 2 000 Years Transmission of Mathematical Ideas, eds Y. DOLD-SAMPLONIUS, J. W. DAUBEN, M. FOLKERTS and B. VAN DALEN. Stuttgart 2002, p. 237–88.

[2] BURNETT, Charles: Antioch as a Link between Arabic and Latin Culture in the Twelfth and Thirteenth Centuries. In: Occident et Proche-Orient: contacts scientifiques au temps des croisades, eds A. TIHON, I. DRAELANTS, and B. VAN DEN ABEELE. Louvain-la-Neuve 2000, p. 1–78 (see p. 61–6: Appendix II: The Latin Alphanumerical and Mixed Numerical Notation in the Works Associated with Stephen the Philosopher in Antioch); IDEM: Latin Alphanumerical Notation and Annotation in Italian in the Twelfth Century: MS London, British Library, Harley 5 402. In: Sic itur ad astra. Festschrift für den Arabisten Paul Kunitzsch zum 70. Geburtstag, eds M. FOLKERTS and R. LORCH. Wiesbaden 2000, p. 79–90.

[3] KING, David A.: The Ciphers of the Monks. A Forgotten Number-Notation of the Middle Ages. Stuttgart 2001.

[4] Discussed several times by Emmanuel POULLE and Richard LEMAY: see LEMAY, R.: L'Introductorium in astronomiam d'Abou Ma'shar de Balkh. In: XIIe Congrès international d'histoire des sciences, Paris, 1968, Actes, I A, textes, p. 101–123 (at p. 111–113), and I B, discussion des rapports (LEMAY and POULLE), p. 105–108; LEMAY, Richard: The Hispanic Origin of Our Present Numeral Forms. Viator 8 (1977) 435–62 and rubrics to illustrations; IDEM: Arabic Numerals. In: Dictionary of the Middle Ages, ed. J. R. STRAYER, I (New York 1982). 382–98; IDEM: Roman Numerals. In: Dictionary of the Middle Ages, VIII (New York 1988) 470–3; IDEM: Abū Ma'šar. Liber introductorii maioris ad scientiam judiciorum astrorum. 9 vols. Naples 1995–6 (henceforth 'LEMAY'), IV, p. 323–4.

IV

in space than additive systems: e. g. .iiii. and .viiii. Sometimes quite bizarre combinations of Roman numerals, initial letters, additive and subtractive systems appear, such as in an Angers manuscript of RAYMOND OF MARSEILLES'S astronomical tables, where 'Xol' is used for '19' $(10+8+1)$.[5] The different choices of numeration found in different twelfth-century texts[6] help one to associate works and authors. The 'Eastern Forms', for example, are distinctive of Latin texts associated with the teaching of ABRAHAM IBN ᶜEZRA in the 1140s; alphanumerical notation is found in works by, or associated with, 'STEPHEN THE PHILOSOPHER' in Pisa and Antioch in the 1120s. This article deals with yet another kind of numeral form, namely the ligature of 'x' and 'l' for the number 40.

In this symbol, the Roman numeral 'x' has a small capital 'L' (often reduced to a hook or cup-shape) attached to its top right-hand arm. This 'L'-element can easily be confused with the termination of the right-hand arm itself, and the symbol becomes difficult to distinguish from a simple 'x'. The appearance of 'ten' where 'forty' would be expected is often evidence that, at any earlier stage in the transmission of the text, the xl-ligature had been used. The occurrence of this ligature in Latin astronomical texts has been noted by Emmanuel POULLE[7] and RICHARD LEMAY.[8] On the other hand, this symbol has been recorded regularly by palaeographers working on documents written in the Iberian peninsula, by whom it is sometimes called 'x aspada'.[9] It is a characteristic of Visigothic script, appearing especially on epigraphic monuments and in charters from the sixth century onwards.[10] In the twelfth century it is a Visigothic feature taken over into Caroline script in Spain (see Plate 1), and it continues to be used in Galicia until well into the fourteenth century. We may presume that scribes who were accustomed to writing diplomatic documents found it natural to use the same ligature when they copied mathematical texts. The purpose of this short article is to list the texts in which the ligature appears and to draw some tentative conclusions from this evidence.

1) Manuscripts of the translations of John of Seville (fl. 1120s and 1130s)

i) ALBUMASAR (787–886), *Great Introduction to Astrology* (probably translated 1133). The xl-ligature is used in Cambridge, University Library, Kk.I.i (beginning/middle of 13th cent.; Lemay, IV, p. 58), Cambridge, Trinity College, O.8.34 (second half of 12th cent./

---

[5]   The tables, from RAYMOND OF MARSEILLES' *Liber cursuum*, occur among the fly-leaves of Angers, Bibliothèque municipale, 280: see RAYMOND DE MARSEILLE. Opera omnia, vol. 1, Traité de l'astrolabe, Liber cursuum planetarum, ed. and trans. Marie-Thérèse D'ALVERNY, Charles BURNETT and Emmanuel POULLE. Paris 2008. The abundance of variations in number forms in this manuscript is in no way an aid to use, and is more likely to be for the sake of show.

[6]   'Texts' rather than 'manuscripts' because scribes tend to copy the numerical forms without changing them (see Conclusions below).

[7]   POULLE, Emmanuel: Le traité d'astrolabe de Raymond de Marseille. *Studi medievali,* 3rd series 5 (1964) 866–900 (see p. 872–3).

[8]   LEMAY, IV, p. 323.

[9]   See MILLARES CARLO, Agustín: Tratado de Paleografía Española, 3rd edition, 2 vols, Madrid 1983, p. 273–4.

[10] See MALLON, Jean: Pour une nouvelle critique des chiffres dans les inscriptions latines gravées sur pierre. *Emerita* 16 (1948), 14–45, who gives two examples from the sixth century and one from the ninth.

early 13th cent.; Lemay, VI, p. 549), Paris, Bibliothèque nationale de France, lat. 14 704 (12th cent.; Lemay, IV, p. 41), Vatican, Vat. lat. 5 713 (12th cent.; Lemay, IV, p. 135).[11] In other manuscripts this ligature has been copied as '.x.': in British Library, HARLEY 3 631 (late 12th/early 13th cent.) '.x.' is written for '40', but an annotator who wrote the Indian numerals above the Roman forms correctly restored 40 were necessary;[12] in the colophon to the *Great Introduction* in Vienna, Österreichische Nationalbibliothek, 5 478, fol. 99r, '.m.cc.xviii.' has been written instead of '.m.cc.xlviii.'[13]

ii) ALCABITIUS (10th c.), *Introduction to Astrology* (date uncertain). The ligature is used in Oxford, Merton College, 259 (late 12th cent./early 13th cent.), Oxford, Bodleian Library, Bodley 430 (late 12th cent.) and Vatican, Reg. lat. 1 285 (early 13th cent.). In *Introduction* II [5], where all three manuscripts use the *xl*-ligature followed by 'iii' for '43', the scribe of the Vatican manuscript has added 'in al. xliii' ('in another <manuscript> 43'). Several manuscripts of the *Introduction* show the misinterpretation of the ligature as '.x.': these include Oxford, Bodleian Library, Ashmole 369, Berlin, Staatsbibliothek-Preussischer Kulturbesitz, lat. fol. 307, Bernkastel-Kues, Bibliothek des St.-Nikolaus-Hospitals, 208, Paris, Bibliothèque nationale de France, nouv. acq. lat. 3 091, Vatican, Barberini 236 and Wrocław, Biblioteka Uniwersytecka, Ac. IV F.[14] Other early manuscripts of the work, however, give the correct value in Roman or Hindu-Arabic numerals; these include Kraków, Biblioteka Jagiellońska, 578 (late 12th cent.) which was written in the Iberian peninsula.

iii) The *xl*-ligature occurs in Oxford, Bodleian Library, Digby 51 (mid 12th cent.), fol. 137vb within a set of three horoscopes datable to 19 November 1131, 22 July 1130 and 27 February 1121, following JOHN OF SEVILLE'S translation[15] of MESSEHALLAH'S *In rebus eclipsis Lune et in coniunctionibus planetarum*.

2) Manuscripts of the Toledan Tables (12th cent.)[16]

Of the many manuscripts of the Toledan Tables several early ones include the *xl*-ligature: Oxford, Merton College, 259 (late 12th cent./early 13th cent.; PEDERSEN, p. 153 and 793), Cambridge, Trinity College, O.8.34 (second half of 12th cent./early 13th cent.; PEDERSEN, p. 98 and 783), Cambridge, Gonville and Caius College, 456/394 (early 13th cent.; PEDERSEN, p. 97 and 783), Cambridge, University Library, Kk.I.i (beginning/middle of 13th cent.; PEDERSEN, pp. 103–4 and 783) and Munich, Bayerische Staatsbibliothek, Clm. 18 927 (late 12th/early 13th cent.). The *De differentiis tabularum* attributed to both

---

11 In this manuscript the ligature is surmounted with a little 'a' – the last letter of 'quadraginta'– which helps to distinguish it from '.x.' (see also Plate 1).

12 LEMAY: The Hispanic Origin. Fig. 3 (*pace* LEMAY, there is no difference in the manuscript between the forms of 'x' used for '10' and '40').

13 For a further example, from Paris, Bibliothèque nationale de France, lat. 16 204 (13th cent.) and Vatican, Barberini, lat. 340 (14th c.), see LEMAY, vol. VI, p. 549.

14 For these manuscripts see AL-QABĪṢĪ: The Introduction to Astrology. Eds Charles BURNETT, Keiji YAMAMOTO and Michio YANO. London and Turin 2004, p.157–191.

15 In this manuscript the text ends with a prayer for PLATO OF TIVOLI ('Tu, domine, miserere Platoni'), the presence of whose works is conspicuous in this manuscript (see below).

16 I owe these references to Fritz Saaby PEDERSEN. In the following paragraph 'PEDERSEN' refers to PEDERSEN, Fritz S.: The Toledan Tables. 4 vols (continuous pagination). Copenhagen 2002.

'IOHANNES ISPANUS' and 'IOHANNES HYSPALENSIS' (fl. 1120 – 1140),[17] includes, in Madrid, Biblioteca nacional, 10 053, fol. 87va, values which coincide with those in the Toledan Tables, but with 'x' written for '40' (PEDERSEN, p. 195, n. 2) and 'xvi' for '46'; in each case the 'x' has extensions that might indicate that the ligature is intended. [18]

3) Manuscripts of the works of Raymond of Marseilles (fl. 1141)

i) *Liber cursuum*. The *xl*-ligature is extant in the twelfth-century copies of RAYMOND'S *Liber cursuum*, i. e., the Tables for Marseilles (with their canons): Paris, Bibliothèque nationale de France, lat. 14704 (12th cent.; see Plate 2) and Angers, Bibliothèque municipale, 280 (12th cent.), while it can be inferred from the confusion of '10' and '40' in the remaining manuscripts of the work: Cambridge, Fitzwilliam Museum, McClean 165 (late 12th cent.), Oxford, Corpus Christi College, 243 (late 12th cent.) and Admont, Stiftsbibliothek, 318 (13th cent.). The use of the ligature in the date of composition of the Tables (m° c° x'° i°), had led earlier scholars to infer that they were composed in 1111. The 'tabula universorum annorum planetarum' (a table of the greatest, great, middle and small years of the planets) from the *Liber cursuum* appears separately in Oxford, Bodleian Library, Digby 51, fol. 131r, where the ligature is used.

ii) *Liber iudiciorum*. The *xl*-ligature appears in the fifteenth-century manuscript of this astrological text copied by ARNAUD DE BRUXELLES (15th c.): Paris, Bibliothèque nationale de France, lat. 10252 (see fol. 79r), and is implied by the error of 13 for 43 in the copy in Madrid, Biblioteca nacional, 10009 (fol. 139ra) and, again, by the writing of the 'present date' as 'millesimi. centesimi. x. i.' for '1141' in Paris, Bibliothèque nationale de France, lat. 16208, fol. 18va.

iii) *De astrolabio*. The sole manuscript of this text, Paris, Bibliothèque nationale de France, lat. 10266, was also copied by ARNAUD DE BRUXELLES, apparently from an early exemplar, which must have contained the *xl*-ligature, since '15' is read for '45' on fols 108v and 112v.[19]

iv) The mean motion planetary tables in Darmstadt, Hessische Landes- und Hochschulbibliothek, 765, fols 204r–209v (first half of 13th cent.; PEDERSEN, pp. 104 – 5 and 1211), bear some relationship to the tables of RAYMOND and use the *xl*-ligature.

4) Manuscripts of the translations of Plato of Tivoli (fl. 1134 – 1145)

i) AL-BATTĀNĪ († 929), *Opus astronomicum*. In Oxford, Bodleian Library, Digby 51, a late twelfth-century manuscript which includes a particularly large number of texts by PLATO OF TIVOLI,[20] the ligature is found within PLATO'S translation of AL-BATTĀNĪ'S *Opus astronomicum* on fols 1ra, 11r, 15r and 17vb.

---

[17] The text is published by MILLÁS VALLICROSA, José Maria: Una obra astronómica desconocida de Johannes Avendaut Hispanus. *Osiris* 1 (1936) 451 – 75 ('Avendaut' is MILLÁS' addition; it does not occur in the manuscripts).

[18] I owe this information to Fritz PEDERSEN.

[19] POULLE: Le traité d'astrolabe de Raymond de Marseille, p. 873.

[20] See the description of contents in LÉVY, Tony; BURNETT, Charles: *Sefer ha-Middot*: A Mid-Twelfth-Century Text on Arithmetic and Geometry Attributed to Abraham Ibn Ezra. *Aleph* 6 (2006), 57 – 238 (see p. 70 – 73).

ii) ABRAHAM BAR ḤIYYA (12th c.), *Liber embadorum.* The translation of the work is described as having been made 'anno Arabum .dx.' (A.H. 510 = A.D. 1116) both at the beginning and at the end of the text, but the horoscope given at the end shows that this is a mistake for '.dxl.' (A.H. 540 = A.D. 1145).[21]

5) Manuscripts of the translations of Gerard of Cremona (1114–1187)

Up to now, the only instance of a confusion probably caused by the *xl*-ligature in a translation by GERARD OF CREMONA is the writing of '.xlv.' as '.xv.' and '.xlix.' as '.xix.' in his translation of PTOLEMY'S *Almagest* in Madrid, Biblioteca nacional, 10 113, fol. 76r.[22]

Conclusions

Different schemes of numerals are specific to texts not to scribes. In Oxford, Bodleian Library, Digby 51, for example, in which the same scribal hands can be recognized throughout the volume, the Latin version of a text attributed to ABRAHAM IBN ᶜEZRA uses the 'Eastern Forms' of Hindu-Arabic numerals (fols 38v–42v), whilst an abacus treatise uses the abacus-forms of the Hindu-Arabic numerals (fol. 36v). It is most likely that the use of the *xl*-ligature in translations of mathematical and astronomical works reflects the usage of the wider society to which the respective translators belonged, rather than their desire to introduce a more convenient symbol for enumerating and calculating. It is unusual for Roman numerals to be combined into one ligature.[23] In the manuscripts of RAYMOND OF MARSEILLES cited above, the *xl*-ligature is used within the Roman numeral system, often alongside initial-letter abbreviations for the numerals ('q' for 4, 'o' for 8 etc.), but its origins are different: initial-letter abbreviations are not found in the charters written in Visigothic or Caroline script that exhibit the *xl*-ligature. In Munich, Bayerische Staatsbibliothek, Clm. 18 927, on the other hand, the ligature accompanies Hindu-Arabic numerals (see Plate 3).

The strongest evidence for the use of the *xl*-ligature in mathematical texts is in the works of JOHN OF SEVILLE, the Toledan Tables and the works of RAYMOND OF MARSEILLES. Of course, in the absence of autographs, one cannot strictly go further than to say that the ligature was used by the copyists of these texts, either because they belonged to a society

---

21 CURTZE, Maximilian: Urkunden zur Geschichte der Mathematik im Mittelalter und der Renaissance, I, *Abhandlungen zur Geschichte der Mathematischen Wissenschaften mit Einschluss ihrer Anwendungen,* 12. Leipzig 1902, p. 10 and 182. The discrepancy between the horoscope and the date was noticed by HASKINS, Charles H.: Studies in the History of Mediaeval Science. 2nd edition. Cambridge, Mass. 1927, p. 11.

22 See Claudius PTOLEMÄUS, Der Sternkatalog des Almagest. Die arabisch-mittelalterliche Tradition. Ed. Paul KUNITZSCH. 3 vols. Wiesbaden 1986–1991, II, p. 168–9. I am grateful to Paul KUNITZSCH for searching through the text on my behalf. The isolated mistake of '10' for '40' in Parma, Biblioteca Palatina 719 (14th cent.) in: *Der Sternkatalog,* p. 120, no. 637 ('8 deg. 10 min.' for '8 deg. 40 min.') is insufficient evidence for the existence of the ligature in the archetype of this manuscript.

23 'v' and 'i' were ligatured in Visigothic texts: see MALLON: Pour une nouvelle critique, Fig. 14. That the Western Form of the Hindu-Arabic '6' derives from from this ligature, as LEMAY (The Hispanic Origin, p. 457) claims, is less likely.

IV

Ten or Forty? A Confusing Numerical Symbol in the Middle Ages

in which the ligature was current, or because they scrupulously copied the forms of numerals they found in their exemplars (most of the manuscripts which preserve the ligature are not, in fact, Spanish and none are written in Visigothic script). Nevertheless, the fact that the *xl*-ligature occurs more frequently in the earliest manuscripts of the texts than in later copies, points towards it being in the authors' original copies. Moreover, the common use of the ligature in the three contexts mentioned above is backed up by their close relationship. It is very probable that JOHN OF SEVILLE was responsible for one version of the Toledan Tables and their canons.[24] RAYMOND OF MARSEILLES, on the other hand, provides the earliest example of the use of JOHN's translations,[25] whilst his Tables for Marseilles (*Liber cursuum*) are adapted from the Toledan Tables.[26] Cambridge, University Library, Kk.I.i and Trinity College, O.8.34, and Oxford, Merton College, 259, all contain both JOHN's translation of ALBUMASAR's *Great Introduction* and the Toledan Tables, while Paris, Bibliothèque nationale de France, lat. 14704 brings the latter text together with RAYMOND OF MARSEILLES' *Liber cursuum*. Moreover, a note in this last manuscript ends with greetings from a certain 'magister R', whom it is tempting to identify with RAYMOND.[27]

More problematic is whether the *xl*-ligature and other distinctive numerical forms which RAYMOND uses reflect in any way at all, local custom at Marseilles. RAYMOND is proud to have written astronomical tables for his native city, and one might presume from his words that he lived there.[28] But I have found no evidence that Visigothic script and/or the *xl*-ligature which is distinctive of it, was known in Marseilles in his time.[29] The main

---

[24] PEDERSEN, while exercising extreme caution, gives the evidence in *The Toledan Tables*, p. 194–5.

[25] In *Liber iudiciorum* §152 RAYMOND refers to the book 'quem Abumasar de annorum revolutionibus scripsit', probably referring to the book *In revolutione annorum mundi*, which forms part of the ALBUMASAR corpus (probably all translations of JOHN OF SEVILLE) in Paris, BNF, lat. 16204 (p. 302–333). In ibid., §214 he refers to the authority of both ALBUMASAR and ALCABITIUS: 'Sed Albumasar atque Adila ceterique quorum auctoritas potior habetur...' ('Adila' is a truncated form of Alcabitius's more complete name: 'Abdilazis Alcabitius'). The lots are the subject of the eighth book of the *Great Introduction* and the fifth book of ALCABITIUS' *Introduction*, respectively. ALCABITIUS, in particular, is followed closely in several paragraphs of the *Liber iudiciorum*: *Liber iudiciorum*, §23–27 = ALCABITIUS 1.16; §28 = 1.20–21; §35 = 1.18.

[26] It cannot be shown, however, that the canons to his tables are based on any of the extant Latin canons (PEDERSEN, *The Toledan Tables, p.* 754).

[27] LEMAY, IV, p. 35. Note on p. 224: 'Rogo vos ut faciatis transcribi librum quam citius poteritis, quia <eum> cum magno labore et difficultate habui. Sciatis quod ego sum sanus et incolumis per Dei gratiam, et magister R. similiter.'

[28] See the *laudatio* in the verse preface of RAYMOND's *Liber cursuum* (RAYMOND DE MARSEILLE. Opera omnia, vol. 1 (n. 5 above), vv. 38–41):

> Et Domini nostri Jhesu Christi super annos
> Massiliamque super nos hunc componere librum
> Sentiat ; est illic quia nostre gentis origo
> Natalemque locum nostro de numero clarum
> Fecimus.

The coincidence that both RAYMOND and WILLIAM THE ENGLISHMAN, 'a citizen of Marseilles', in the *Astrologia Marsiliensis* of 1220, knew the Toledan tables, which they ascribed to AZARCHEL, may be significant here: see PEDERSEN, *The Toledan Tables,* p. 754 and 1577 and JACQUART, Danielle: William English. Oxford Dictionary of National Biography, vol. 18 (Oxford 2004) 458–9.

[29] NEBBIAI-DELLA GUARDA, Donatella: La bibliothèque de l'abbaye Saint-Victor de Marseille (XIe–XVe siècle). Paris, 2005, makes no mention of the ligature, or of any other Visigothic features.

religious foundation in Marseilles, the Benedictine Abbey of St Victor, retained close links with centres in Catalonia, from which the scribal practices used in its scriptorium are said to have been derived.[30] There are instances of the *lx*-ligature in manuscripts containing the translations of PLATO OF TIVOLI, who worked exclusively in Barcelona. But Caroline script had replaced Visigothic in Catalonia much earlier than elsewhere in Spain, and by the twelfth century all traces of Visigothic features had disappeared.[31] Thus the use of the *xl*-ligature in the writings of RAYMOND OF MARSEILLES and PLATO OF TIVOLI is more likely to be due to the contacts of their authors, or of early scribes, with JOHN OF SEVILLE and his circle. PLATO'S relationship with this circle still requires investigation; the mention of his name at the end of a translation by JOHN OF SEVILLE in Oxford, Bodleian Library, Digby 51 (see n. 15 above) already hints at a connection. Other Arabic-Latin translators, working in Aragon and Navarre in the first half of the twelfth century, have not been observed to use the *xl*-ligature: e. g., PETRUS ALFONSI, HERMANN OF CARINTHIA, ROBERT OF KETTON, and HUGO OF SANTALLA (all 12th c.).

In the course of the twelfth century Hindu-Arabic numerals gradually became more common amongst astronomers. It seems that JOHN OF SEVILLE himself turned from using Roman numerals in his translation of ALBUMASAR'S *Great Introduction* to using Hindu-Arabic numerals when he translated the same astrologer's *Great Conjunctions*.[32] When GERARD OF CREMONA came to translate the *Almagest* (before 1175) he appears still to have been using Roman numerals,[33] though the evidence that he used the *xl*-ligature is sparse. When Hindu-Arabic numerals finally prevailed amongst mathematicians, the ligature disappeared altogether.

---

[30] See AMARGIER, P.: Les "scriptores" du XIe siècle à Saint-Victor en Marseille. *Scriptorium* 32 (1978) 213–220. See p. 214: 'De Catalogne les leçons de graphie sont arrivées à Saint-Victor'.

[31] MILLARES CARLO, Agustín: Tratado de Paleografía Española (n. 9 above), p. 153–7.

[32] The use of Hindu-Arabic numerals in both versions of this text is investigated in BURNETT, Charles: The Strategy of Revision in the Arabic-Latin Translations from Toledo: The Case of Abu Ma'shar's On the Great Conjunctions. In: Les Traducteurs au travail: leurs manuscrits et leurs méthodes, ed. J. HAMESSE. Turnhout 2002, p. 51–113, 529–40 (see p. 66–68).

[33] This is the conclusion of KUNITZSCH, *Sternkatalog* (n. 22 above), II, p. 8–9.

Ten or Forty? A Confusing Numerical Symbol in the Middle Ages

Plates

Plate 1:   A charter written in Caroline minuscule, dated 12 November 1157 (Spanish era 1195). Madrid, Archivo Histórico Nacional. Clero. Carpeta 899, núm. 7. Reproduced with transcription in MILLARES CARLO, Agustín: Tratado de Paleografía Española, 3rd edition, 2 vols, Madrid 1983, no. 150. Note the *xl*-ligature in the date, with a suprascript 'a' (= quadraginta).

Plate 2:   RAYMOND OF MARSEILLES, *Liber cursuum*. Paris, Bibliothèque nationale de France, lat. 14 704, fol. 125r.

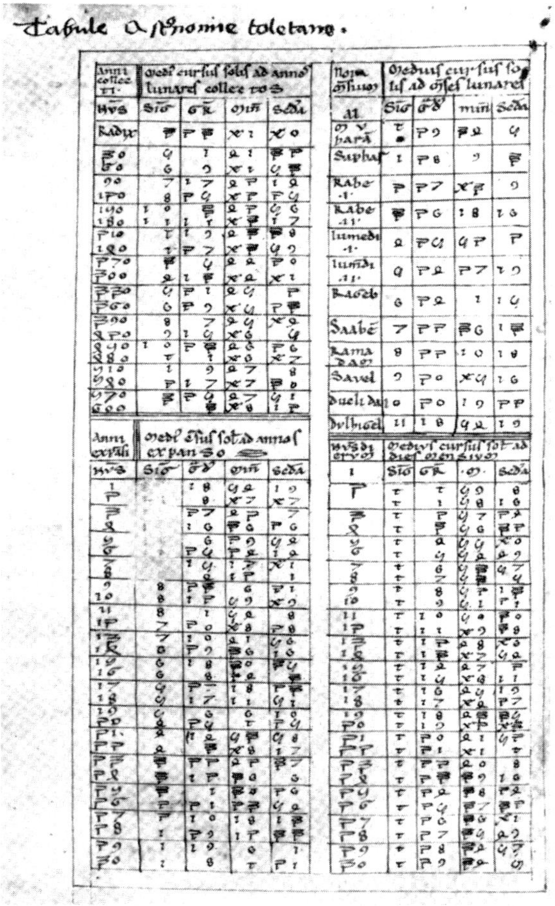

Plate 3: The Toledan Tables, in Munich, Bayerische Staatsbibliothek, Clm. 18927, fol. 1v. Note the *xl*-ligature alternating with Hindu-Arabic 4 in the place of the tens in the fourth column from the left.

V

Indian Numerals in the Mediterranean Basin in the Twelfth Century, with Special Reference to the "Eastern Forms"

The Arabic numerals which are now used universally are of ultimately Indian origin and were called "Indian" by Arabic, Greek and Latin scholars of the Middle Ages. The time, place and context of their introduction and development in all three language cultures are still obscure. This article seeks to illuminate, in particular, the use of one kind of Indian numeral (the "Eastern forms") which was shared by Arabic, Greek and Latin mathematicians in Italy and the Eastern Mediterranean in the twelfth century. By the early thirteenth century this kind had been displaced in Latin contexts by the "Western forms" which are the ancestors of our present Arabic numerals.

Mathematical notation is independent of language; it is symbolic, and does not represent sounds. Therefore, there is no need for different notations to be used in different languages, even when they are written in different scripts. Nowadays, the same mathematical notation is used and understood throughout the world, by Chinese, Arabic, Russian and American mathematicians. There was a potential for this to happen in the Middle Ages too. The Indians had invented a symbolic notation for the nine digits and the zero, which was taken over by Syrian and Arabic writers and eventually passed to Western Europeans. Scholars writing in Syriac, Arabic, and Latin alike referred to these symbols as "Indian figures," and they all participated in the same, distinctive, method of calculating with them. Thus a common mathematical language was shared by mathematicians in Bath and Baghdad, in Roskilde and Marrakesh. However, unlike the situation in the modern world of global communication in which systematization has become the norm, in the Middle Ages, symbols inevitably changed and diversified as they travelled from place to place, and as a result of the passage of time. One change which had great consequences for the history of Indian numerals was that which took place somewhere on the Western fringe of the Mediterranean World, which resulted in the forms of the numerals that prevailed among Latin scholars and have eventually been adopted universally by mathematicians (our "Arabic numerals"). What I would like to draw attention to in this paper are certain forms of Indian numerals shared for a while by Latin, Greek and Arabic scholars, but which, in the end, were *not* adopted by Western scholars, with the result that there is now a split between the printed forms of numerals used in the Western world and those used in most parts of the Islamic world.[1]

---

1 For the Arabic side of the story, see Kunitzsch in [Folkerts 1997], and [Kunitzsch 2002]. I am most grateful for Paul Kunitzsch's advice. I am also indebted to Menso Folkerts for the generous and prompt loan of microfilms, to Nigel Wilson for a careful reading of the article,

The Eastern and Western Forms of Indian Numerals[2]

The most obvious differences between the numerals which became the norm in Latin Europe (henceforth the "W[estern forms]") and forms closer to the printed (Eastern) Arabic shapes (henceforth, the "E[astern forms]") are as follows:[3]

W: 2 and 3 tend to be "upright," giving the impression of being the cursive representation of two or three horizontal lines. Sometimes the vertical orientation of the form is emphasised by being terminated with a straight descender.

E : 2 and 3 look as if they are "on their backs," giving the impression of being the cursive representation of two or three vertical lines.

W: In the case of 4, the Arabic form was made up of a hook and loop. In W, the loop predominates over the hook, which disappears.

E : The hook predominates over the loop, which disappears.

W: 5 is a cup-shape terminating on the right with a vertical descender.

E : 5 resembles a capital "B," sometimes turning into a figure-of-eight, at other times into a circle either crossed by a horizontal line or squeezed by a belt round its waist. In Arabic, the capital "B" is usually reversed.

W: 6 is a circle or spiral terminating on the left with a vertical ascender, which is sometimes bent over into a horizontal plane, or continues the curve of the spiral.

E : 6 is a cup terminating on the right with a vertical descender. Sometimes the curve-and-descender is replaced by a stepped- or zigzag-shape.

and to Jeremy Johns, Michael Matzke, Bernd Michael, Fritz Saaby Pedersen, Julien Veronese and Clare Woods. All the examples of Eastern forms in Latin manuscripts mentioned in this article are given in the Table on pages 265–267, alongside representative examples of other forms, and of forms found in Arabic and Greek manuscripts. Note that the following terminology is used:

algorism   A text describing how to use Indian numerals in arithmetical calculations deriving ultimately from al-Khwārizmī's *Indian Arithmetic*.

Arabic   Written in Arabic. Therefore, "Arabic numerals" are numerals (of any kind) written by Arabic scribes, and not "Arabic numerals" written by Latin scribes.

Indian numerals   The symbols for numerals known now in the West as "Arabic," but referred to by Arabic and Latin scholars as "Indian." They derive from Indian symbols and are characterised by having place value.

row   The numerals 1 to 9 set out in a line.

"standard"   The Western forms that became most widespread in the Middle Ages, from the early thirteenth century onwards.

2 Scholars have often referred to these forms respectively as *hindī* ("Indian") and *ghubārī* ("dust"), but the inappropriateness of these terms has been demonstrated in [Kunitzsch 2002]. For the forms of the Indian numerals in Indian scripts see [Renou & Filliozat 1985: 702–708]; their diffusion is described in [Ifrah 1981: 460–490].

3 Occasionally Latin scribes change the *direction* in which the numerals face, but usually the numerals are not turned round in respect to their Arabic forms: see [Burnett 2001a].

W: The two branches of 7 make the shape of a gallows (gnomon) or a lamda.

E : The two branches of 7 make the shape of a "v."

W: 8 is formed from two circles, one on top of the other.

E : 8 is a lamda shape in which the left-hand branch is sometimes tucked in to form a bow or a circle.

W: 9 is a circle extended downwards on the right in either a straight or a curved line.

E : The form is the same as in W except that sometimes the mirror-image is substituted, perhaps to differentiate it from 8.

Both W and E use a small circle for 0, sometimes substituting the astronomical symbol "t."[4]

It is not the purpose of this article to explore the origin of these differences. Suffice to say that it is more plausible to suppose that all the Western and Eastern forms derive ultimately from the same source, rather than that, in some cases, symbols from a completely different kind of source have been substituted. For example, it has been suggested that the Western form of 5 is simply the adoption of the way of writing the Roman "v" in Visigothic Spain [Lemay 1977: 452–453]. However, already in Arabic contexts the loops of the reversed-"B" form of 5 often fall away from the straight ascender, which becomes abbreviated; if one imagines this process continuing until they fall into a horizontal plane and the last curve becomes straight, the Western form of 5 would result. Also, the spiral of the Western form of 6 could derive from the cup of the Eastern form: the spiral is still attached to a vertical descender in several versions of the abacus numerals of the eleventh and twelfth centuries (see [Folkerts 1970, Table I] and the *Liber Floridus*, facsimile edition, Gent 1968, fol. 85v). But the ease in which purely hypothetical conjectures like this can be made shows the danger of spending too much time speculating on origins.

Variant Forms Given in al-Khwārizmī's *Indian Arithmetic*

Before looking at the regional development of different numeral forms, it would seem natural to consider the transmission of the essential text on calculating in the Indian way (*al-ḥisāb al-hindī*), which promoted the use of Indian numerals: al-Khwārizmī's *On Indian Calculation* (ca. 820). Although the original Arabic text has been lost, it gave rise to a whole genre of works, in Arabic, Latin and Greek, which convey more or less of the original text, and are known generically as "algorisms," after the name of the first author.

---

4   Hence zero is called "circulus" in early algorisms (*Dixit Algorismi* and *Liber Alchorismi*) and in the first version of Abū Maʿshar's *On the Great Conjunctions* (see below, p. 242). The term "dāʾira ṣaghīra" ("small circle") is also used for "zero" in Arabic.

Already in the Arabic algorisms different ways of writing certain numerals are mentioned (see [Kunitzsch 2002]). In a copy of an algorism by Ibn al-Yāsamīn of Morocco (d. ca. 1204) a row of Western forms is given first, followed by a row of Eastern forms (Plate 1). Ibn al-Yāsamīn writes:

> These are the shapes which are called "al-ghubār" (followed by the Western forms), and they are also like this (followed by the Eastern forms).[5]

The differences in certain forms are mentioned (but not illustrated) in the Latin text which represents al-Khwārizmī's work most closely, known from its incipit as *Dixit Alchoarizmi* (*DA*):

> Est quoque diversitas inter homines in figuris earum. Fit autem hec diversitas in figura quinte littere ac .VI., .VII. quoque et octave [Folkerts 1997: 28, commentary 111–112].

In the *Liber Alchorismi* (*LA*), a similar passage can be found:

> Est autem in aliquibus figurarum istarum apud multos diversitas. Quidam enim septimam hanc figuram representant Ꮞ, alii autem sic Ɣ, vel sic Ꮭ; quidam vero quartam sic Ᵹ [Allard 1992: 69].

Of these alternatives, Ɣ is clearly the Eastern form of 7, while Ᵹ is the hook-and-loop form of 4 which underlies both the Eastern and Western forms.[6] In one manuscript that contains the hybrid version of the *Liber pulveris* and *Liber Alchorismi* — Vat. Pal. lat. 1393 — the row of Eastern forms are added after the phrase "Quidam etiam sic scribebant figures." In MS Dresden C 80, similarly, two rows of forms are given, and the author seems to be saying that one is more genuinely "Indian" than the other (see p. 257 below); the Eastern forms are clearly recognizable, and 4, consisting of two curves, one on top of the other, like an open cursive "e," resembles that of Vat. Pal. lat. 1393. In another fifteenth-century manuscript of the algorism, Paris, BNF, lat. 10252, the two rows are given, along with the abacus names, but the numerals have been copied so badly that it is difficult to recognise the Western or Eastern forms in either row.

The fact that the numeral forms vary from one algorism to another, and that different numerals are picked out as having alternate forms in different algorisms, suggest that the differences reflect not the original text of al-Khwārizmī, but rather the practices current at the time and in the locality of the author or scribe of the

---

5 The photograph of MS Rabat, Maktaba al-ʿāmma, k 222 has been reproduced in [Abū Fāris 1973: 232] and [Kunitzsch 2002], and the text is discussed in [Köbert 1975]. This manuscript itself is written in the Eastern Arabic script and is of a later date than Ibn al-Yāsamīn, but it is interesting to note that the 5 in the row of Eastern Numerals has its loops on the right, just as in the Palermitan Arabic sources discussed below (p. 243).

6 Form Ꮭ is similar to our present 7, which is also the shape of the "tironian *et*," and belongs to the tradition of the Western forms. I have not found Ꮞ, which is like an upside-down lower-case "h" with a curved ascender, among the Eastern or Western forms, but it is curious to observe that occasionally it appears, in an upright form, for 7, in the fifteenth-century copy of an early manuscript of Raymond of Marseilles's *Liber iudiciorum* in Paris, BNF, lat. 10252 (see fols. 88r and 89v).

algorism. The algorisms did not determine which forms to use, but rather reproduced the forms in use; nor were they the sole means by which Indian numerals were transmitted. It is to the temporal and regional differences in the numerals that we should now turn.

The Western Forms

A consistency can be observed in the Western forms, from their earliest examples in a Latin manuscript of Isidore of Seville's *Etymologiae* written in the monastery of Albelda in the Rioja in 976, through their stylised representation on the *apices* (counters) of the "Gerbertian" abacus (where they are also used to number the columns of units, tens, hundreds, etc.), to the forms found in most of the twelfth-century texts on the algorism (exceptions have been mentioned above), until, by the time they were used in the early thirteenth-century manuscripts of translations made in Toledo, they became the standard Medieval forms of Indian numerals. The tenth-century examples and the abacus texts already show all the distinctive features of the Western forms, except that the 4 still retains both its hook and its loop.[7] The earliest form of the Latin algorism — *Dixit Alchoarizmi* — preserves forms (for 2, 3 and 4) that are strikingly similar to those of the Albelda MS; the rotation of the forms of some of the numerals (5, 7, 9) in the context of the abacus is attributable to the non-directionality of the *apices* (see [Beaujouan 1948]). To identify a specific time and place for the origin of these distinctive features is more difficult. It is possible that they arose among mathematicians in Islamic Spain, from which they were brought to the monasteries of León and Old Castile by Christian refugees in the ninth and tenth centuries, and they remained in use after the Reconquista.

A Latin manuscript written in Bavaria in the late twelfth century — Munich, Clm 18927 — compares different forms of Indian numerals by setting them out in three rows: the first is described as "toletane f⟨igure⟩" ("Toledan figures"), the second as "indice f⟨igure⟩" ("Indian figures"), the last is not described, but is that which the scribe of the manuscript uses himself (Plate 3). This last row, of which the most distinctive element is the vertical tails extending below the 2 and 3, is characteristic of some manuscripts written in England, France and Germany in the twelfth century, e.g. Cambridge, Trinity College, O 7 41, Paris, BNF, lat. 14704 and lat. 16208 (both manuscripts of the works of Raymond of Marseilles), and Munich, Clm 13021 (written by Sigisboto of Prüfening (?), between 1163 and 1168). The Munich manuscript also gives 3 in the form of a cleft stick, which occurs too in a multiplication table added to a manuscript written by Wolfger of Prüfening in 1143 (MS Vienna, ÖNB, 275), and in year-values in the Annals of

---

7   To the examples listed in [Folkerts 1970, Table I] and [Ifrah 1981: 506], one may add Vat. Reg. lat. 1308, copied in the eleventh century in France by an Italian scribe who numbered the quires with Arabic numerals and their names (see [Bischoff 1990: 23, n. 27]). For an aberrant example see [Gibson & Newton 1995]. See also [King 2001: 309–317].

Regensburg for the period 1152 to 1197 (MS Munich, Clm 14733 written by Hugo von Lerchenfeld).[8] These forms may have been transmitted with the early manuscripts of the *Liber Ysagogarum* and Toledan Tables, both of which are included in Clm 18927 itself.[9]

What is more difficult to ascertain is the significance of the description of the first row of numerals as "toletane f⟨igure⟩." The 2 and 3 do not have the vertical tails. The non-tailed form is normal for the numerals in later manuscripts of translations made in Toledo (see below p. 255), and eventually became standard. That Indian numerals were known to Latin scholars in Toledo would seem to be guaranteed by their presence in the *Liber Alchorismi* of "magister Iohannes" which accompanies Toledan translations and precedes a calendar relating to Toledo and computus tables for 1143–1159 in its best manuscript, Paris, BNF, lat. 15461.[10] However, that Indian numerals were used by the principal translator of mathematical works in Toledo, Gerard of Cremona (1114–1187), has been questioned [Kunitzsch 1990, II: 8–9]. Two pieces of evidence could be significant in assessing this question. In the earliest copy of Gerard's major mathematical translation — Ptolemy's *Almagest* in Paris, BNF, lat. 14738 — the scribe, who is probably French, but using Spanish parchment, is evidently copying from a manuscript which used Indian numerals, for he starts by transcribing them (with obvious difficulty), and then gives up after the first few pages, and substitutes Roman numerals.[11] In the best manuscripts of a revision (possibly by Gerard, and certainly made in Toledo) of the translation of Abū Ma'shar's *On the Great Conjunctions*, two year-values have been copied in a "fossilised" form, presumably because they were not understood by the copyist (Plates 2 and 4).[12] They show the early form of 4 consisting

8   Examples of these German manuscripts are given in [Menninger 1958, II: 238–239; von Fichtenau 1937; Arrighi 1968; Lemay 1977].

9   The earliest Latin versions of the Toledan Tables were not necessarily written at Toledo; for they owe their name to the fact that they derive from tables made for the meridian of Toledo by al-Zarqāllu in ca. 1080; the first evidence of Latin knowledge of them is in Aragon [North 1995] and Marseilles (Raymond of Marseilles's adaptation of the Tables to the meridian of Marseilles in 1141). Associated with this form is the writing of a compendium of .xl. . ⟩ᴄ. for "40" which is found in the works of Raymond and in Clm 18927, and which is of Spanish origin.

10 For other references to the algorism by Toledan scholars see [Allard 1992: xx; Burnett 1994: 428–429].

11 I am grateful to Patricia Stirnemann for her advice concerning the provenance of this manuscript. Further examples from early copies of Gerard's translations are discussed in [Jacquart 2001: 215–219], to which may be added the copy of Gerard's translation of the *Liber ad Almansorem* in Cambridge, University Library, Additional 9213, where 2 has a tail, but 3 has not (as in Bodleian, Digby 51).

12 The passage containing these year-values does not occur in the original translation of *On the Great Conjunctions* (apparently by John of Seville); for the two Latin versions of the text see [Burnett 2001b]. It is possible that a gloss close to these forms (but without a corresponding reference mark in the text) refers to these strange forms: "hanc litteram nec hunc numerum intellexi" ("I have not understood this letter or this number"). In this case, the reviser himself (who was also responsible for the gloss: see [Burnett 2001b: 55–56]) seems to have been confused by the numeral forms.

of a hook and a loop, and a version of the Western form of 6 that occurs only in the earlier Latin tradition and in Arabic.[13] These forms may retain the shapes of the numerals written in the original Arabic manuscript. That the Arabic manuscripts of *On the Great Conjunctions* included Indian numerals is certain, since Abū Maʿshar writes a very high number (of days in a long period of years) first in words, and then "in the Indian form" (*aṣ-ṣūra al-hindīya*, translated "figura indica / figure indice"). It is curious that, aside from in the two cases mentioned above, the Latin manuscripts (none of which are before the very end of the twelfth century) give the Indian numerals in their standard Western form, and the fossilised forms too have been glossed with these standard forms.

Thus, though it is likely that Indian numerals were used in Toledo, even outside the context of the algorism, the fact remains that we cannot yet, as far as I am aware, point to a manuscript definitely written in Toledo in the twelfth century, in which the scribe shows himself familiar with Indian numerals.[14] Moreover, no Arabic manuscript has been found written in Spain (or any part of the North-West African realm of the Almoravides or the Almohades) which uses the Indian numerals, in any form, until after 1284 [Kunitzsch 2002].

The Palermitan Forms of the Court of Roger II (1130–1154)

For Sicily we are more fortunate. For there is a manuscript written in an Arabic hand in which Indian numerals occur, and which can be dated and located quite accurately. The manuscript is London, British Library, Harley 5786, which consists of a Psalter written in three languages: the Greek text on the left, the Latin text in the centre, and the Arabic text on the right (Plate 5). The scribes of each language are different from each other and are evidently professional. Each psalm is numbered in the numeral system appropriate to the language. Thus alphabetical numeration is used for the Greek, and Roman numerals for the Latin. For the Arabic, Indian numerals are used, and among these we encounter both the Western form 6 with a straight ascender (as in the Toledan translation of Abū Maʿshar's *On the Great Conjunctions*), and a 4 with a hook and loop. Moreover, 8 is written in its Western form, with two circles. It must be noticed, however, that 2, 3, 5 and 7 have the characteristics associated with the Eastern forms.

13 6 with a straight ascender is found in the Albelda manuscript of 976 and in the abacus texts, but is also used in the Arabic text in Harley 5786 discussed in the next section.

14 The best candidate for such a manuscript is Paris, Bibliothèque de l'Arsenal, 1162, written in Spain, with Spanish decoration, in 1143, which contains the copy of the translations of texts on Islam (including the Qurʾān) commissioned for Petrus the Venerable, abbot of Cluny. The last text is a translation by "Peter of Toledo," and the whole collection has been called the "Collectio Toletana," but most of the translations were made by Hermann of Carinthia and Robert of Ketton (see below, p. 248), and Peter the Venerable's itinerary took in only the valley of the Ebro and Old Castile. The Indian numerals, which are used to number the folios on the *verso* on fols. 1–17, are, however, close to the "toletane f⟨igure⟩" in form and look later than the text. See [d'Alverny 1948: 77–80, 108–109].

244

MS Harley 5786 was written before 1153, for this date has been written on the last folio: "anno incarnationis dominice .m.c.liii. ind. ⟨i⟩ mensis ianuarii die octavo die mercurii."[15] The nature of the work makes the most likely place of origin the Palermo of Roger II. For King Roger astutely promoted the interests of the three language communities of his Sicilian kingdom, and set up a separate chancery for Latin, Greek and Arabic documents. Jeremy Johns has shown that his Arabic chancery (or "Dīwān") was formed on the model of that of Fatimid Egypt, and Roger most probably brought in scribes trained in writing the script from Egypt or another East Mediterranean Islamic chancery.[16] The Arabic script of the Psalter is typical of the Eastern Arabic script used in the Royal Dīwān, but concedes to local custom by writing the Arabic letters *fā'* and *qāf* in the *maghribī* way, rather than as the Eastern Arabs did. This may account, too, for the substitution of Western 6 and 8 for the predominantly Eastern forms of the numerals.

There is a striking similarity between the Indian numerals in the Psalter and those used in the tables in one manuscript of the translation of Ptolemy's *Almagest* made in Sicily in ca. 1165 [Haskins 1927: 157–162; Murdoch 1966]. The translation (as we are informed in the preface, written by the unnamed translator), was made from a Greek manuscript which was brought from Constantinople to Palermo as a present from the Greek emperor to the Sicilian king. Although the translation was made from Greek, it also includes some material from the Arabic astronomer, al-Battānī,[17] thus nicely combining Greek, Arabic and Latin culture, as does the Harleian Psalter. The most accurate, and probably the earliest, of the four known manuscripts of this translation is Vatican, Pal. lat. 1371, and it is this manuscript that contains the numerals which are very similar to those of the Harleian Psalter, when allowance is made for the fact that one is written by an Arabic scribe, the other by a Latin one (Plate 6). Particularly striking is the resemblance of the form for 5, which looks like a "B" with the ascender extended upwards and which is reversed in respect to the usual form found in Arabic manuscripts.[18] The only observable differences are in 4, in which the loop in the Arabic has been

15 The date (discussed in [Haskins 1927: 184]) is now very faint (even under ultraviolet light), due to the rubbing of the last folio, and the indiction number has disappeared.

16 In his detailed study [2002] Johns has shown that the script of the Dīwān is found only in the royal palace and court, and differs from the variety of scripts used in private documents from Sicily of the period, which exhibit strong *maghribī* (Western Arabic) features. Roger's "Great Emir" ("magnus ammiratus"), George (d. 1151), who had ultimate charge of the chancery, was from Antioch. I am most grateful to Jeremy Johns for his advice, and for sending me material in advance of its publication.

17 [Lemay 1987: 466] quoting MS Vat. pal. lat. 1371, fol. 67v: "Hoc fuit necesse, ut ait Albategni, tabulas diversitatum apponere."

18 In the Arabic context of the Harleian Psalter it is clear that the scribe, writing from right to left, has executed the form in one stroke, ending with a sweep of the pen upwards. In the Latin context, because the scribe was writing the form in the opposite direction, he uses two strokes of the pen, the first a single downward stroke, the second, the two bows of the "B." The resulting forms, however, are very similar to each other (I owe the confirmation of this to Clare Woods).

simplified to a right-angle, and in 6, in which the straight ascender has been bent over into a vertical plane.

The agreement of the Indian numerals in the Harleian Psalter and a manuscript of a translation of Ptolemy's *Almagest* made in Sicily, suggests that both texts were written in the same milieu. This is probably Palermo, to which the Greek manuscript of the *Almagest* had been brought, and where, according to a marginal note, the translation was made.[19] That Indian numerals had a privileged position in Palermo is indicated by the fact that Roger II himself used them on his Arabic coins, and is the first Western European — or indeed Arabic — ruler known to have done this. From the evidence of the two dated coins surviving,[20] the same forms of Indian numerals are used as in the Harleian Psalter; in particular, both sources give a 5 with the ascender extended upwards and reversed in comparison with the usual Arabic form. It is likely, then, that the translator of the Sicilian *Almagest* himself used the Indian numerals, which would have been familiar to the scholars of his circle who would have been associated with the royal household. These Palermitan forms, however, do not seem to have been known widely amongst Latin scholars, since they have not been identified in any other Latin manuscript. Therefore, when copies of the translation were made for distribution outside this circle, more familiar forms would have been substituted; hence the Roman numerals used in Florence, Biblioteca Nazionale, Conventi Soppressi, A. 5. 2654 (ca. 1300). This suggests that, in this instance, the use of Indian numerals *preceded* that of Roman numerals in the history of the transmission of the text.

The Eastern Forms: Hugo of Santalla and Hermann of Carinthia

As we have seen above, the second row of numerals in Clm 18927 is described as "indice f⟨igure⟩" ("Indian figures"). These are the Eastern forms of the Indian numerals, to which the rest of this article will be devoted.[21] While we are badly informed about the use of Western forms in Arabic manuscripts in the Western part of the Islamic world, there are several examples of the use of Eastern forms

---

19 Vat. Pal lat. 1371, fol. 41r: "Translatus in urbe Panormi tempore Roggerii per Hermannum de greco in latinum." For the attribution to "Hermann" see below, p. 248.

20 For a "folles" of 533 (= 1138/39) on which the *hijra* date is given in Indian numerals see [Travaini 1995: 53, 284 (no. 193), Pl. 12], in which the illustration is taken from [Spinelli 1844: Pl. VI, no. 32]. Another coin was struck in Messina with the date 543 (= 1148/49; [Travaini 1995: 300–302, no. 247; Grierson 1998: 626]). Travaini concludes (p. 302) that "questi due follari sono forse, a quanto pare, i più antiche esempi nella numismatica di data con l'anno in cifre arabe, e tale aspetto merita ulteriori indagini." I owe these references to Jeremy Johns. See Plates 7 and 8.

21 The only parallel I have found for singling out the Eastern forms as "Indian" is in a commentary by Ḥusayn ibn Muḥammad al-Maḥallī to a work on arithmetic by al-Sakhāwī (14th century), cited in [Ifrah 1981: 502] in which the Western forms are called *ghubārī* and the Eastern, *hindī*; but see also [Kunitzsch 2002].

in Arabic manuscripts written before 1200 (see [Irani 1955/56; Kunitzsch 2002]). What has received less attention is the use of Eastern forms in Latin and Greek manuscripts in the same period. One of the few scholars who have commented on the Latin diffusion of these forms, Richard Lemay, sees Hermann of Carinthia (fl. 1138–1143) as a key figure [Lemay 2000]. While his arguments are based on dubious attributions, it is instructive to start our investigation by looking at the evidence of the copies of works by Hermann of Carinthia and his close associate Hugo of Santalla.[22]

A rare testimony to the origin of an Arabic manuscript used by a Latin translator is provided by Hugo of Santalla (fl. 1145–1179), a *magister* in the service of Michael, bishop of Tarazona (1119–1151) and his successors, who states, in the preface to his translation of the commentary by Ibn al-Muthannā on the *Zīj* (Astronomical Tables) of al-Khwārizmī, that his patron got his manuscript from the "rotense armarium." This has been identified as the library of Rueda de Jalón, the stronghold to which the Banū Hūd kings of Islamic Zaragoza retreated after the fall of their kingdom in 1118 [Haskins 1927: 75]. Two kings of the Banū Hūd dynasty were renowned for their mathematical prowess [Hogendijk 1986], and it is likely that at least part of the library of mathematical texts that they possessed survived the removal to Rueda. Since they were aware of recent mathematical works written in the East, such as the *Optics* of Ibn al-Haytham, it is possible that they were familiar with the Eastern forms of the numerals. No extant Arabic manuscript has been identified as belonging to their library. But it may be possible to pick up hints of what these manuscripts looked like from the earliest manuscripts of Hugo's translations.

Four manuscripts take us close to the activity of the translator himself. Two of these (Oxford, Bodleian, Digby 159 and Cambridge, Caius College, 456 [= C]) belong to what must originally have been a collection of the translations of Hugo, possibly put together within his life-time and written, for the most part, by an English scribe.[23] The other two (Bodleian, Arch. Seld. B 34 [= A] and Digby 50) were also written in the same hand — that of a professional Italian scribe — and are probably a little later than the first two. In all these manuscripts Roman numerals are normally used, but the Eastern forms occur in two contexts. The first is that of two illustrative tables in Ibn al-Muthannā's commentary on al-Khwārizmī (the very text that Hugo mentions as coming from the library of Rueda) in MSS A and C (see Plates 9 and 10). It is clear from both manuscripts that in the exemplars from which the numerals were copied the Eastern forms were used. However, the scribe of C was evidently unfamiliar with the numerals and did not understand

22 I refer to Lemay's attributions to Hermann of the Sicilian translation of the *Almagest* (discussed above p. 244 and below p. 248) and the *Liber Mamonis* (discussed below, p. 251). It is *prima facie* unlikely that Hermann of Carinthia would have translated the *Almagest* from Arabic as he claimed to be intending to do) and from Greek *and* would have written a book — the *Liber Mamonis* — which uses the terminology of yet a third translation of the work, again from Arabic: see [Burnett 2000b: 10–13].

23 At least one further volume must have existed when a copy was made of the collection into MS Bodleian Savile 15, written in the fifteenth century: see [Burnett & Pingree 1997: 9–10].

how they functioned; for he reverses the order of the numerals and the direction they face (thus providing a mirror image of the table), and misinterprets the symbol for 4 as the Roman numeral xxxvi or lxxvi. In MS A, on the other hand, the numerals have been filled in by a hand other than the professional scribe: presumably by a scholar, and a scholar who was familiar with Indian numerals.[24] The forms copied by the scribe of C apparently differed a little from those used by A. The two scribes share a distinctive form of 7 with a loop at the bottom, but A's 5 is a figure-of-eight, whereas B's 5 is the "B" shape. More perplexing is the 4: while A clearly draws the hook-and-loop shape, C evidently misinterpreted quite a complex symbol in his exemplar in writing xxxvi or lxxvi for the single Arabic digit.[25]

That the scribe (as distinct from the annotator) of MS A was unfamiliar with the Eastern forms is indicated by their appearance in the other text copied by him: Oxford, Bodleian, Digby 50, which contains Hugo's translation of an Arabic text on geomancy, the *Ars geomancie*.[26] In a sequence of chapters following Hugo's text (beginning "Notandum est quod Leticia"; fols. 93–106), it is evident that the scribe was copying from a text that used the Eastern forms, but was not himself used to writing these forms (Plate 11). His first attempts (on fol. 94r–v) are rather crude, and include only the numbers 1, 2, 3 and 4. From fols. 99r onwards, however, he gives systematic lists of each of the sixteen geomantic figures in each of the twelve astrological places. Hence we can see what the scribe does as he repeats the numbers 1 to 12 sixteen times. He copies the Eastern forms throughout for the first three times. The first time his numerals are a little clumsy, with the vertical strokes of 3 separated, and the curve of 6 exaggeratedly rounded; for the second and third times the numbers are written more evenly and confidently. For the fourth time, however, he writes Roman numerals for 1, 2 and 3 and (inadvertently?) realises the symbol for 6 as "etiam."[27] The next time all the numerals are Roman except 9; and after this the scribe only uses Roman numerals. It is as if he, having at last understood the significance of the pattern of recurring symbols, transcribed them into the symbols that he knew best, namely Roman numerals.

---

24 The same hand has provided, in the margins of fol. 20r, a revised translation of a large part of the chapter headed "Quare alhoarizmi maius attadil…" [Millás Vendrell 1963: 111]: he crosses out the passage printed at [Millás Vendrell 1963: 112, line 9–113, line 6] and rewrites it with different terminology and different transliterations of the Arabic words. He also comments, on the same folio, that the geometrical figures are in the wrong order. It is possible that this scholar has filled in the values in the tables from a copy of Abraham ibn Ezra's *Book on the Foundations of the Astronomical Tables*, which quotes this section of Ibn al-Muthannā's text: see [Millás Vallicrosa 1947: 152] and below p. 249.

25 The scribe of MS C fails to fill in the second table; in MS A, on the other hand, almost the same material is provided in two tables, thus giving three tables altogether.

26 For an account of this text see [Charmasson 1980: 95–109].

27 The Eastern form of 6 does, in fact, resemble a common form of the "tironian *et*," though, strictly speaking, this should only be realised as "etiam" if there is a bar on top of it; but bars are often found over Indian numerals; e.g. in MS Harley 3631 of Abū Ma'shar's *On the Great Conjunctions*.

Once again we have an indication that Indian numerals precede Roman numerals in the transmission of a text.

The forms of the numerals that the scribe was attempting to imitate differ in at least two cases from those of the manuscripts of Ibn al-Muthannā: 4 is written as a "z" with its base-line extended in a downwards curve; 5 is the "crossed-zero" version turned through 90 degrees.

A third work with which Hugo of Santalla was associated was the *Book of the Three Judges*.[28] In two out of the three manuscripts that contain this compendium of judicial astrology there is a dedication to "bishop Michael" ("antistes Michael"), Hugo's patron in Tarazona, and the style of writing is that of Hugo.[29] In a third manuscript the dedication is to "mi karissime R." who is likely to be Robert of Ketton, archdeacon of Pamplona, the close friend of Hermann of Carinthia. It is probable that Hermann and Hugo, both of whom were in the valley of the Ebro in the 1140s, collaborated in the compilation of the *Book of the Three Judges*. In the manuscript with the dedication to "mi karissime R." — British Library, Arundel 268 — the Eastern forms are used exclusively (Plate 12). Moreover, only Eastern forms are used throughout the codex which contains the *Book of the Three Judges*, and they can therefore be assumed to be the forms favoured by the scribe himself. Some of the numerals resemble those found in Digby 50: the 8 is identical, while the 6, with its left-hand stroke almost forming a circle, resembles one of the forms of 6 in the Digby manuscript. The 4, however, consists only of the hook, and the rounded form of 9 is used.

An early thirteenth-century copy of another text possibly by Hermann of Carinthia occurs in a manuscript written with the Eastern forms: namely, the translation of the *Centiloquium* of Pseudo-Ptolemy with the incipit "Mundanorum ad hoc et illud mutatio" (which incorporates part of Plato of Tivoli's translation of the same work), whose copyist in MS Berlin, SBB-PK, Hamilton 557, has numbered the *Verba* in the margin away from the body of the text.[30] The presence also of Western and mixed forms in this manuscript may indicate that it was written when the Eastern forms were already defunct.

As we have seen (n. 19 above), a "Hermann" is also named as the translator of the Sicilian *Almagest*. While it is not *prima facie* likely that this is Hermann of Carinthia, it is interesting to note that the same translator was responsible for the major part of a version of Euclid's *Elements* made from Greek [Murdoch 1966; Busard 1987: 2–3]. The scribe of the earliest manuscript — MS Paris, BNF, lat. 7373 (Plate 13) — uses Eastern forms similar to those of Arundel 268, and both scribes wrote all the texts in their manuscripts in the Eastern forms; they were

---

28 Note that in the *Biblionomia* of Richard of Fournival the *Book of the Nine Judges* (an expanded version of the *Book of the Three Judges*) was in the same manuscript as Hugo's geomancy ("In uno volumine liber IX judicum, geomancia que vocatur Rerum opifex" [Delisle 1868–81, III: 67, no. LVI-17]).

29 The evidence for Hugo's authorship is assessed in [Burnett 1977: 67–70]. To the two manuscripts mentioned in this article should be added Dublin, Trinity College, MS 368 (fols. 43r–137v).

30 Richard Lemay first drew attention to this manuscript; see [Lemay 1995–96, VII: 141].

evidently the symbols they were most used to using. Moreover, the Eastern forms numbering the theorems of Euclid are placed in the margin away from the body of the text in the same way as in MS Berlin, SBB-PK, Hamilton 557. The Paris manuscript was written in Tuscany [Avril 1980–84, I: 56], and the version of Euclid's *Elements* that it contains, as [Busard 1987: 18–20] has shown, was known to Fibonacci who was working in Pisa, who may have added an appendix to the translation. The association of this translator's work with that of "Hermann" may be due to the fact that his translations were copied in the same environment as those of Hermann of Carinthia and Hugo of Santalla. That this environment was Tuscany will be explored below (see p. 251).

Abraham ibn Ezra

Another scholar who translated the commentary by Ibn al-Muthannā (this time into Hebrew) and came from the same area as Hugo and Hermann, was the Jewish polymath, Abraham ibn Ezra, who was born in Tudela between 1089 and 1092. It is in the twelfth-century Latin works associated with this scholar that we have the most sustained use of the Eastern forms. These occur in four twelfth-century manuscripts of his *Book on the Foundations of the Astronomical Tables* (written in 1154) — Erfurt Q 381, British Library, Cotton Vespasian A II (Plate 15), Cambridge, Fitzwilliam McClean 165 and Oxford, Bodleian, Digby 40 (Plate 16) — and in his text on the astrolabe also contained in the Cotton manuscript. In another manuscript — British Library, Arundel 377 — the astrolabe text accompanies an introduction to the Pisan Tables attributed to Abraham. The two texts in this manuscript are written with the Western forms, but a list of Eastern forms with the Western forms written on top of them is included (Plate 17); this suggests that, at an earlier stage in its diffusion, it was written in the Eastern forms. These Eastern forms occur also in an anonymous text on arithmetic and geometry which has been copied out into the important late-twelfth-century collection of Arabic-Latin translations made by scholars working in North East Spain (especially Plato of Tivoli in Barcelona): Oxford, Bodleian, Digby 51. It has recently been shown, from a comparison with a Hebrew version of the same text, that this work too is by Abraham (see Plate 14 and [Lévy 2001]).[31]   *

In all the texts explicitly attributed to Abraham the numeral forms are very similar to each other, with 6 acquiring a zigzag form, 8 written as a curve and 9 written with a straight back and facing right; only in the case of 4 is there variation, but the hook with an extended tail (but without a loop) seems to underly all the versions. The forms in Digby 51 diverge a little in that 4 takes the form that we also meet in Digby 50, the manuscript associated with Hugo of Santalla.

---

31   The presence of the Eastern forms in the earliest manuscript of the *Ysagoge* and *Liber*   *
     quadripartitus attributed to Iohannes Hispalensis (Venice, Biblioteca Marciana, Lat. Z. 344)
     confirms the close relationship of this text with the astrological works of Abraham ibn Ezra,
     which is the subject of a forthcoming monograph by the present author.

So far, the picture that is building up is of the employment of the Eastern forms in works written by writers on astronomy who can all be associated with the North East of Spain and the South of France. Hugo of Santalla was translating works for Michael, bishop of Tarazona from 1119 to 1151, and was possibly still in the service of one of his successors in 1179. Hermann of Carinthia probably collaborated with Hugo over the compilation of the *Book of the Three Judges* and its sequel, the *Book of the Nine Judges*, and is attested in León (1138), "on the banks of the Ebro" (1141), and in Toulouse and Béziers (both in 1143). But Hermann's principal collaborator and inseparable friend was Robert of Ketton, successively archdeacon of Pamplona, and canon of Tudela (1157). Abraham ibn Ezra was born in Tudela, and this was his base until the early 1140s, when he started to travel. His *Foundations of the Astronomical Tables* and *Astrolabe* in the Cotton manuscript sandwich another work on the astrolabe written in 1144, again in Béziers, by Hermann's pupil, Rudolph of Bruges. Plato of Tivoli worked in Barcelona, and collaborated with Abraham bar Ḥiyya, whose astronomical writings were known to Abraham ibn Ezra. Both Plato and Rudolph addressed texts to a certain "John David," whose mathematical prowess they praise.

Both geographical vicinity and the combinations of their works in manuscripts show that there were close associations between these scholars. It is possible that they had some acquaintance with the Western forms of Indian numerals.[32] But it is equally possible that they worked in an environment in which the Eastern forms were used. This is particularly the implication of the evidence of the numerals in Hugo's translation of Ibn al-Muthannā, whose Arabic text was found in the library of the Banū Hūd. What is more clear is that the Eastern forms were used in Italy. For it is preeminently copies of these authors' works written by *Italian* scribes that exhibit these forms: Digby 50, Arch. Seld. 34, Hamilton 557, and Arundel 268 are all written in Italian hands on Italian parchment. The master scribe of Digby 51 is also Italian; he writes most of the annotations, including a diagram which he labels in a competent Arabic script (fol. 88v), and could be a key player in our story. Moreover, while Abraham ibn Ezra came from Spain, in the early 1140s he was in Lucca, and his *Foundations* and explanation of the use of astronomical tables presuppose the use, specifically, of the tables of Pisa, which he implies he composed himself.[33]

32 The only text in Digby 51 in which the Western forms are used is Plato's translation of al-Battānī, and the master scribe and another hand use the Western forms in their annotations to this text. However, Roman numerals are used in another twelfth-century copy of al-Battānī — Digby 40. The Western forms are also used in Ibn Ezra's *Sefer ha-Mispar* ("book of arithmetic"), but MS Paris, BNF, ébreu 1052, adds the Eastern forms in the margin. However, these manuscripts are of the fifteenth century and it is unwise to draw sharp conclusions from their evidence.

33 The *Foundations* are referred to, by Henry Bate and Nicholas of Cusa, as "De motibus et opere tabularum super Pisas" [Birkenmajer 1950]. Abraham's responsibility for composing the Pisan tables rests on the presumption that "the tables of aṣ-Ṣūfī" and those drawn up for the meridian of Pisa are the same; see [Millás Vallicrosa 1947: 87]: "Proinde omnium aliorum tabulis omissis, tabulas medii cursus Solis secundum Azofi composui ... Et he tabule composite sunt secundum meridiem Pisanorum."

Pisa and Lucca

The Pisan tables were drawn up in 1149, or soon after, since 1149 is the epoch from which they are calculated. Their earliest manuscript — Berlin, SBB-PK, lat. fol. 307 (Plate 18) — provides the most sustained example of the use of the Eastern forms, employing them throughout, both in the instructions and in the twelve pages of tables. They differ from those found in the manuscripts of Abraham's works in that the "s" form of 4 is used, that 6 is no longer a zigzag, and 8 has a straight back, whilst 9 is curved. In 1160 someone added some annotations to MS British Library, Harley 5402 in Lucca, some 15 miles from Pisa, and gave the date as "1160" in Roman numerals; the only Indian numerals used are in the Eastern form, of which the most distinctive is 6. At about the same time "Stephen the Philosopher," who, for several reasons, should be identified with "Stephen of Pisa," wrote a cosmology which he called the *Liber Mamonis*; in its only surviving manuscript (Cambrai, Médiathèque municipale, 930), the Eastern forms are used for high numbers.[34] In some cases these manuscripts show forms closer to the (supposed) original Arabic than in Abraham's works, such as the "B"-shaped 5 (in Berlin, SBB-PK, lat. fol. 307) and 4 with a hook and a loop (in Cambrai 930).

The Pisan tables are those for which the two Latin works on astronomical tables bearing Abraham ibn Ezra's name (the *Foundations* in several manuscripts, and the introduction to the tables in Arundel 377) were written. But there are at least two more instructions for the use of the tables which do not mention Abraham, and which rather appear to be written by anonymous Christian scholars: those in Berlin, SBB-PK, lat. fol. 307 and Bodleian, Selden Supra 26 (Plate 19). All these instructions use the Eastern forms, as would be natural if they are introducing tables written in those forms (the tables are lacking in all the manuscripts except that of Berlin). The use of Eastern forms in the Latin texts associated with Abraham ibn Ezra is probably due, therefore, not so much to Abraham himself as to his Latin associates, who were using the tables of Pisa. The combined testimony of these manuscripts strongly indicates that the Eastern forms were being used in Pisa and Lucca in the mid-twelfth century. It has already been noted that MS Paris, BNF, lat. 7373 was written in Tuscany and that the text it contains was known to Fibonacci in Pisa (p. 249 above). This manuscript, in fact, shares with the annotator of MS Harley 5402 the unusual spelling "pacta" for "epact."

It would be expected that these forms rather than the Western forms were used in Pisa, since, in the mid-twelfth century the orientation of Pisa was towards the cultural centres of the Eastern Mediterranean, both the Greek/Arabic centre of Antioch (where the Pisans had a quarter, and where Stephen the Philosopher made his translations), and the Greek centre of Constantinople. It is quite plausible that the Pisan tables themselves came directly from an Eastern source, possibly through the agency of Stephen; see [Burnett 2000b: 14–15].

---

34 In this manuscript the scribe appears to be unfamiliar with the numerals and writes them as the letters that they most nearly resemble; see [Burnett 2000b: 64–65].

The Greek Connection

A Latin scholar working in the Crusader States could easily have had access to Arabic manuscripts with Indian numerals written in the Eastern form. That these forms were actually adopted by Christian scholars is proved by the Greek evidence. For, already by the twelfth century, Greek mathematicians had begun to use the Eastern forms.[35] These are found in marginal notes to Euclid's *Elements* which, on palaeographical grounds, can be assigned to the twelfth century. Among these are those of MSS Oxford, Bodleian, Auct. F.6.23, d'Orville 301 (see Plates 20 and 21), and Paris, BNF, gr. 2466. The forms found in these manuscripts are remarkably similar to each other, and are written with an ease that suggests that they were in normal use among certain Greek mathematicians. We find both the curved and the straight-backed 9, a 6 sometimes tending towards a zigzag, and a curved 8, like that in Ibn Ezra's Latin works. Most striking is 4, which is an "s" shape as in the Pisan Tables, sometimes with an added terminal line, which is a vestige of the loop.[36] The main difference is 5, which is written as an oval tipped slightly forward, though this form is given a "belt" which squeezes it into a kind of figure-of-eight in an alternative form 5 written above a row of numerals in the d'Orville manuscript.[37] At least one of the Greek manuscripts with Indian numerals in its twelfth-century glosses was possibly known in a Latin context. For MS Bodleian d'Orville 301 has been claimed by John Murdoch to be the very manuscript that the translator of the Sicilian *Almagest* used as his principal text for his translation of Euclid's *Elements*, and from which he incorporated some glosses into his translation [Murdoch 1966: 260–263].[38]

Pisa, in particular, was an important centre of translation from Greek, thanks to its close connections with Constantinople, and the activity of some outstanding local scholars, such as Richard Burgundio, Leo Tuscus and his brother Hugo Etherianus. It is quite plausible that Pisan mathematicians were using the Eastern forms that they found in Greek manuscripts. But what we may rather be seeing in

---

35 This had been made clear, on palaeographical grounds, in [Wilson 1981], but seems to have been overlooked by historians of mathematics. I am very grateful to Dr. Wilson for sending me an offprint of this article, to which the following paragraph is much indebted. The other manuscripts with notes dating from the twelfth century (aside from those discussed here) are Paris, BNF, gr. 2344 and Vienna, ÖNB, phil. gr. 31. A different style of Eastern forms is used in one manuscript (Vat. gr. 211) of the late-thirteenth-century astronomical tables of Chioniades, translated from the Arabic tables known as the *Zīj al-ʿAlāʾī* [Pingree 1985–86, Part 2: 11, 15, etc.]: these presumably have been reintroduced directly from the Arabic models, but are sufficiently close to those found in the twelfth-century Euclid glosses to be immediately recognizable. Western forms appear in Greek for the first time in an anonymous Greek algorism written in 1252 — the *Psēphēphoria* — and dependent on Leonardo of Pisa's *Liber abbaci*: see [Wilson 1981; Allard 1976, 1977].

36 This alternative form of 4 is given above the "s" form in the well-known row of Arabic numerals in D'Orville 301, fol. 32v (now thought to be considerably later than the main text of the ninth century, and probably post-twelfth century [Wilson 1981: 401]).

37 See Table III, Greek b, below.

38 However, [Busard 1987: 8–9] expresses reservations about this claim.

the mid-to-late twelfth century is a "common language" of numerical symbols shared by the mathematicians of Greek Byzantium, Latin Tuscany and the Arabic Middle East, and characteristic especially of those places where two or three of these cultures coexisted, such as Pisa, Constantinople, Antioch and Tripoli. Aside from the common use of the Eastern forms in the three language cultures, one can point to another feature shared across these language boundaries: namely, the use of a mixed system, in which lower values (in principle, values which do not exceed 360, the number of degrees of the ecliptic) are represented by letters of the alphabet, the higher values by Indian numerals. This is a feature common to Eastern Arabic astronomical tables and the *Liber Mamonis* of Stephen the Philosopher, and is found in the thirteenth-century Greek tables of Chioniades; see [Burnett 2000b: 65–66].

Conclusions

The Tables of Pisa had a moderate diffusion in the twelfth century, and were adapted to the meridians of Angers, Winchester and London. The extant instructions (including those of Ibn Ezra) were written in these places rather than in Pisa itself, and the various local adaptations of the numerals in which the tables were written may explain the variation in their forms. The Berlin manuscript of the tables was copied in the Ile de France — perhaps in Paris itself —, MSS Fitzwilliam McClean 165 and Arundel 377 are in English hands, both including copies of Adelard of Bath's *De opere astrolapsus*.[39] The Eastern forms have been expertly copied in these manuscripts, but Arundel 377, and also Digby 40 (whose place of copying is unknown), provide a key to them in the same hand as the manuscripts, which suggests that they were not current forms in the places where they were copied, and needed interpretation. In Italian manuscripts the Eastern forms are used without an accompanying key. Similarly, when a scribe copies a work in a context in which the Eastern forms of the numerals are no longer current, he is likely to use other forms of numerals in other texts: e.g. the same scribe (probably English?) copied Raymond of Marseilles's *Liber iudiciorum* in Erfurt, Amplon. Q 365 and Abraham ibn Ezra's *Foundations* in Erfurt, Amplon. Q 381, but he uses Western forms in the former. The master-scribe of Digby 51 uses the Western forms when annotating the one text which uses these forms (e.g. the work of al-Battānī), and uses Eastern forms only when copying the text by Abraham ibn Ezra. When a scribe copies a variety of texts using the same numeral system it is reasonable to suppose that this is the system that is normal for him: e.g. in Arundel 268, the numbers are always written in their Eastern forms whether the scribe is copying the *Book of the Three Judges* of Arabic provenance or the *Aratea* of the Latin tradition. A similar copying of more than one work in the Eastern forms can be observed in Berlin, lat. fol. 307, Paris, BNF, lat. 7373, and Oxford, Bodleian, Selden supra 26, though

---

39   An Anglo-Norman hand also accompanies the Italian hand in MS Digby 51.

the last also includes a key. Of these manuscripts, Arundel 268, Paris, BNF, lat. 7373 and (possibly) Selden supra 26 are in Italian hands.

The evidence for the active use of the Eastern forms in Italy is strong. They were used in Pisa and Lucca in the mid-twelfth century. They seem also to have been adapted to local usage in Sicily at the same time, and were used in some glosses in a Greek manuscript which may have been in Italy. The places of writing of the Italian manuscripts of Hugo of Santalla's and Hermann of Carinthia's works, and of the Italian hand in Digby 51, are not known, but the works of these authors are associated with the Latin works of Ibn Ezra, who moved from Spain to Tuscany.[40] If the Eastern forms were used by scholars in Italy, then, *a fortiori*, one would have expected them to have been used by Latin scholars in Byzantium and in the Crusader States, but no manuscripts with Indian numerals have yet been identified as having these provenances. All one can say is that the evidence points to the spread of the numerals to Northern European centres from Pisa and Lucca, together with the Pisan tables, and that eventually they became unintelligible in these areas.[41]

But the manuscript evidence, on its own, may lead to an underestimation of the currency of the Eastern forms. The clearest examples of their current use in both Greek and Latin manuscripts are in glosses and annotations rather than in texts: i.e., in the writing of scholars rather than in the writing of scribes. Among these scholars are the annotators of Harley 5402, Arch. Seld. B 34 and Soest 24,[42] as well as the Greek manuscripts listed by Wilson. We can therefore assume that the Eastern forms are likely to have been more current among scholars, who found them convenient to use and who understood each other's mathematical jargon. Such annotations are not so likely to have survived.

The question to be asked is why did Italian scholars not continue to use the Eastern forms? Part of the answer may lie in the developments of the early thirteenth century, when the principal successors to the Toledan translators of the twelfth century moved to Northern Italy. Among these were the compilers of the most authoritative collection of Toledan mathematical translations, which is comprised in three surviving manuscripts copied by the same scribe in Northern Italy in the early thirteenth century: Paris, BNF, lat. 9335 and 15461 and Vatican City, BAV, Ross. lat. 579, and in another manuscript copied by a closely related hand in British Library, Harley 3631.[43] Paris, BNF, lat. 15461 includes the *Liber Alcho-*

---

40 For the possibility that the annotator of one of the Italian manuscripts of a translation by Hugo used a text of Ibn Ezra, see n. 24 above.

41 A curious relic of these numerals is in an illustration in a manuscript in the monastery of Heiligenkreuz (MS 226), where they adorn the book of a "clericus" — perhaps intending to convey the obscure symbols used by scholars (see Plate 22).

42 In all three of these manuscripts the Eastern forms appear only in annotations: for the annotations in Soest, in which the Western form of 5 is found among otherwise Eastern forms see the thorough study in [Becker 1995].

43 For these manuscripts see [Burnett 2001b]. [Avril 1980–84, II: 5] assigns the decoration of Paris, BNF, lat. 9335 to the Venice / Padua region. Forms very similar to those found in these manuscripts occur in the the notarial documents of Raniero da Perugia, which are a valuable testimony because they can be dated and located to Perugia between 1184 and 1206: see

rismi of "magister Iohannes," the most advanced and complete algorism of the twelfth century, and the numeral forms given in this text are found throughout the collection; they are similar to the "toletane f⟨igure⟩" of Clm 18927, and they were to become the standard forms of Medieval Europe.

One of these scholars was Michael Scot, who left Toledo for Bologna sometime between 1217 and 1220. In 1228, Leonard of Pisa (Fibonacci), working in Pisa, dedicated a revised version of his *Liber abbaci* to Michael. As we have already seen (p. 249 above) Fibonacci knew the Greek-Latin version of Euclid's *Elements*, whose earliest manuscript uses the Eastern forms. But Fibonacci chose to use the Western forms of the Arabic numerals in his mathematical works. He would have known these numerals from learning the art of calculation in Bougie (in Algeria) in his youth [Boncompagni 1857–62, I: 1].[44] But his adoption of these forms is also likely to have been due to his association with the successors of Gerard of Cremona,[45] and the eclipsing of other traditions of Arabic-Latin scientific learning by that stemming from Toledo.

Appendix: Latin Manuscripts Containing the Eastern and Palermitan Forms of Indian Numerals

Included here are all the Latin manuscripts known to contain examples of the Eastern and Palermitan forms of Indian numerals. In each case the following information is provided (where possible): the manuscript number, the relevant folios, and their date and place of writing; the text in which the numerals occur, the place and date of the writing of the original text. Reference is given to the illustrations of the manuscripts and their numerals in the Plates and Table included in this article. Unless the manuscript itself provides a date of copying, the dates of manuscripts can only be approximate. The variations in numeral forms described

[Bartoli Langeli 2000] and Table I, entry o. Raniero inserted the numerals at the beginning of documents, indicating the number of lines that the document contained, perhaps as a kind of private code. The Greek-Latin translators of Northern Italy in the twelfth century, James of Venice and Burgundio of Pisa, were also notaries and legal scholars.

44 One variant of his numerals, taken from the fourteenth-century MS Florence, Biblioteca nazionale centrale, fondo Magliabechiano, conv. sopp. Scaffale C. Palchetto 1, no. 2616 by [Boncompagni 1852: 103], looks remarkably similar to Western *Arabic* versions of the Indian numerals, and could plausibly have been picked up by Fibonacci during his studies in Bougie. But the other variant recorded by Boncompagni, and attributed to Fibonacci in [Allard 1976] is the "standard" Western forms also found in the copies of the Toledan works.

45 The close connection of Fibonacci with the successors of Gerard of Cremona is suggested by his use of Gerard's translation of the *Algebra* of al-Khwārizmī (see [Miura 1981: 59–60]) and his reference to a work well-known to Gerard's *socii* ("students"): the *Liber de proportione et proportionalitate* of Aḥmad ibn Yūsuf ibn al-Dāya ("Ametus filius Iosephi"), which Fibonacci refers to in a way which can only be interpreted as a truncated form of the *Latin* name of the author and his work: "Ametus filius ... in libro quem de proportionibus composuit" [Boncompagni 1857–62, I: 119]; see also [Rashed 1994: 148, 160].

in this article can, themselves, help in determining the date and provenance of a manuscript.

1) Berlin, Staatsbibliothek zu Berlin-Preußischer Kulturbesitz, lat. fol. 307 (Rose 956), s. XII^ex, Ile de France (cf. fol. 1r: "nos sumus Parisius existentes"). The normal usage of the scribe is the Eastern forms which appear on fol. 1r, as the chapter numbers in his copy of al-Farghānī's *Rudimenta* (fols. 19r–21r), and, most conspicuously, in the Tables of Pisa and their instructions (giving the meridian of Angers) on fols. 27, 30, 28, 31–34 (ancient fols. 103–108). Only in the copy of John of Seville's translation of al-Qabīṣī's *Introduction to Astrology* do the Western forms appear in one table (fol. 29v); in another table, however, Eastern and Western forms are given on alternate rows (fol. 35r), and a third table (also on fol. 35r) uses Eastern forms only. [Rose 1893–1919: 1177–1185; Leonardi 1960: 9–10; Folkerts 1981: 65]. I am very grateful to Bernd Michael for information on this manuscript. See Plate 18 and Table IIIc.

2) Berlin, Staatsbibliothek zu Berlin-Preußischer Kulturbesitz, Hamilton 557, s. XIII (the year 1202 is mentioned on a horoscope on fol. 8r; the year 1221 on another on fol. 15r), Italy, fols. 1–15v. A mixed version of Pseudo-Ptolemy, *Centiloquium*, combining the translations of Plato of Tivoli and Hermann of Carinthia (?). The Eastern forms of the numerals are used to number the 100 *verba* in the margin, and in a table of the planetary terms (fol. 14r). The same scribe, in the text, retains some of these forms (3, 4, and 5) but replaces others with Western forms (2, 6, 7 and 8); see fols. 8r and 14r; in the other major text in the manuscript (Albumasar, *Flores*) he uses Western forms entirely. Cited wrongly as Berlin, Hamilton 16 in [Lemay 1995–96, IV: 141], but correctly in [Lemay 2000: 382–383; Boese 1966: 273–274]. Table IIIn.

3) Cambrai, Médiathèque municipale 930 (829). Stephen the Philosopher, *Liber Mamonis*. The Eastern forms are used for high numbers (on fols. 27v–28r) alongside Roman numerals and Latin alphanumerical notation. [Burnett 2000b]; for illustrations see [Lemay 2000: 391–392; Burnett 2000b: 73]. Table IIIs.

4) Cambridge, Fitzwilliam Museum, McClean 165, s. XII, fols. 48v–49r contain a portion of Abraham ibn Ezra's *Book on the Foundations of the Astronomical Tables* ("Nunc artem de fructu Almagesti sumptam communem trademus … sed scriptore errasse" = [Millás Vallicrosa 1947: 145–147]). The Western forms are used in another introduction to astronomical tables associated with Pisa [Mercier 1991, Text II] on fols. 67r–80v. Table IIId.

5) Cambridge, Gonville and Caius College, 456/394, s. XII^2, in St Augustine's Monastery, Canterbury, by the fourteenth century. The Eastern forms occur in tables (fol. 73r) within Hugo of Santalla's translation of Ibn al-Muthannā's commentary on the tables of al-Khwārizmī. [Burnett & Pingree 1997: 9]. Plate 9 and Table IIIm.

6) Dresden, Sächsische Landesbibliothek, C 80, fols. 156v–157r, s. XV. Although texts on the algorism occupy several folios of this manuscript, the Eastern

forms first occur within a chapter headed "De abaci Sarracenici multi(plica)-t(i)on(e)" beginning: "De aliis abacis illud breviter dico, quod quicumque ⟨...?⟩ Sunt autem ibi caracteres alii quam in latino, quos duobus ⟨...?⟩ non enim illos dicunt (?) indeos (?) quos rectius sic indeus (?) " (there follows a row of Eastern forms and a row of Western forms) "ternarius sic etiam fit ℇ, octonarius Ϸ." The examples of calculation are entirely in the Eastern forms, and the excerpt ends on fol. 157r with the triangle of multiplication accompanying the text of the *Liber Ysagogarum* [Allard 1992: 27]. The manuscript is exceedingly difficult to read, but special photographs are being made for Menso Folkerts who is preparing an edition. Table IIIx.

7) Erfurt, Wissenschaftliche Bibliothek der Stadt, Amploniana Q 381, s. XII^{ex}, fols. 1–34 (a separate codex, probably from a larger MS). Abraham ibn Ezra, *Book on the Foundations of the Astronomical Tables*. [Schum 1887: 638–639; 1882 no. 13] and Table IIIh.

8) Heiligenkreuz, Stiftsbibliothek 226, fol. 129r. Burgundy or Lorraine (?), s. XII^{ex}.[46] Five numerals in their Eastern forms have been added to the open book being read by the "clericus" on the frontispiece of Hugh of Fouilloy's *Aviarium*. No other manuscripts of the *Aviarium* has these numerals in the book. [Clark 1992: 283 and Fig. 1a]. Plate 22 and Table IIIu.

9) London, British Library, Arundel 268, fols. 75r–103v, s. XII^{ex}, a paper manuscript written in Southern Italy (apparently the earliest paper manuscript of a Latin Classical text). The beginning of the *Book of the Three Judges* (fols. 81rb–84rb), an edition of Germanicus's *Aratea* with scholia (85rb–92v and 96r–103v) and a cento from Virgil's *Aeneid* (fols. 93r–95v). [Lemay 2000: 387–390; Burnett 1977: 73, n. 39 (here the hand is erroneously said to be English) and 78–97; Reeve 1980: 511–515 (gives a s. XIII date)]. Plate 12 and Table IIIp.

10) London, British Library, Arundel 377, fol. 36r, s. XII^{ex}, Ely. A row of Eastern forms underneath a row of Western forms and some alternative forms, opposite the end of Abraham ibn Ezra's *Introduction to the Pisan Tables* [Mercier 1991: Text III]. Plate 17 and Table IIIe.

11) London, British Library, Cotton Vespasian A II, fols. 27–40 (a separate codex, incompletely copied), s. xii. An acephalous copy of Abraham ibn Ezra, *Book on the Foundations of the Astronomical Tables*; Rudolph of Bruges, *On the Astrolabe*, and Abraham ibn Ezra, *On the Astrolabe*. The two texts by Abraham ibn Ezra use the Eastern forms throughout; Rudolph's text uses only Roman numerals. Plate 15 and Table IIIg.

12) London, British Library, Harley 5402, fols. 60r–v, 1160 A.D., Lucca. Some notes on how to find the position of the Moon in 1160, followed on fol. 80v with instructions from tables made "super civitas (*sic*) luce" in the same hand. Aside from Roman numerals, the notes include the Indian numerals 0 and 6, the last in its Eastern form. [Burnett 2000a] and Table IIIr.

---

46   I am grateful to Baudouin van den Abeele for drawing my attention to this manuscript.

13) London, British Library, Harley 5786, ca. 1153, Palermo. A Psalter in Greek, Latin and Arabic, in which the Arabic psalms are numbered in Indian numerals which share characteristics of both Eastern and Western forms. The Arabic numerals are written adjacent to the Arabic text, but in the margin (as in Paris, BNF, lat. 7373), at first at the same time as the text, then (from fol. 86r) after the writing of the main text, but probably at the same time as the addition in the margin (in Arabic) of the times for the recitation of the psalms. Plate 5 and Table IIb.

14) Madrid, Biblioteca nacional, 10009, between 1267 and 1286, Italy? (the MS includes the notes of Alvaro of Toledo, who was in Italy with his patron, Gonzalo Gudiel, archbishop of Toledo), fols. 23va–38va. 'Alī ibn Aḥmad al-'Imrānī's *Elections* "interpreted" by Abraham bar Ḥiyya. Eastern forms are used throughout this text, except in the date in the colophon (here the scribe writes "1124" where the other manuscripts of the work give "1134," suggesting that the scribe had misinterpreted a numeral, probably because it was written in a form that he was not used to). 2 and 3 face right and left respectively, presumably to avoid confusion between 2 and 6, as in Arundel 268. The same scribe has copied other texts in the manuscript, including Hugo of Santalla's version of the *Centiloquium*, using Western forms. [Millás Vallicrosa 1942: 171–172]. Plate 23 and Table IIIv.

15) Munich, Bayerische Staatsbibliothek, Clm 18927, s. XII^2, Southern Germany, fol. 1r. Three rows of numerals of which the first is labelled "toletane f⟨igure⟩," the second (the Eastern forms) "indice f⟨igure⟩" and the third is unnamed (a species of the Western forms). The *Liber Ysagogarum* and Toledan Tables in this manuscript are written in the third form. [Lemay 1977, Fig. 1a]. Plate 3 and Table Ic, Im and IIIa.

16) Oxford, Bodleian Library, Arch. Seld. B 34, 13r–63v, s. XIII^{in}, written in the same Italian hand as Digby 50. The Eastern forms occur in the tables on fols. 32v and 33r, within Hugo of Santalla's translation of Ibn al-Muthannā's commentary on the tables of al-Khwārizmī (fols. 11–62v). [Millás Vendrell 1963]. Plate 10 and Table IIII.

17) Oxford, Bodleian Library, Digby 40, fols. 52–88 (a separate codex), s. XII^2, English? Abraham ibn Ezra, *Book on the Foundations of the Astronomical Tables*. Written throughout with the Eastern forms. A later hand has added Western forms above the Eastern forms, but makes mistakes, e.g. in interpreting the Eastern 8 as a 7. A key to the numerals is given on fol. 88v. [Birkenmajer 1950; Millás Vallicrosa 1947: 32 (illustration), 69]. Plate 16 and Table IIIi.

18) Oxford, Bodleian Library, Digby 50, s. XIII^{in}, written in the same Italian hand as Arch. Seld. B 34. Eastern forms used only within the supplement to Hugo of Santalla's *Ars geomancie*, on fols. 94r–101r; from 100v onwards they are progressively replaced by Roman numerals which eventually are used exclusively. Plate 11 and Table IIIo.

19) Oxford, Bodleian Library, Digby 51, s. XII^2, fols. 38v–42v. The whole manuscript appears to be written on Italian parchment; these folios are written by

the Italian hand which is that of the master-scribe. Abraham ibn Ezra, *Arithmetic and Geometry*. Although the hand of a single master-scribe is found throughout the manuscript, the Eastern forms are used only in this text. Elsewhere are examples of the abacus numerals (fol. 36v) and the Western forms (used by an Anglo-Norman hand and the Italian annotator on fols. 5–9, 12–15, 17). The scribe becomes more confident in using the Eastern forms as he progresses, and starts off with more complex forms: e.g. 5 is a double "bow", developing into a circle with a horizontal line through it; 4 is like a "z" with the lower line curved round, but on fol. 38v there is still a vestige of a detached loop to the lower curve; 6 begins like the Arabic 6, with the right-hand line sometimes sharply curved, at other times straight, making it indistinguishable from a tironian *et*; it later becomes a zigzag shape. [Hunt & Watson 1999: 25–26; Lévy 2001]. Plate 14 and Table IIIj.

20) Oxford, Bodleian Library, Selden supra 26, fols. 96–100 (I), 106–121 (II) and 122–129 (III) (original a single codex, written in one hand, with I apparently following II and III and breaking off incomplete), s. XII^{ex}–s. XIII^{in}, of unknown provenance; brought to St Augustine's Canterbury by William de Clara in 1277, probably immediately from Paris. Fols. 96–121v, *Algorismus* (a hybrid of the *Liber pulveris* and the *Liber Alchorismi*); fols. 122r–129v, Instructions for the use of the Pisan tables ([Mercier 1991, Text II] = Cambridge, Fitzwilliam McClean 165, fols. 68v ff.). The Eastern forms are used throughout, alongside Roman numerals. On fol. 96r a later hand has written the Western forms above the Eastern forms. A key giving the Western forms above the Eastern forms appears on fol. 106r. Bruce Barker-Benfield (letter to Allard); [Allard 1992: xxxviii–xxxix]. Plate 19 and Table IIIf.

21) Paris, Bibliothèque nationale de France, lat. 7373, 1181 A.D., Tuscany. Euclid's *Elements* (the translation from Greek made by the scholar who translated the "Sicilian *Almagest*"; see Vatican, Biblioteca apostolica Vaticana, Pal. lat. 1371 below). The original numbering of the propositions in the first four books is in the Eastern forms written on the extreme edges of the pages. In Books 5–9 Roman numerals take the place of the Eastern forms; thereafter no numerals are visible (in some cases they may have been cut off when the margins were trimmed). The numbering of the theorems is repeated in the Western forms for Books 1–5, and is continued in these forms in Book 6. Fol. 176r is the "figure of rhetoric" from the *Ars notoria*. Fols. 176v–178r consist entirely of computus tables for the church calendar written in the Eastern forms. The headings of the columns on fol. 176v are "luna Ianuarius; feria termini; pentecostes; rogatio⟨nis⟩; termini .xl.; termini .lxx." Those on fol. 177v are: "numerum aureo (*sic*); quantum habet luna in kalendis (?); in qua feria termini fiunt (?); claves terminorum; terminus pentecost.; terminus rogationis; terminus hebreorum; terminus quadrag.; terminus septuag.; pacta: circulus solis; pacta lune; circulus lune." The date "mclxxxi" (1181) is written in the margin of fol. 177r. [Murdoch 1966: 249–250; Avril 1980–84, I: 56, no. 95]; information from Julien Veronese. Plate 13 and Table IIIq.

22) Paris, Bibliothèque nationale de France, lat. 10252, s. XV, fol. 70r (within a miscellany of chapters on the algorism including some — e.g. on fol. 69r — that correspond with chapters in British Library, Egerton, 2261, fols. 226ra–b). The abacus names are followed by two rows of badly copied Indian numerals. The second row (after "vel aliter") was originally, in all likelihood, the Eastern forms, some of which appear to have been rotated: 5 and 9 are upside down, 4 is turned through 90 degrees. This work is immediately followed by Raymond of Marseilles's *Liber iudiciorum*, which has evidently been copied from an early manuscript. [Poulle 1963: 33]. Table IIIw.

23) Soest, Stadtarchiv, 24, s. XII, fol. 32v, between 15 October 1185 and 5 April 1186, North French? A note on how to convert between *anni domini*, the era of Nebuchadnezzar and the *hijra*, using the present date (15 October 1185) as an example, added to a copy of Firmicus Maternus, *Mathesis*, books II–IV. [Becker 1995]. Table IIIt.

24) Vatican City, Biblioteca apostolica Vaticana, Pal. lat. 1371, fols. 41r–97v, s. XII, Italy. Ptolemy's *Almagest*, translated from a copy brought to Palermo before 1160. Roman numerals are used in the text and in the first tables

* (fols. 46r–47r, 48v). The distinctive "Palermitan" forms are used in the remaining tables (fols. 53v–54, 57v–60v, 66v, 69r–70r, 81v, 86v, 89r–90r) with the exception that, on fol. 63r, the Western forms are used (a later addition? note that some tables are left unfilled: fol. 70r, 75v, 95v). [Lemay 1987: 468–470]. Plate 6 and Table II.

25) Vatican City, Biblioteca apostolica Vaticana, Pal. lat. 1393, s. XIII[mid], fols. 1–60. *Liber Alchorismi / Liber pulveris* (the same hybrid version as in Selden supra 26), including a row of Eastern forms as an alternate to the Western forms which are otherwise used throughout the text. [Allard 1992: xl]. Table IIIb.

26) Venice, Biblioteca Marciana, Lat. Z. 344 (1878; Valentinelli, Cl. XI, 104), s. XIII[in], Italian. Fols. 1–30 contain a copy of "Iohannes Hispalensis," *Ysagoge* and *Liber quadripartitus* in which the Eastern forms are used. In the bottom margins of both fols 1r and 2r keys to these forms are given in the form of three rows of numerals: Roman, Western forms and Eastern forms respectively; over the first key the scribe has written "figure algorismi alterius quam utamus" (*sic*!). On the first folio a scribe has written Western forms over erasures; thereafter he occasionally writes the Western forms above the Eastern forms. Table IIIk.

The crossed-zero form of the Eastern "5" has survived in at least two manuscripts of al-Kindī's *De mutatione temporum*, ch. 4 ([Bos & Burnett 2000: 273]; the manuscripts are Paris, BNF, lat. 16204, p. 374, and *ibid.*, lat. 7316). There are also examples where a mistake in a number may have arisen from a misreading of an Eastern form: e.g. al-Kindī, *De mutatione temporum* ("180" read as "150"; [Bos & Burnett 2000: 274]); Hugo of Santalla, *Liber Aomaris* ("138" read as "130"; [Burnett 1977: 91]); and Hermann of Carinthia, *De essentiis*, 70vA ("120" read as "150"; [Burnett 1982: 166]).

Bibliography

Abū Fāris 1973. Dalīl jadīd ʿalā ʿurūbat al-arqām al-mustaʿmala fī al-maghrib al-ʿarabī. *Al-Lisān al-ʿarabī* 10: 232–234.

Allard, André 1976. Ouverture et résistance au calcul indien. *Colloques d'histoire des sciences: I (1972) et II (1973)*, pp. 87–100. Louvain: Université de Louvain.

—— 1977. Le premier traité byzantin de calcul indien: classement des manuscrits et édition critique du texte. *Revue d'histoire des textes* 7: 57–107.

——, Ed., 1992. *Muḥammad ibn Mūsā al-Khwārizmī, Le Calcul Indien (Algorismus).* Paris / Namur: Blanchard.

d'Alverny, Marie Thérèse 1948. Deux traductions latines du Coran au Moyen Age. *Archives d'histoire doctrinale et littéraire du Moyen Age* 16: 69–131.

Arrighi, Gino 1968. La numerazione "arabica" degli *Annales Ratisponenses* (codex Monacensis lat. 14733 olim Sancti Emmerammi G. 117). *Physis* 10: 243–257.

Avril, François, *et al.* 1980–84. *Manuscrits enluminés d'origine italienne*, 2 vols. Paris: Bibliothèque nationale.

Bartoli Langeli, Attilio 2000. I notai e i numeri (con un caso perugino, 1184–1206). In *Scienze matematiche e insegnamento in epoca medioevale: Atti del convegno internazionale di studio, Chieti, 2–4 maggio 1996*, Paolo Freguglia, Luigi Pellegrini, & Roberto Paciocco, Eds., pp. 225–254. Naples: Edizioni scientifice italiane.

Beaujouan, Guy 1948. Etude paléographique sur la "rotation" des chiffres et l'emploi des apices du xᵉ au xiiᵉ siècle. *Revue d'histoire des sciences* 1: 301–313.

Becker, Wilhelm 1995. *Frühformen indisch-arabischer Ziffern in einer Handschrift des Soester Stadtarchivs.* Soester Beiträge zur Geschichte von Naturwissenschaft und Technik. Soest: Uni-GH Paderborn, Abt. Soest.

Birkenmajer, Alexander 1950. A propos de l'Abrahismus. *Archives internationales d'histoire des sciences* 11: 378–390.

Bischoff, Bernhard 1990. *Latin Palaeography: Antiquity and the Middle Ages*, Dáibhi Ó Cróinín & David Ganz, Trans. Cambridge: Cambridge University Press.

Boese, Helmut 1966. *Die lateinischen Handschriften der Sammlung Hamilton zu Berlin.* Wiesbaden: Harrassowitz.

Boncompagni, Baldassarre 1852. *Della vita e delle opere di Leonardo Pisano matematico del secolo decimoterzo.* Atti dell'Accademia Pontificia de'Nuovi Lincei, anno V, sessioni I, II, III (1851–1852). Roma: Tipografia delle Belle Arti.

—— 1857–62. *Scritti di Leonardo Pisano matematico del secolo decimoterzo*, 2 vols. Rome: Tipografia delle scienze matematiche e fisiche.

Bos, Gerrit, & Burnett, Charles 2000. *Scientific Weather Forecasting in the Middle Ages: The Writings of al-Kindī.* London / New York: Kegan Paul International.

Burnett, Charles 1977. A Group of Arabic-Latin Translators Working in Northern Spain in the mid-Twelfth Century. *Journal of the Royal Asiatic Society*: 62–108.

—— 1982. Hermann of Carinthia, *De essentiis. A Critical Edition with Translation and Commentary.* Leiden: Brill.

—— 1994. Magister Iohannes Hispanus: Towards the Identity of a Toledan Translator. In *Comprendre et maîtriser la nature au moyen âge. Mélanges d'histoire des sciences offerts à Guy Beaujouan*, pp. 425–436. Paris: Droz.

—— 2000a. Latin Alphanumerical Notation, and Annotation in Italian, in the Twelfth Century: MS London, British Library, Harley 5402. In *Sic itur ad astra. Studien zur Geschichte der Mathematik und Naturwissenschaften. Festschrift für den Arabisten Paul Kunitzsch zum 70. Geburtstag*, Menso Folkerts & Richard Lorch, Eds., pp. 76–90. Wiesbaden: Harrassowitz.

—— 2000b. Antioch as a Link between Arabic and Latin Culture in the Twelfth and Thirteenth Centuries. In *Occident et Proche-Orient: contacts scientifiques au temps des croisades. Actes du colloque de Louvain-la-Neuve, 24 et 25 mars 1997*, Isabelle Draelants, Anne Tihon, & Baudouin van den Abeele, Eds., pp. 1–78. Turnhout: Brepols.

—— 2001a. Why We Read Arabic Numerals Backwards. In *Ancient and Medieval Traditions in the Exact Sciences. Essays in Memory of Wilbur Knorr*, Patrick Suppes, Julius M. Moravcsik, & Henry Mendell, Eds., pp. 197–202. Stanford: Center for the Study of Language and Information.

—— 2001b. The Strategy of Revision in the Arabic-Latin Translations from Toledo: The Case of Abū Ma'shar's *On the Great Conjunctions*. In *Les Traducteurs au travail: leurs manuscrits et leurs méthodes*, Jacqueline Hamesse, Ed., pp. 51–113. Turnhout: Brepols.

Burnett, Charles, & Pingree, David, Eds. 1997. *The Liber Aristotilis of Hugo of Santalla*. London: Warburg Institute.

Busard, Hubertus L.L. 1987. *The Mediaeval Latin Translation of Euclid's* Elements *Made Directly from the Greek*. Stuttgart: Steiner.

Charmasson, Thérèse 1980. *Recherches sur une technique divinatoire: la Géomancie dans l'Occident médiéval*. Geneva: Champion / Paris: Droz.

Clark, Willene B., Ed., 1992. *The Medieval Book of Birds: Hugh of Fouilloy's* Aviarium. Binghampton, NY: State University of New York Press.

Delisle, Léopold 1868–81. *Le Cabinet des manuscrits de la Bibliothèque impériale [nationale]*, 3 vols. Paris: Imprimerie impériale [nationale]; reprinted Hildesheim: Olms, 1978.

Fichtenau, Heinrich von 1937. Wolfger von Prüfening. *Mitteilungen des Österreichischen Instituts für Geschichtsforschung* 51: 313–357.

Folkerts, Menso 1970. *"Boethius" Geometrie II. Ein mathematisches Lehrbuch des Mittelalters*. Wiesbaden: Steiner.

—— 1981. Mittelalterliche mathematische Handschriften in westlichen Sprachen in der Berliner Staatsbibliothek. Ein vorläufiges Verzeichnis. In *Mathematical Perspectives. Essays on Mathematics and its Historical Development*, Joseph W. Dauben, Ed., pp. 53–93. New York: Academic Press.

—— 1997. *Die älteste lateinische Schrift über das indische Rechnen nach al-Ḫwārizmī. Edition, Übersetzung und Kommentar, unter Mitarbeit von Paul Kunitzsch*. Abhandlungen der Bayerischen Akademie der Wissenschaften, philosophisch-historische Klasse, Neue Folge 113. Munich: Bayerische Akademie der Wissenschaften.

Gibson, Craig A., & Newton, Francis 1995. Pandulf of Capua's *De Calculatione*; an Illustrated Abacus Treatise and Some Evidence for the Hindu-Arabic Numerals in Eleventh-Century South Italy. *Mediaeval Studies* 57: 293–335.

Grierson, Philip, & Travaini, Lucia 1998. *Medieval European Coinage. With a Catalogue of the Coins in the Fitzwilliam Museum, Cambridge*, Vol. 14. Cambridge: Cambridge University Press.

Haskins, Charles Homer 1927. *Studies in the History of Mediaeval Science*, 2nd edition. Cambridge, MA: Harvard University Press.

Hill, George Francis 1915. *The Development of Arabic Numerals in Europe, Exhibited in Sixty-Four Tables*. Oxford: Clarendon Press.

Hogendijk, Jan P. 1986. Discovery of an 11th-Century Geometrical Compilation: The Istikmāl of Yūsuf al-Mu'taman ibn Hūd, King of Saragossa. *Historia Mathematica* **13**: 43–52.

Hunt, R. W., & Watson, A. G. 1999. Notes on Macray's Descriptions of the Manuscripts. In *Bodleian Library Quarto Catalogues*, Vol. IX: *Digby Manuscripts* (reprint of the 1883 catalogue by William Dunn Macray). Oxford: Clarendon Press.

Ifrah, Georges 1981. *Histoire universelle des chiffres*. Paris: Seghers.

Irani, Rida A. K. 1955/56. Arabic Numeral Forms. *Centaurus* **4**: 1–12; reprinted in E.S. Kennedy *et al.*, *Studies in the Islamic Exact Sciences*. Beirut: American University, 1983: 710–721.

Jacquart, Danielle 2001. Les manuscrits des traductions de Gérard de Crémone. In *Les Traducteurs au travail: leurs manuscrits et leurs méthodes*. Jacqueline Hamesse, Ed., pp. 207–220. Turnhout: Brepols.

Johns, Jeremy 2002. *The Royal Dīwān: Arabic Administration in Norman Sicily*, in press.

King, David A. 2001. *The Ciphers of the Monks. A Forgotten Number-Notation of the Middle Ages*. Stuttgart: Steiner.

Köbert, Raimund 1975. Zum Prinzip der *ġurāb* [read: *ġubār*]-Zahlen und damit unseres Zahlensystems. *Orientalia* **44**: 108–112.

Kunitzsch, Paul 1986–91. *Der Sternkatalog des Almagest*, 3 vols. Wiesbaden: Harrassowitz.

—— 2002. The Transmission of Hindu-Arabic Numerals Reconsidered. In *New Approaches to Islamic Science*. Jan P. Hogendijk & Abdelhamid I. Sabra, Eds. Cambridge, MA: MIT Press, in press.

Lemay, Richard 1977. The Hispanic Origin of Our Present Numeral Forms. *Viator* **8**: 435–462.

—— 1987. De la scolastique à l'histoire par le truchement de la philologie. In *La diffusione delle scienze islamiche nel medio evo europeo (Roma, 2–4 ottobre 1984): convegno internazionale*, Biancamaria Scarcia Amoretti, Ed., pp. 399–535. Rome: Accademia dei Lincei.

—— 1995–96. Abū Ma'šar al-Balḫī [Albumasar], *Liber introductorii maioris ad scientiam judiciorum astrorum*, 9 vols. Naples: Istituto Universitario Orientale.

—— 2000. Nouveautés fugaces dans des textes mathématiques du XIIe siècle. Un essai d'abjad latin avorté. In *Sic itur ad astra. Studien zur Geschichte der Mathematik und Naturwissenschaften. Festschrift für den Arabisten Paul Kunitzsch zum 70. Geburtstag*, Menso Folkerts & Richard Lorch, Eds., pp. 376–392. Wiesbaden: Harrassowitz.

Leonardi, Claudio 1960. I codici di Marziano Capella [II]. *Aevum* **34**: 1–99.

Lévy, Tony 2001. Hebrew and Latin Versions of an Unknown Mathematical Text by Abraham ibn Ezra. *Aleph* **1**: 295–305.

Menninger, Karl 1958. *Zahlwort und Ziffer. Eine Kulturgeschichte der Zahl*, 2nd revised edition, 2 vols. Göttingen: Vandenhoeck & Ruprecht.

V

264

Mercier, Raymond 1991. The Lost *Zīj* of al-Ṣūfī in the Twelfth Century Tables for London and Pisa. In *Lectures from the Conference on al-Ṣūfī and Ibn al-Nafīs, 5–8 October 1987*, pp. 38–71. Beyrut and Damascus: Dār al-Fikr.

Millás Vallicrosa, José Maria 1942. *Las traducciones orientales en los manuscritos de la Biblioteca Catedral de Toledo.* Madrid: Consejo superior de investigaciones científicas.

—— 1947. *El libro de los fundamentos de las Tablas astronómicas de R. Abraham ibn ʿEzra.* Madrid / Barcelona: Consejo superior de investigaciones científicas.

Millás Vendrell, Eduardo 1963. *El comentario de Ibn al-Muṭannā' a las Tablas Astronómicas de al-Jwārizmī.* Madrid / Barcelona: Consejo superior de investigaciones científicas.

Miura, Nobuo 1981. The Algebra in the *Liber Abaci* of Leonardo Pisano. *Historia scientiarum* **21**: 57–65.

Murdoch, John E. 1966. Euclides Graeco-Latinus: A Hitherto Unknown Medieval Latin Translation of the *Elements* Made Directly from the Greek. *Harvard Studies in Classical Philology* **71**: 249–302.

North, John D. 1995. "Aragonensis" and the Toledan Material in Trinity O.8.34. *Cahiers du moyen âge grec et latin* **65**: 59–61.

Pingree, David, Ed. 1985–86. *The Astronomical Works of Gregory Chioniades*, Vol. I: *The Zīj al-ʿAlā'ī*, 2 Parts. Amsterdam: J.C. Gieben.

Poulle, Emmanuel 1963. *La Bibliothèque scientifique d'un imprimeur humaniste au XV siècle.* Geneva: Droz.

—— 1964. Le traité d'astrolabe de Raymond de Marseille. *Studi medievali*, 3rd series **5**: 866–900.

Rashed, Roshdi 1994. Fibonacci et les mathématiciens arabes. *Micrologus* **2**: 145–160.

Reeve, Michael D. 1980. Some Astronomical Manuscripts. *Classical Quarterly* **30**: 508–522.

Renou, Louis, & Filliozat, Jean 1985. *L'Inde Classique. Manuel des études indiennes*, 2nd edition, 2 vols. Paris: École française d'Extrême Orient.

Rose, Valentin 1893–1919. *Verzeichnis der lateinischen Handschriften der Königlichen Bibliothek zu Berlin*, 3 vols. Berlin: Asher / Behrend.

Schum, Wilhelm 1882. *Exempla codicum amplonianorum erfurtensium saeculi IX.–XV.* Berlin: Weidmann.

—— 1887. *Beschreibendes Verzeichnis der amplonianischen Handschriften-Sammlung zu Erfurt.* Berlin: Weidmann.

Spinelli, Domenico 1844. *Monete cufiche battute da principi Longobardi, Normanni, e Svevi nel Regno delle Duo Sicilie.* Naples; reprinted Sala Bolognese: Forni, 1977.

Travaini, Lucia 1995. *La monetazione nell'Italia normanna.* Rome: Istituto storico italiano per il medio evo.

Wilson, Nigel G. 1981. Miscellanea Palaeographica. *Greek, Roman and Byzantine Studies* **22**: 395–404.

Table

In the following table, numeral forms have been grouped together according to their similarities with each other. All manuscripts are twelfth or early thirteenth century, unless otherwise indicated. While only a representative selection of Western forms is given, a complete list of Eastern forms in Latin is provided. An asterisk (*) indicates that the forms in question are not the forms usually used by the scribe, or have been copied by him with obvious difficulty. A double asterisk (**) indicates that the whole of the relevant manuscript has been written in Eastern forms.

I Western Forms

Arabic		9	8	7	6	5	4	3	2	1	0
a)	Rabat, Maktaba al-'āmma, k 222 (Ibn al-Yāsamīn)	ꝯ	8	7	6	ɣ	ʄ	3	2	1	
b)	Univ. of Tunis, 2043, 1611, (ash-Sharīshī)	ꝯ	8	1	6	ɣ	ﻉ	ع	ح	1	o
Latin											
a)*	Escorial, d.I.2 (from Albelda, 976 A.D.)	9	8	1	b	Ч	ʄ	ع	ح	I	
b)	British Library, Harley 3595, s. XI^mid (Abacus forms)	9	8	ſ	ⱶ	Ч	ﻉ	ᵮ	Ꮿ	1	
c) ·	Munich Clm 18927 (unnamed row)	ꝯ	8	7	6	ɣ	ʒ	ʒ	ꝑ	1	oꞇ
d)	Cambridge, Trinity O.7.41	9	8	ʌ	ᴳ	ч	४	ʒ	ꝑ	1	
e)	Paris, BNF, lat. 16208 (Liber ysagogarum)	ꝯ	8	ꝩ	ᴳ	S	ʒ	ᵮ	ꝑ	1	oꞇ
f)	Vienna 275 (Liber ysagogarum)	ꝯ	8₈	7	ᴳ	ч	ʒ	ᵥ	ꝑ	1	O
g)	Bodleian, Digby 51 (al-Battānī)	ꝯ	ʒ	1	ᴳ	ч	ʒʒ	ʒʒ	ꝑz	1	ꝧₒ
h)	Hispanic Society of America, HC 397/726 (Dixit Algorismi)	9	8	ꝗ	ᴳ	ч	ﻉ	ʒ	z	J	o
i)	Erfurt, Q 351 (John of Seville)		8		6	ч	ꝩꝩ	ʒ	ʒ	J	o
j)	Erfurt, Q 365 (Raymond of Marseilles)			ʌ	ᴳ		ꝩ	ʒ		1	o
k)*	Paris, BNF, lat. 16204 (On the Great Conjunctions)				b		ꝺꝺ			1	
l)*	Paris, BNF, lat. 14738 (Gerard of Cremona)	9	ʒ	ʌ	6	ч	ꝩ	ʒʒ	ᶻᴢ	1	o
m)*	Munich, Clm 18927 ('toletane f<igure>')	ꝯ	8	ꜩ	6	ч	४	ʒ	ʒ	1	
n)	Paris, BNF, lat. 15461 (Liber Alchorismi)	9	ƀ	ʌ	ᴳ	ɣ	ꝗ	ʒ	z	1	o

o) Raniero	⅃	8	7	ᕲ	ᶌ	ʌ	3	z	ι	o
p)* Florence, Bibl. naz. centrale, Magliabech. (Fibonacci I)			6	ʏ	ᵏ	Ƹ	ᴢ		ι	o
q) Florence, Bibl. naz. centrale, Magliabech. (Fibonacci II)	꟫	8	7	6	ᶘ	9	3	z	ι	
Greek										
Vatican, gr. 184, s. XIII (*Psēphēphoria*)	9	8	ᐱ	6	ᴟ	ᴚ	3	ᴢ	ι	o

II Palermitan Forms

Arabic										
a) Paris, BNF, ar. 2457	꟫꟫	ᐱ	V	ᴟ	ᴇ	ᵍᶜ	ᴦᴣ	ᴨ	ι	٥
b) British Library, Harley 5786, c.1153 A.D. (Harleian Psalter)	꟫	8	V	ᒪ	ᒪᴣ	ꟼᶜ	ᴘ	ᴦ	ι	٥
c) Roger II's coins, 1138 and 1148 A.D.					ᴣᴊ		ᴘ			
Latin										
Vat. Pal. lat. 1371 (Sicilian *Almagest*)	ꟼ	ꟹ	V	ᴕ	ᒪ	ᵏᶜ	ᴘ	ᴦ	ι	٥

III Eastern Forms

Arabic										
a) Rabat, Maktaba al-'āmma, k 222 (Ibn al-Yāsamīn)	ꟼ	ʌ	V	ᶘ	ᒪᴣ	ᴦᶜ	ᴦ	ᴦ	ι	
b) Bodleian, Or. 516, 1082 A.D. (al-Bīrūnī)	ꟼ	ᐱ	V	ᴟ	ᴂ	ᵍᶜ	ᴘ	ᴦ	)	ᵞ٥
c) Chester Beatty, 3910, 1177 A.D. (al-Bīrūnī)	ꟼ	ʌ	ᴜ	ᴟ	ᶒ	ᵠᶜ	ᴘ	ᴦ	/	٥
Latin										
a)* Munich, Clm 18927 ('indice f<igure>')	ᴘ	ꟼ	ᵛ	ᴟ	ꟼ	ᴦᶜ	ᴘ	ᴘ	ι	
b)* Vat. lat. 1393 (*Liber Alchorismi/Liber pulveris*	ᶘ	ꟼ	ᵞ	ᴟ	ᴃ	ᴦᶜ	ᴊ	ᵞ	ι	
Pisan Tables and Ibn Ezra										
c)**Berlin, lat. fol. 307	ꟾ	ꟼ	ᴠ	ᴟ	ᴃ	ᴩꜱ	ᴣ ᴓ	ᴘ	ι	o
d) Cambridge, Fitzwilliam Museum, McClean 165	ᴘ	ꟼꟼ	ᴠ	h	8	ꟼ	ᴘ	ᴦ	ι	o
e)* British Library, Arundel 377	ᴘ	ꟼ	ᴠ	ᴟᴟ	ᴃᴆ	ᴨꜱ	ᴦ	ᴘ	ι	
f)**Bodleian, Selden supra 26	ꟾ	ꟼ	ᴠ	ᴟ	ᴃᴃᴓ	ᵏꜱ	ᴟ	ᴦ	ι	o
g) British Library, Cotton Vespasian A.II	ᴘ	ꟾ	ᴠ	ᴥ	ᴃ	ᴣ	ᴦ	ᴘ	ꟾ	o
h) Erfurt, Q 381	ᴘ	ꟾ	ᵏᴠ	ᴥ	ᴃᴃ	ᴣᶜ	ᴦ	ᴘ	ι	o

i)	Bodleian, Digby 40	የ	ꝰ₁	⌣	4ꝯ	8	ꞔ	ᴘ	ᴦ	ı	o
j)	Bodleian, Digby 51	p	ꝰ₂	⌵	н	℮₈	33	ᵏᴘ	ᵱ	ı	o
k)*	Venice, Bibl. Marciana, Lat. Z. 344 (*Iohannes Hispal.*)	ꝰ	q	⌵	Ꮁ	8	Ꮭ	ᴄᵱ	ꝓ	ı	ꞇ
Hugo of Santalla and Hermann of Carinthia											
l)	Bodleian, Arch. Seld. B 34			⅄	ꞩ	8	ᵍᶜ	ᴘ		ı	o
m)*	Cambridge, Caius 456			⅄	Ꮁ	Ᏼ	Ьᵡᵥı	ꞮꞮ		ı	o
n)	Berlin, Hamilton 557	ꝑ	q	⌵	Ꮁ	℮℮₈	ᵣᴄ	ᵣᴜᴚ	ᵱ	ı	o
o)*	Bodleian, Digby 50	ꝑ	ꝗ	⌵	Ꭻ℘	Ꝑ	ꝝ3	ᴘ	ᴦ	ı	o
p)**	British Library, Arundel 268	ꝰ	ꝗ	⋁	Ꮯ	℮	ꞓᴜ	ꞮꞮ	ᴘ	ı	o
Other manuscripts											
q)**	Paris, BNF, lat. 7373 (Euclid)	ꝰ	⋀	⌵	Ꮁ	℮	ꞑ	ꞮꞮ	ꞯ	ı	o
r)	British Library, Harley 5402 1160 A.D. (Luccan annotator)				Ꮁ						ꞇ o
s)*	Cambrai 930 (Stephen the Philosopher)	q	ꝰₔ	u	ꙑ	ꝿ	ᵍᶜ	ꞮꞮ	ꝓ	ı	
t)	Soest 24, 1185 A.D. (conversion of eras)	ꝰ	⌵	⋀	ꙑ	Ꮯ	ᴘ	3	ꝑ	ı	o
u)*	Heiligenkreuz 226 (Hugh of Fouilloy, *Aviarium*)				ꙃ	℮	ꝯ		ꝑ		o
v)*	Madrid 10009, 1267-86 A.D. (Abraham bar Hiyya)	ꝰ	Ꮯ	⋁	Ꮁ	8	Ꮭ	ᴄᵱ	ꝓ	ı	ꞇ o
w)*	Paris, BNF, lat. 10252, s. XV (algorism)	G	Ꝗ	⌵	ᴋɗ	ꞕ	ꞕ	F	ꞕ	1	
x)*	Dresden C 80, s. XV (algorism)	ꝗₔ	Ꝓ	⌵	н	⊗	ℰ	ᴦ	ᴦ	ı	o
Greek											
a)	Bodleian, Auct. F.6.23 (Euclid)	ꝰq	⋀	⌵	Ꮁ	ꝍ	Ꮭ	ᴦ	ᴦ	ı	o
b)*	Bodleian, d'Orville, s. XIII (?)	ꝰ	⋀	⌵	ꙑ	ꝏ₀	ᵏꞌꞩ	ᴘ	ᴦ	ı	
c)	Vat. gr. 211, s. XIII (astronomical tables)	ꝗ	⋀	⌵	Ꮁ	Ꞓ	ꞓ	ᴘ	ᴦ	ı	o

Sources and comments

I Western Forms

Arabic **a:** see Plate 1; **b:** from [Ifrah 1981: 503].

Latin **a:** this "Codex Vigilanus" has been reproduced many times: e.g. [Ifrah 1981: 504]; **b:** [Folkerts 1970, Table I; Ifrah 1981: 506]; **c:** see Plate 3; **d:** from a table of numbers on fol. 62v; **e:** the numeral forms on fols. 67r–69v; **f:** the numeral forms used in a table from the *Liber ysagogarum* (Vienna 275, fol. 27r [Menninger 1958, II: 239]); **g:** forms used in the text and notes to al-Battānī's *Opus astronomicum* on fols. 5–9, 12–15 and 17; **h:** fol. 17r from [Folkerts 1997, Table 1]; **i:** fols. 103–130, John of Seville's translation of al-Farghānī's *Rudimenta*; **j:** a version of Raymond of Marseilles's *Iudicia* on fols. 40v–42r; **k:** see Plates 2 and 4; **l:** the Arabic numerals on the first folios of a copy of the *Almagest*; **m:** see Plate 3; **n:** see [Lemay 1977, Fig. 5]; **o:** for the forms used by the notary Raniero, see [Bartoli Langeli 2000]; **p:** the "Arabic-looking" numerals in the table in the *Liber abbaci* in MS Florence, Biblioteca nazionale centrale, fondo Magliabechiano, conv. sopp. Scaffale C. Palchetto 1, no. 2616, fol. 1r, reproduced in facsimile in [Boncompagni 1852: 103]; **q:** the standard numeral forms in the *Liber abbaci* in the same manuscript [Boncompagni 1852: 103].

Greek **a:** the *Psēphēphoria* of 1252 [Allard 1976].

II Palermitan Forms

Arabic **a:** a manuscript of the works of al-Sijzī, commonly thought to have been written in Shiraz between 969 and 972, but probably much later (illustration in [Kunitzsch 2002]), given here because of its "mixed" forms of 2s and 3s and similar appearance to the script of the Harleian Psalter; **b:** see Plate 5; **c:** see Plates 7 and 8.

Latin see Plate 6.

III Eastern Forms

Arabic **a:** see Plate 1; **b:** from [Irani 1955/56: 4]; **c:** from personal inspection.

Latin **a:** see Plate 3; **b:** from [Allard 1992: 69]; **c:** see Plate 18; **d:** from personal inspection; **e:** see Plate 17; **f:** see Plate 19; **g:** see Plate 15; **h:** from [Schum 1882, no. 13]; **i:** see Plate 16; **j:** see Plate 14; **k:** from personal inspection; **l:** see Plate 10; **m:** see Plate 9; **n:** from personal inspection; **o:** see Plate 11; **p:** see Plate 12; **q:** see Plate 13; **r:** see [Burnett 2000a: 86]; **s:** see [Lemay 2000: 391–392]; **t:** from [Becker 1995]; **u:** see Plate 22 (the identification of 2 and 9 is not certain); **v:** see Plate 23; **w:** from personal inspection; **x:** from personal inspection.

Greek **a:** see Plate 20; **b:** see Plate 21; **c:** these are the numerals in [Pingree 1985–86, Part 2: 11, 15, 17, 20–21, 35, 44].

Indian Numerals in the Mediterranean Basin in the Twelfth Century   269

Plates

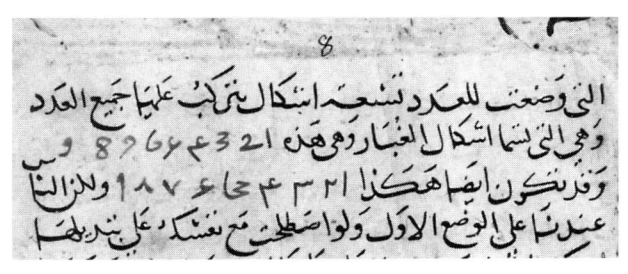

Plate 1. The Western and Eastern forms in the algorism of Ibn al-Yāsamīn; MS Rabat, Maktaba al-'āmma, k 222, fol. 5a.

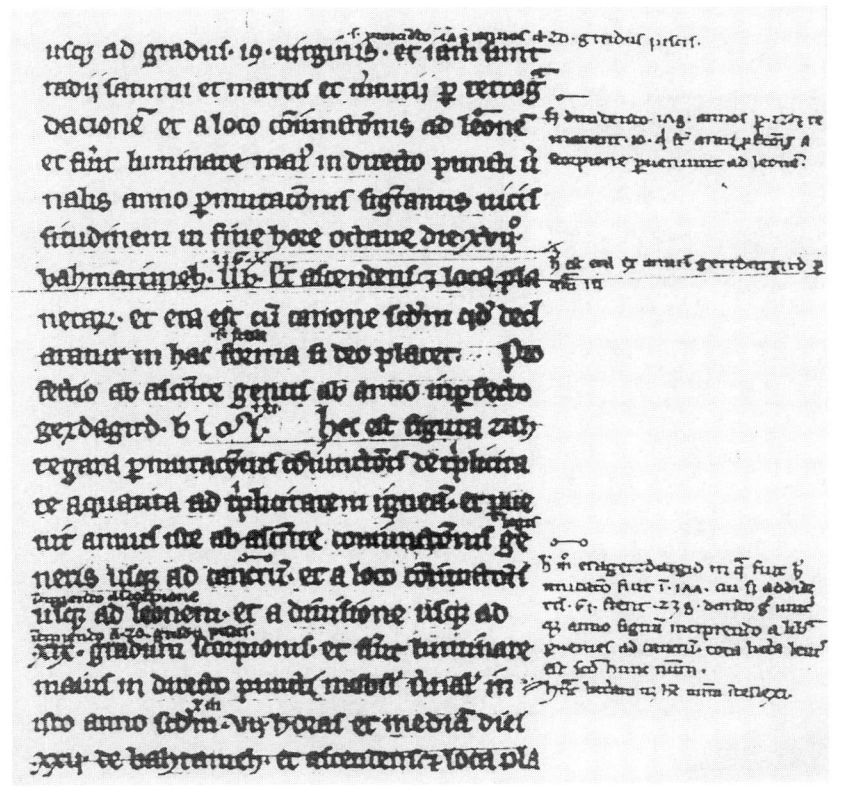

Plate 2. The "fossilised" forms of early numerals (116 in line 7; 141 in line 11) in the revised translation of Abū Ma'shar's *On the Great Conjunctions*, in MS Paris, Bibliothèque nationale de France, lat. 16204, p. 299.

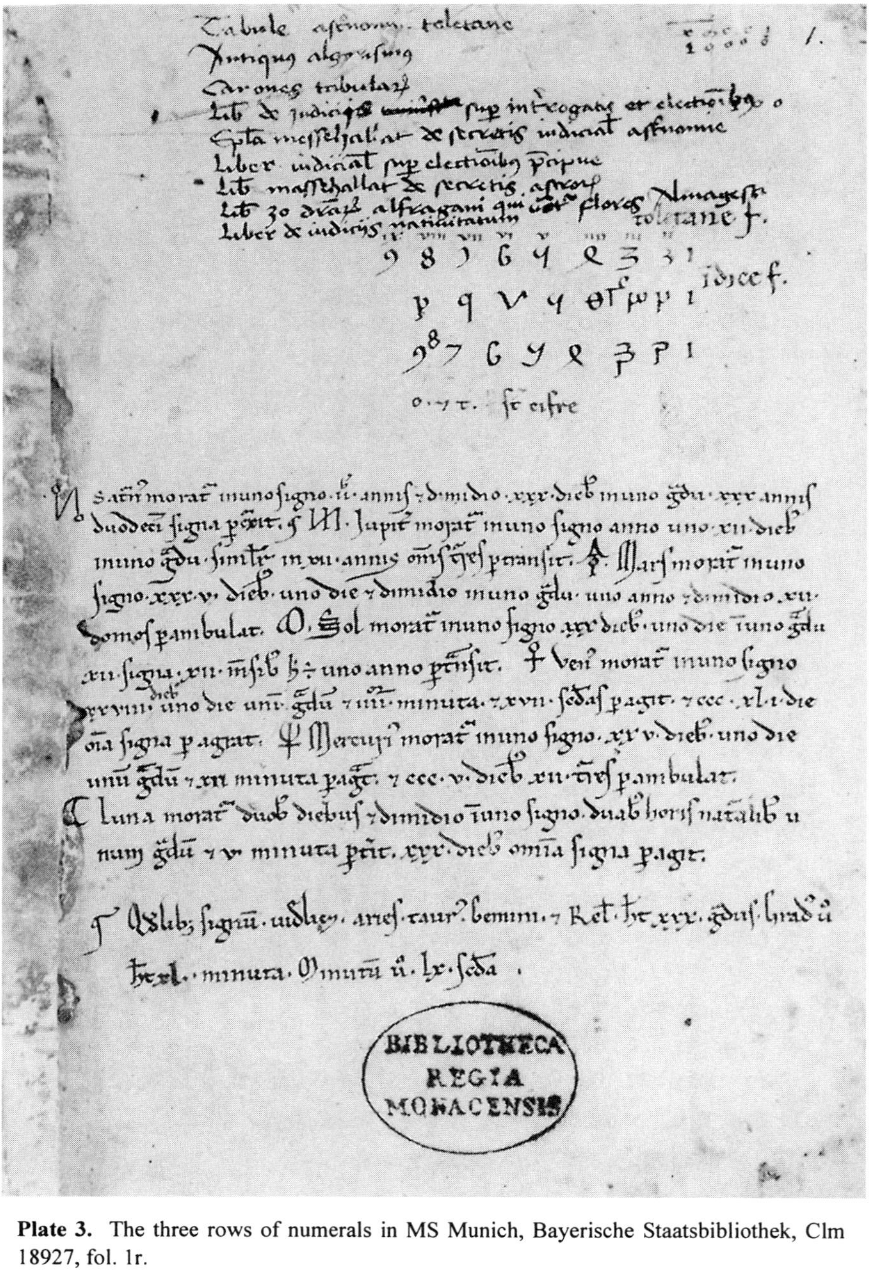

Plate 3. The three rows of numerals in MS Munich, Bayerische Staatsbibliothek, Clm 18927, fol. 1r.

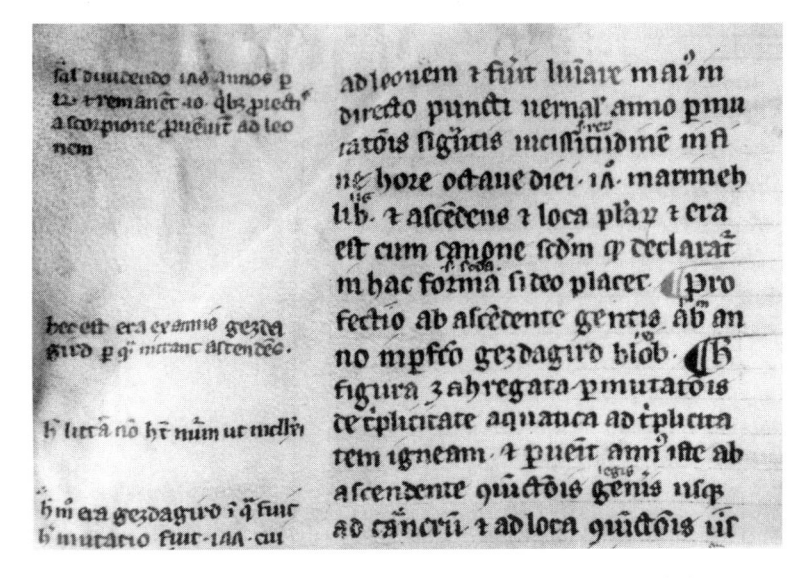

Plate 4. The "fossilised" forms of early numerals in MS Cambrai, Médiathèque municipale, 168 (163), fol. 98v (see lines 5 and 9).

Plate 5. The number "56" in the Harleian Psalter (note that Psalm 57 in Latin is equivalent to Psalm 55 in Greek and 56 in Arabic); MS British Library, Harley 5786, fol. 74r.

Plate 6. The numerals in the Sicilian translation of the *Almagest* in MS Vatican City, Biblioteca apostolica Vaticana, Pal. lat. 1371, fol. 69v.

Plate 7. A "folles" of Roger II of 1138/39, on which the *hijra* date of 533 is given in Indian numerals; from [Travaini 1995, Pl. 12].

Plate 8. A "folles" of Roger II of 1148/49, on which the *hijra* date of 543 is given in Indian numerals; from [Travaini 1995, Table 15].

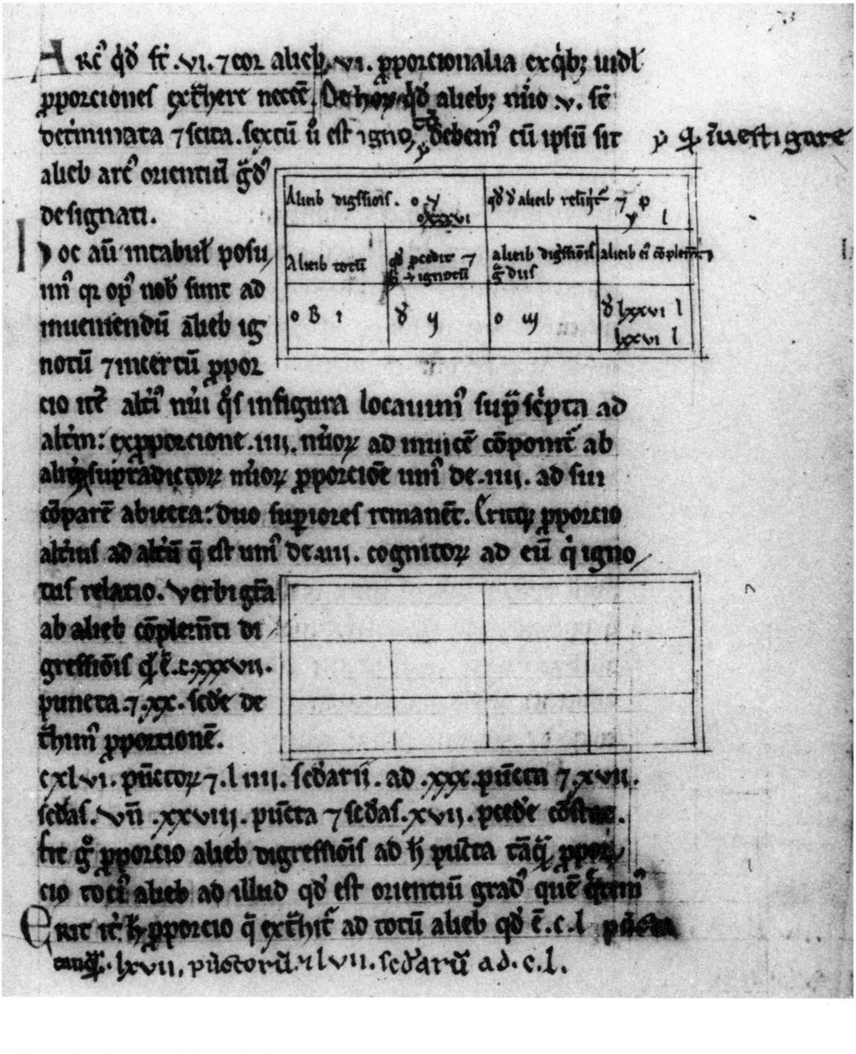

Plate 9. Hugo of Santalla's translation of Ibn al-Muthannā's commentary on al-Khwā-rizmī in MS Cambridge, Gonville and Caius 456/394, fol. 73r. Note the numbers 150, 67, 30 and 147 in the third row of the table.

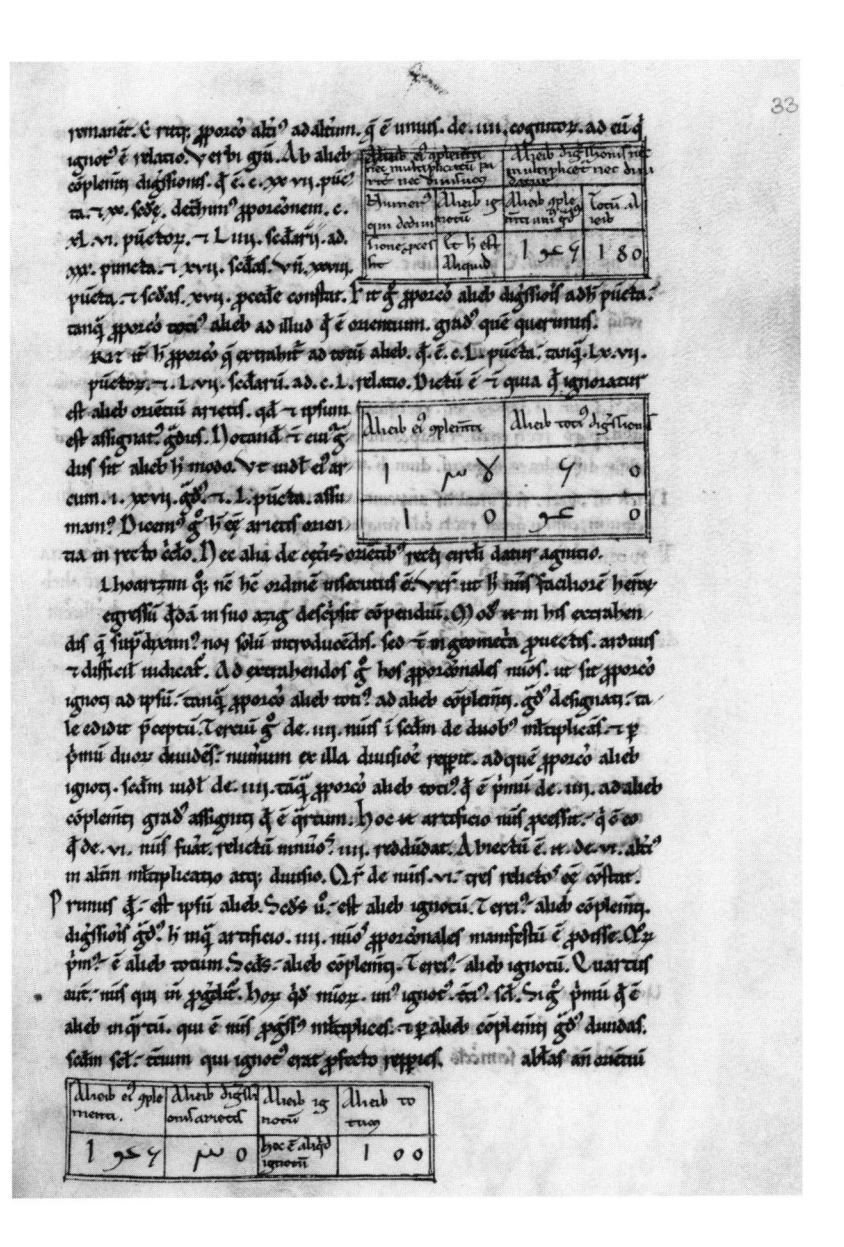

Plate 10. The same in MS Oxford, Bodleian Library, Arch. Seld. B 34, fol. 33r. Note the numbers 137, 60, 10 and 40 in the middle table, and 146, 30 and 100 in the lower table.

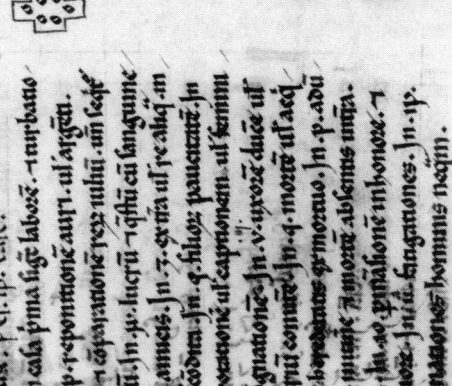

Plate 11. From the supplement to Hugo of Santalla's *Ars geomancie* in MS Oxford, Bodleian Library, Digby 50, fols. 94v (note the number 4 in the second line) and 100r (note the sequence of numbers from 1 to 12).

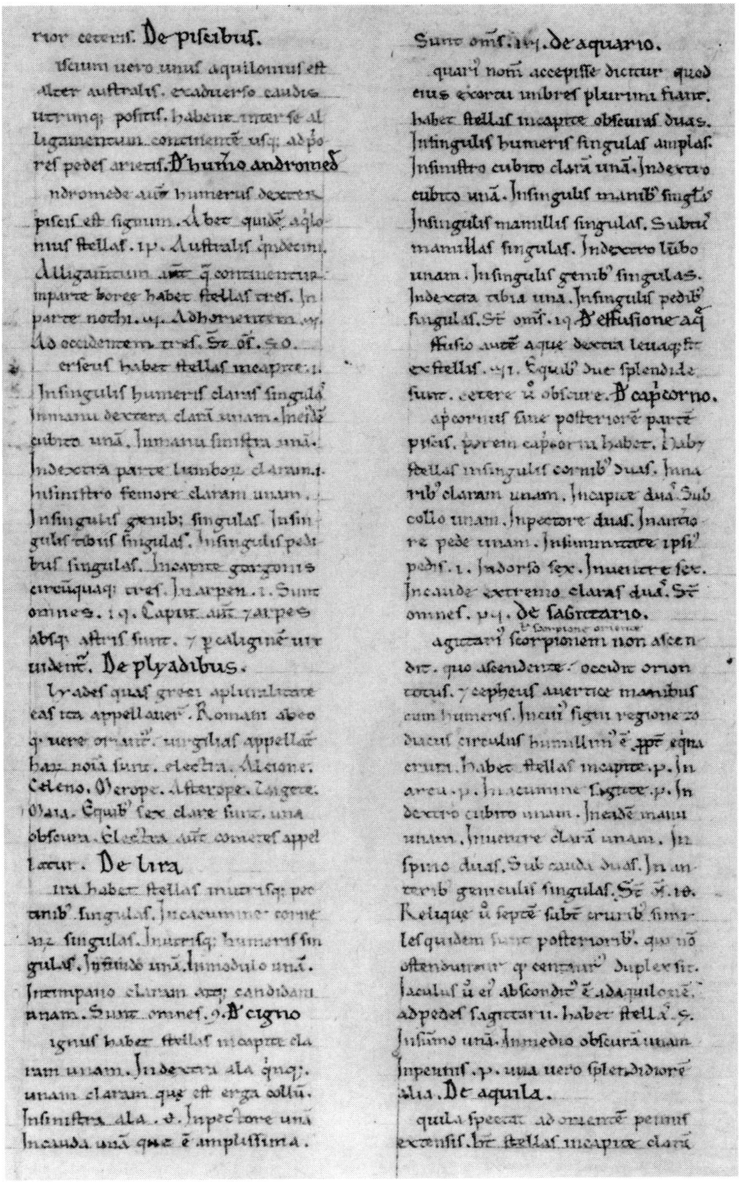

Plate 12. From the *Aratea* in British Library, Arundel 268, fol. 87r. Note the numbers 12, 3 and 40 within the description of Andromeda.

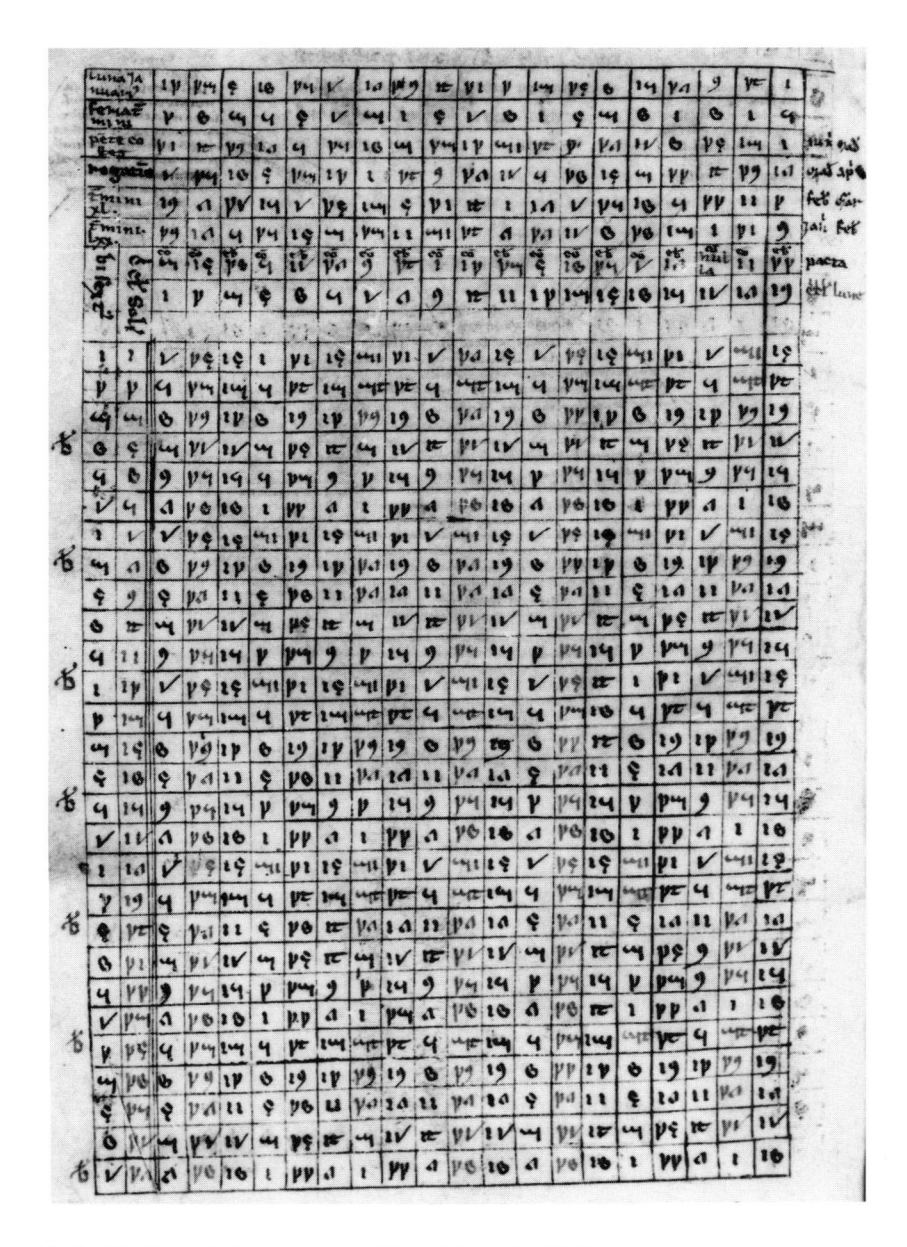

Plate 13. Part of the computus table in MS Paris, Bibliothèque nationale de France, lat. 7373, fol. 176v.

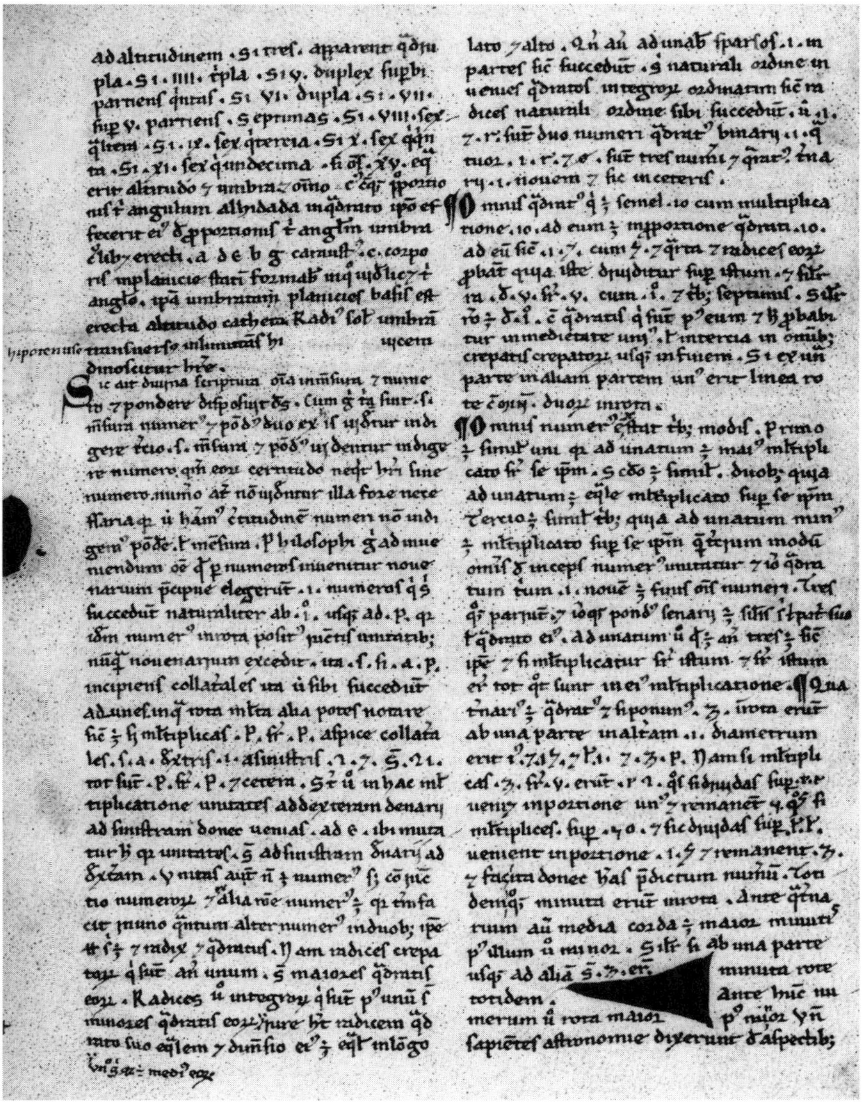

Plate 14a. Abraham ibn Ezra's *Arithmetic* and *Geometry* in MS Oxford, Bodleian Library, Digby 51, fol. 38v.

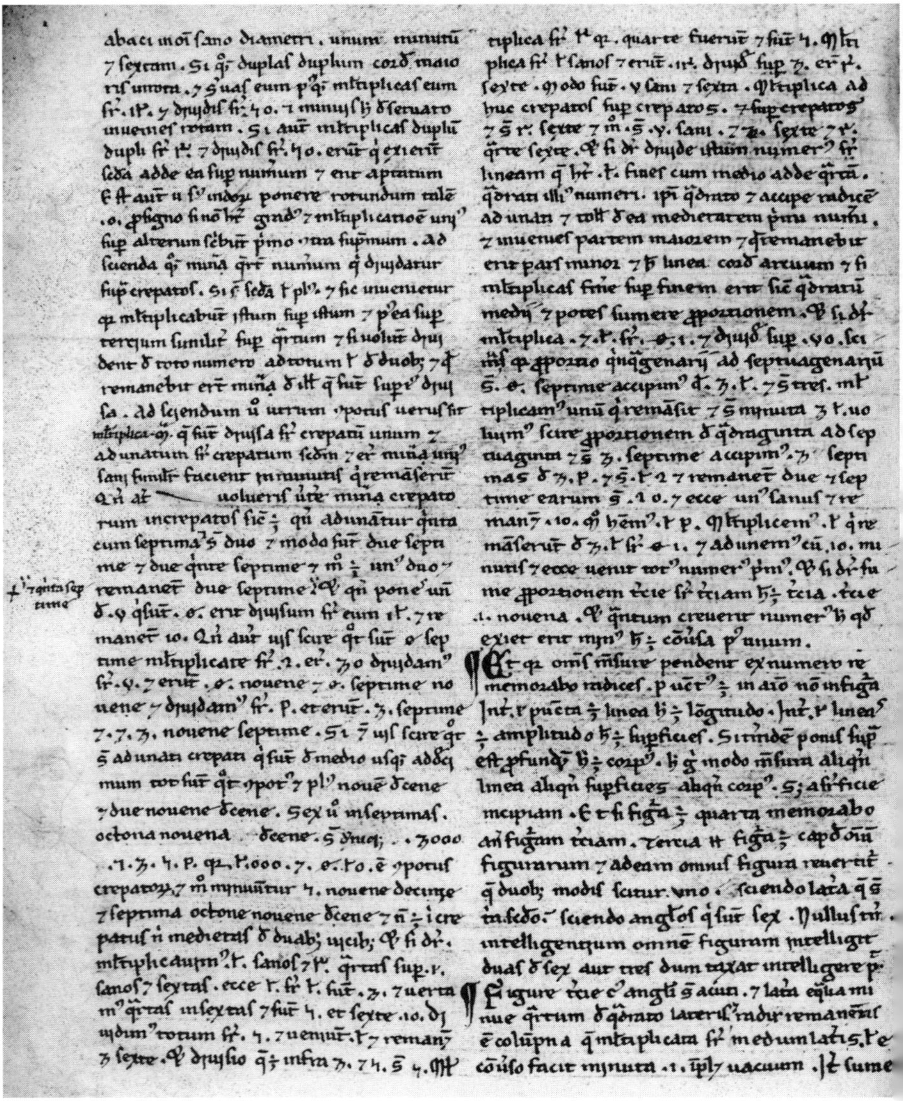

Plate 14b. Abraham ibn Ezra's *Arithmetic* and *Geometry* in MS Oxford, Bodleian Library, Digby 51, fol. 39v.

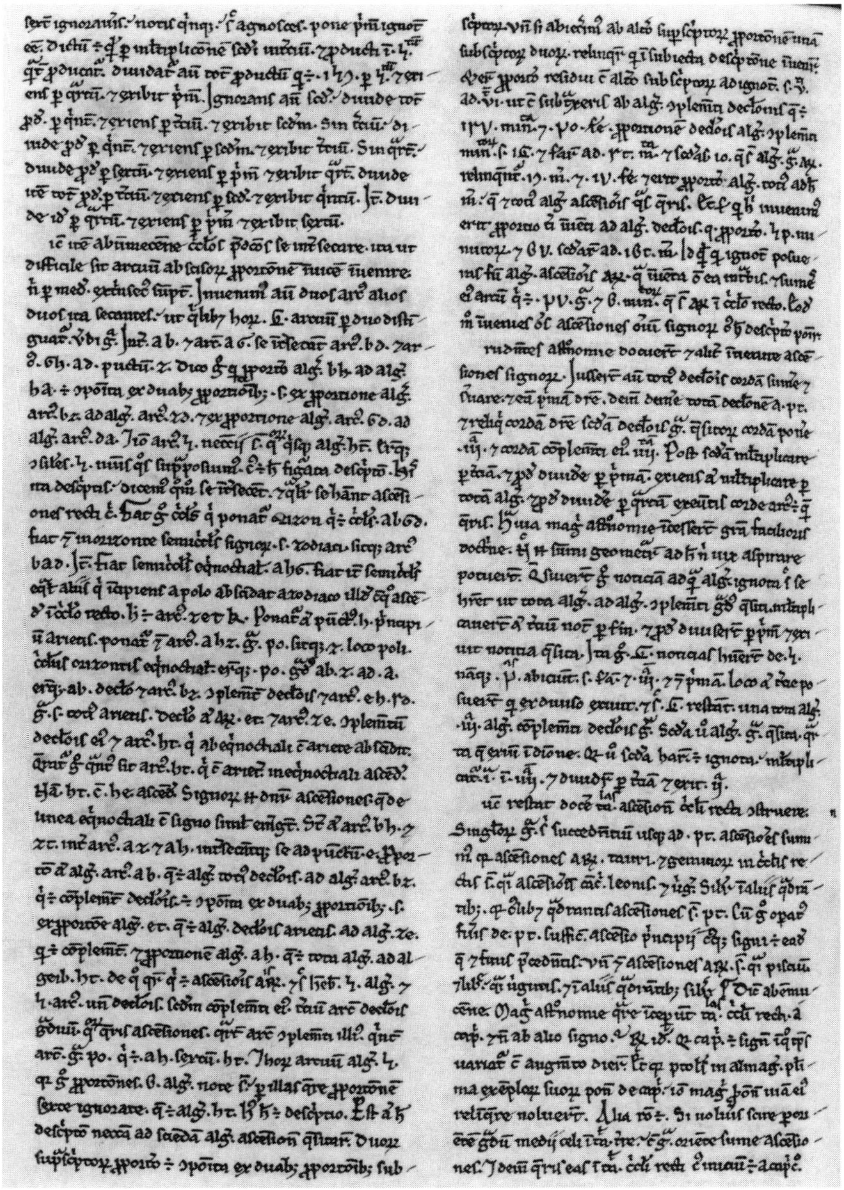

Plate 15. Abraham ibn Ezra's *Book on the Foundations of the Astronomical Tables* in MS British Library, Cotton, Vespasian A II, fol. 29v.

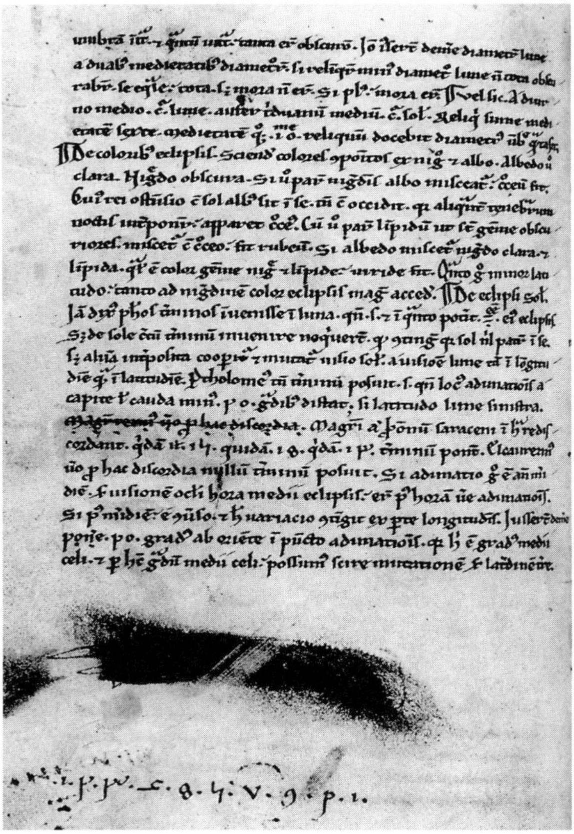

Plate 16. The same in MS Oxford, Bodleian Library, Digby 40, fol. 88v.

Plate 17. The varieties of numerals in MS British Library, Arundel 377, fol. 56r. Note that, in the lower row, the scribe has written two forms of 4 and omitted 5.

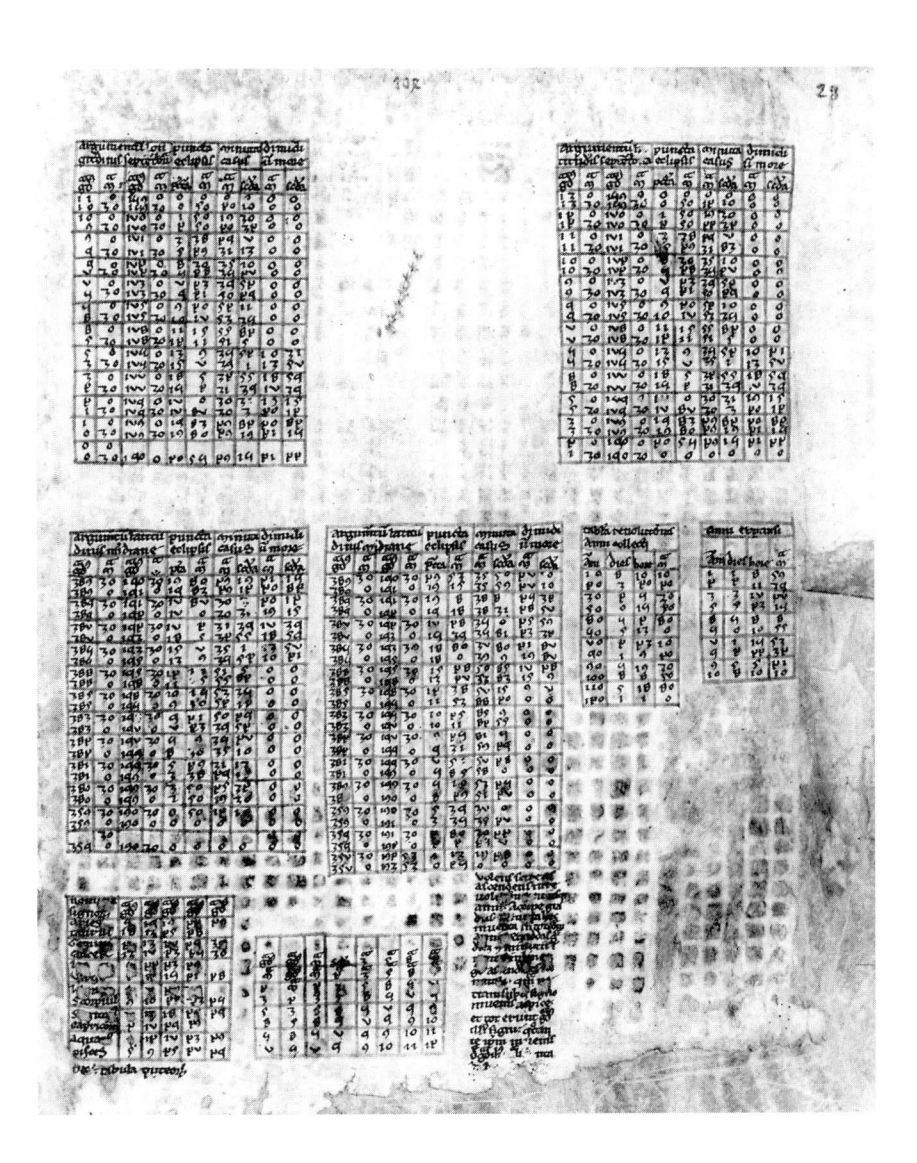

Plate 18. The tables of Pisa in MS Berlin, Staatsbibliothek Preußischer Kulturbesitz, lat. fol. 307, fol. 28r.

V

284

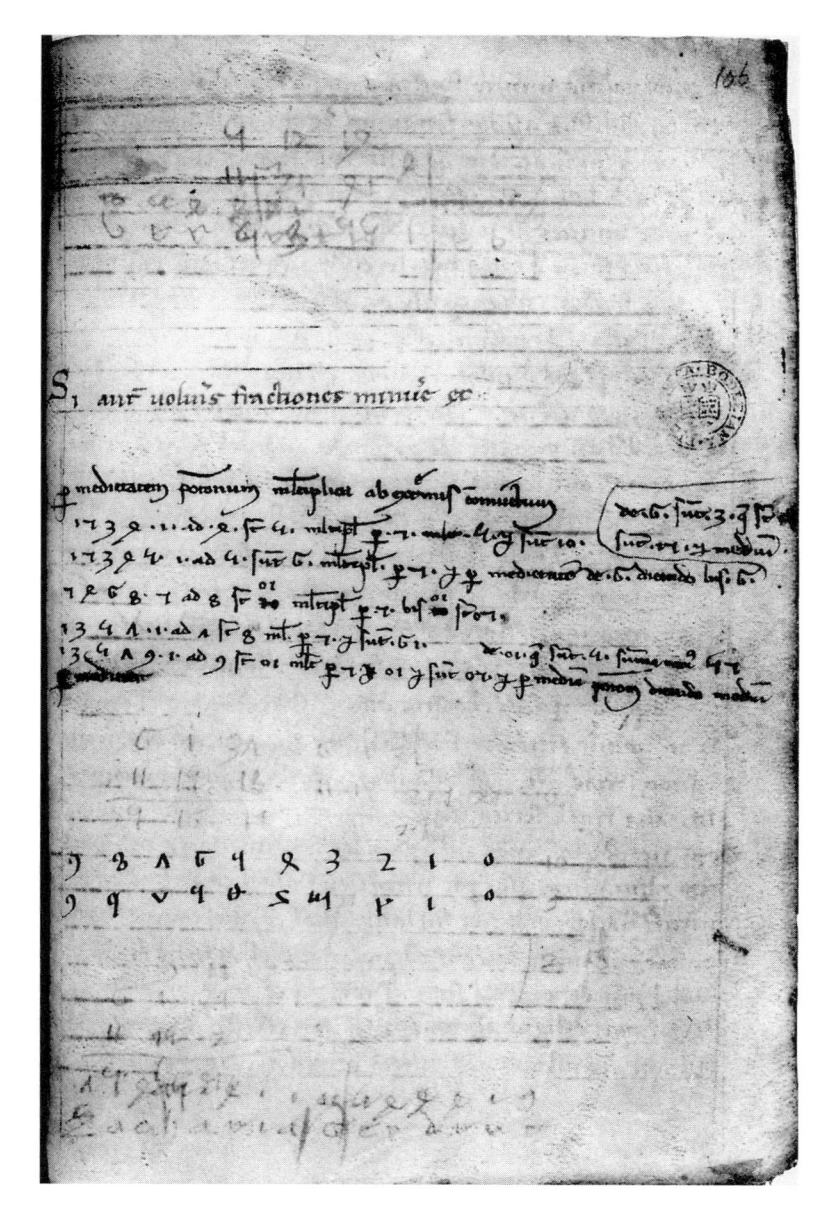

Plate 19a. The *Liber Alchorismi / Liber pulveris* in MS Oxford, Bodleian Library, Selden supra 26, the key on fol. 106r (with some later additions).

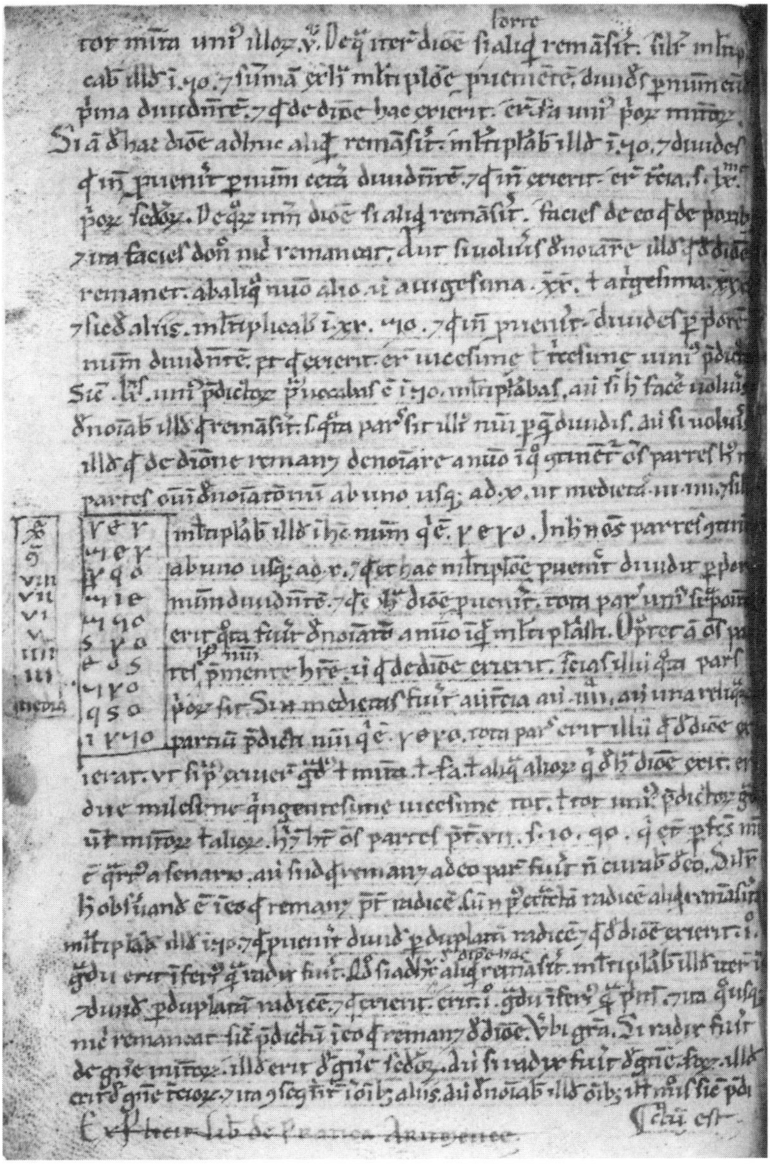

Plate 19b. The *Liber Alchorismi / Liber pulveris* in MS Oxford, Bodleian Library, Selden supra 26, fol. 121v.

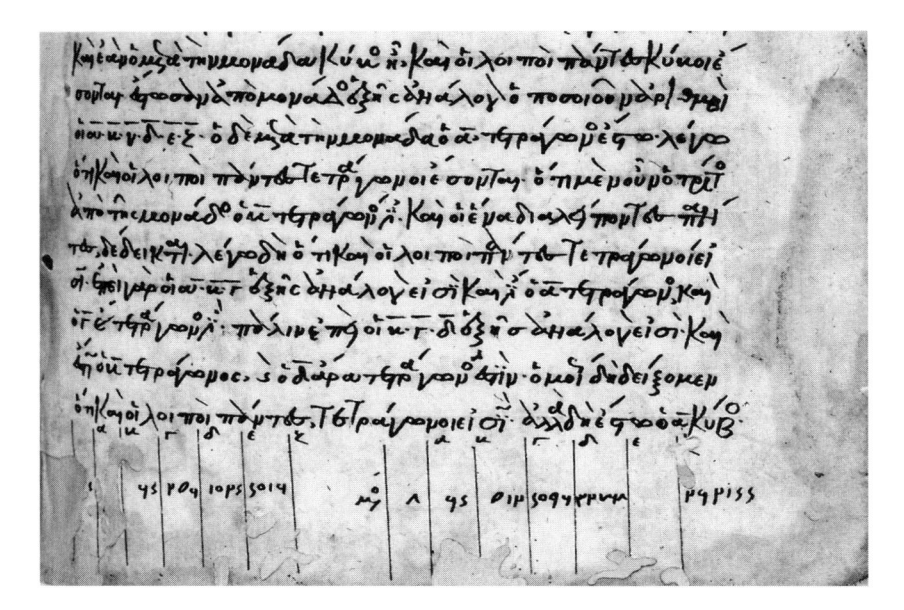

Plate 20. Indian numerals in annotations in MS Oxford, Bodleian Library, Auct. F.6.23, fol. 127r.

Plate 21. The same in MS Oxford, Bodleian Library, D'Orville 301, fol. 1v.

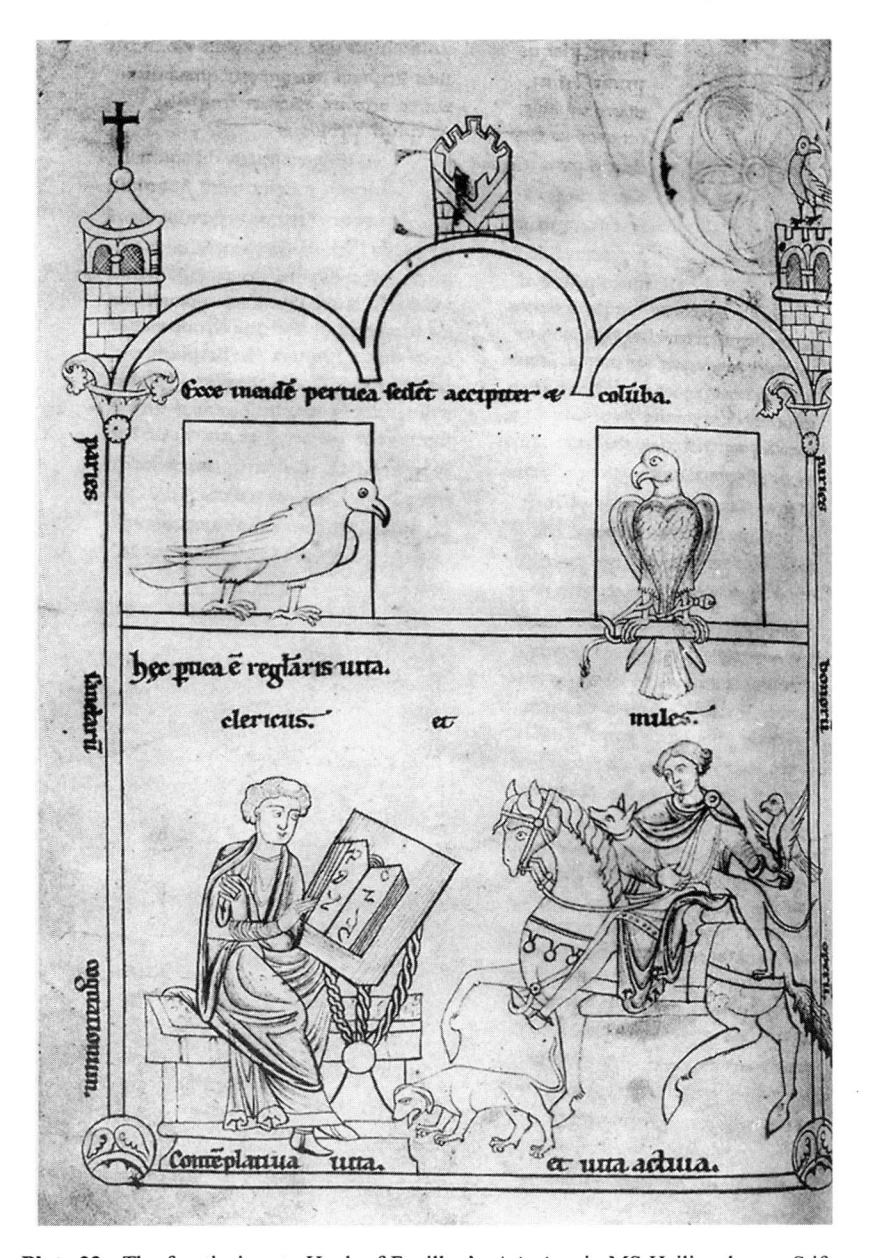

Plate 22. The frontispiece to Hugh of Fouilloy's *Aviarium* in MS Heiligenkreuz, Stifts-bibliothek 226, fol. 129r.

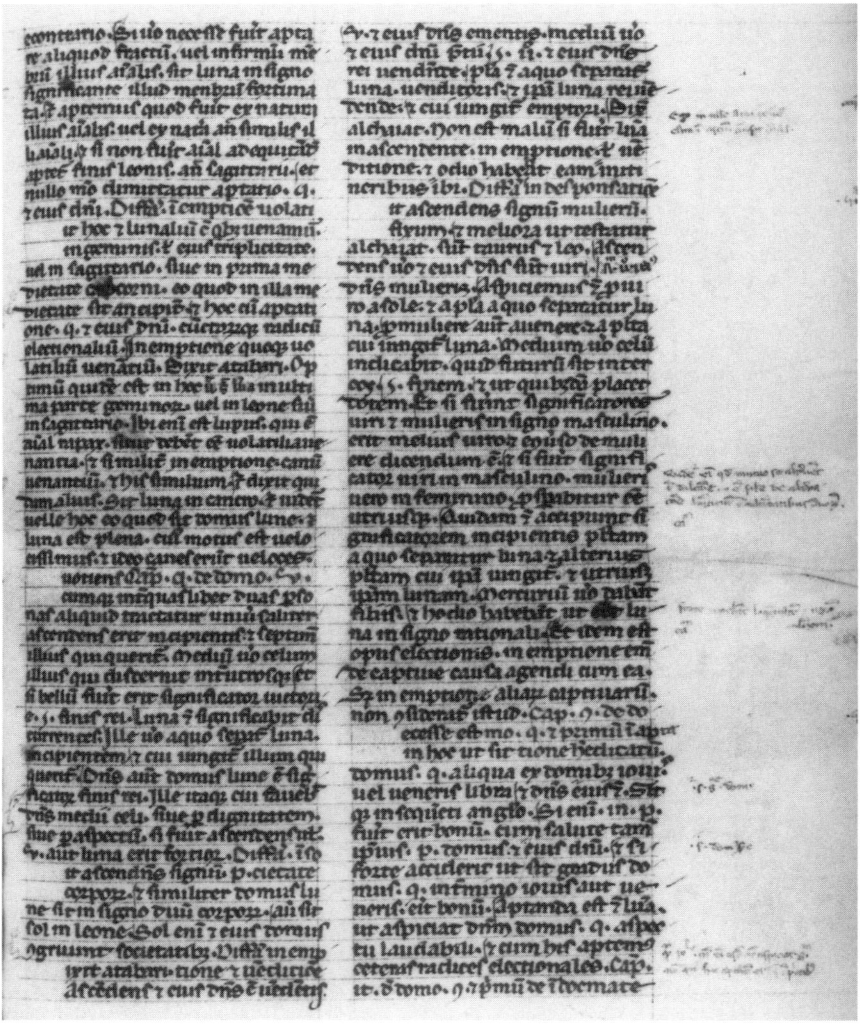

Plate 23. The Eastern forms in the chapter-headings of ʿAlī ibn Aḥmad al-Imrānī's *Elections* in MS Madrid, Biblioteca nacional, 10009, fol. 37r. Note "Cap. 8 de domo 7" and "Cap. 9 de domo 8."

VI

THE USE OF ARABIC NUMERALS AMONG THE
THREE LANGUAGE CULTURES OF NORMAN SICILY

The importance of the three cultures in Norman Sicily for the transmission of science has been recognised at least since the discoveries of Valentin Rose, Charles Haskins and Evelyn Jamison.[1] But these scholars have tended to isolate only individual streams of the transmission: first, translations from Greek into Latin made by Henricus Aristippus, archdeacon of Catania, who translated Plato's *Phaedo* and *Meno*, and the fourth book of Aristotle's *Meteora*; or the still anonymous mathematician (or perhaps two mathematicians) who translated from Greek Euclid's *Elements*, *Data*, *Optica and Catoptrica*, Proclus's *Elementatio physica* (*De motu*) and Ptolemy's *Almagest*;[2] secondly, works written in Arabic, such as the *kitab Rujiri* of al-Idrisi; thirdly, translations from Arabic, such as the Admiral Eugene's translation of the *Optics* of Ptolemy.[3] But Arabic-Latin translations were also made in Spain where no Greek was known, and Greek-Latin translations were made in Constantinople and northern Italy in a non-Arabic context. More recent scholarship has drawn attention to the special situation in Sicily, in which we see the *interweaving* of Greek and Arabic streams.[4]

For there to be reliable translations, there had to be scholars who knew both the subject matter and the languages.[5] This is what we find in Sicily in the twelfth century. Ade-lard of Bath, himself a pioneer in translating mathematical works from Arabic, addresses the dedicatee of his original work, *De eodem et diverso*, William, bishop of Syracuse, "most erudite in all the mathematical arts."[6] The anonymous translator of the *Almagest*, in turn, addresses *his* anonymous dedicatee as "you who have freely soaked your thirsty mind with the doctrine of those arts which Aristotle calls 'the most exact', as if with a draught of living water."[7] The same translator refers to the admiral Eugene as being "a man most skilled both in Greek and in Arabic, and not ignorant of Latin."[8] That Greek and Arabic sources may have been used by the same mathematicians is, therefore, *prima facie*, quite likely.

The translation of Ptolemy's *Almagest* is known to have been made in about 1165 and, according to the rubric in the earliest manuscript, in Palermo but, according to notes in Paris, Bibliothèque de l'Arsenal 1036, in 1183 and 1188 the positions of the stars were being established, also in Palermo, by means of Arabic star tables.[9] On another folio of

---

[1] Valentin Rose, "Die Lücke im Diogenes Laërtius und der alte Übersetzer", *Hermes*, 1 (1866), pp. 367–97; Charles Homer Haskins, *Studies in the History of Mediaeval Science*, 2nd ed., Cambridge, Mass. 1927, pp. 155–93 ("The Sicilian Translators of the Twelfth Century"); Evelyn Jamison, *Admiral Eugenius of Sicily*, London 1957.

[2] On this mathematician see John Murdoch, "Euclides Graeco-Latinus: A Hitherto Unknown Medieval Latin Translation of the Elements Made Directly from the Greek", *Harvard Studies in Classical Philology*, 71 (1966), pp. 249–302; Hubert L. L. Busard, *The Mediaeval Latin Translation of Euclid's Elements Made Directly from the Greek*, Stuttgart 1987, and Ken'ichi Takahashi, Takako Mori and Youhei Kikuchihara, "A Paraphrased Latin Version of Euclid's Optica: a Text 'De visu' in MS Add. 17368, British Library, London", *Sciamus*, 3 (2002), pp. 127–92.

[3] Baldassarre Boncompagni, "Intorno ad una traduzione latina dell' ottica di Tolomeo", *Bullettino di bibliografia e di storia delle scienze matematiche e fisiche*, 4 (1871), pp. 470–92, and *L'ottica di Claudio Tolomeo da Eugenio ammiraglio di Sicilia ridotta in latino*, ed. Gilberto Govi, Turin 1885.

[4] Jeremy Johns, "The Norman Kings of Sicily and the Fāṭimid caliphate", *Anglo-Norman Studies*, 15 (1993), pp. 133–59; idem, "I re normanni e I califfi fāṭimiti. Nuove prospettive su vecchi materiali", in *Del nuovo sulla Sicilia musulmana. Giornata di studio (Accademia dei Lincei)*, Rome 1996, pp. 9–50.

[5] This point was made both by the students (socii) of Gerard of Cremona, quoting words of the tenth-century Arabic mathematician Ahmad ibn Yusuf: "Oportet ut interpres preter excellentiam quam adeptus est ex notitia lingue de qua et in quam transfert, artis quam transfert scientiam habeat" ("It is necessary that the interpreter, in addition to the excellence which he has acquired from the knowledge of the languages from which and into which he translates, should also have knowledge of the subject which he translates"): see Charles Burnett, "The Coherence of the Arabic-Latin Translation Program in Toledo in the Twelfth Century", *Science in Context*, 14 (2001), pp. 249–88 (see pp. 255 and 276); and by Roger Bacon, *Opus maius, Pars tertia, de utilitate grammaticae*, ed. J. H. Bridges, 2 vols., Oxford 1897, vol. 1, pp. 66–69 and 81, and *Opus tertium*, chapter 25, ed. James S. Brewer, London 1859, p. 91.

[6] "Omnium mathematicarum artium eruditissime": Adelard of Bath, *Conversations with his Nephew: On the Same and the Different, Questions on Natural Science and On Birds*, ed. and trans. Charles Burnett, Cambridge 1998, p. 2.

[7] "Earum quas Aristotiles acrivestatas vocat arcium doctrina quasi haustu aque vive animum sicientem liberaliter imbuisti": Haskins (as in note 1), p. 191.

[8] "Virum tam grece quam arabice lingue peritissimum, latine quoque non ignarum": ibid., p. 191.

[9] Paul Kunitzsch, "The Astronomer Abu'l-Husayn al-Sufi and his Book on the Constellations", *Zeitschrift für Geschichte der Arabisch-Islamischen Wissenschaften*, 3 (1986), pp. 56–81, see p. 72: "Currente anno Christi millesimo centesima (sic) octuagesimo tertio, per instrumenta Panormi in Leone gradu .17. minuta .46. Currente anno Christi millesimo centesimo octuagesimo octavo, indictione sexta, inventus fuit locus Cordis Leonis in ipso gradu .17. minuta .47. secunda .24."

the same manuscript the anonymous annotator says "These tables are drawn up for the meridian of Sicily by means of very accurate instruments, and they are those which William (II, 1166–1189) arranged to be made in the city of Palermo in the year of our lord Christ 1188, indictio 6." But the Palermitan astronomer goes on to say that "since these tables take into account the motion of the eighth sphere, it has been decided to call them 'Latin' or 'Constantinopolitan', because the teacher who made the instruments came from that city: they were (originally) made in the year of our lord Christ 1080."[10] The tables, as Paul Kunitzsch has shown, ultimately go back to those drawn up for Toledo by al-Zarqalluh in 1080, but it is curious to see that the instruments (presumably astrolabes) used for checking them, were allegedly brought by a teacher from Constantinople. Now we know that the Greek manuscript of the *Almagest* was also brought from Constantinople to Palermo, as a present from the Greek emperor, by an ambassador of the Sicilian king, Henricus Aristippus himself.[11] Only recently has it been discovered, however, that this translation does not simply follow the Greek text: rather, it includes references to the work of the Arabic astronomer al-Battani.[12] When one turns to the earliest manuscripts of another translation made by this anonymous translator – the *Optics* of Euclid – one sees a similar *mélange*. Next to several of Euclid's enunciations translated from Greek are the translations of the same enunciations from Arabic.[13]

Particularly indicative of this cross-fertilisation is the use of Indian numerals in Arabic and Latin works written in twelfth-century Sicily. I should summarise the history of the introduction of Indian numerals (our 'Arabic numerals') briefly.[14] The forms, which originated as symbols within the Sanskrit alphabet, were introduced from India into the Islamic world, where they were described in Arabic by al-Khwarizmi (ca. 820). They first appear in the West as an addition to the mathematical section of Isidore's *Etymologies*, in a manuscript written in 976 in the monastery of

Albelda in the Rioja. Thereafter they appear on counters on the abacus that was used for teaching arithmetic in the Latin Schools in the eleventh and early twelfth century. Then they were *re*-introduced with the translation of al-Khwarizmi's *On Indian Calculation* in the early twelfth century. The forms of numerals found in these Latin works are the 'Western forms'. These are the forms which have been used by Arabic writers in the Maghreb until modern times. The problem is that the earliest instance of the use of these forms in an Arabic context is in a manuscript dated after 1284. Before that date only Latin manuscripts manifest these forms, and their use in Latin contexts, even among mathematicians, is limited (e.g. most astronomical tables of the twelfth and early thirteenth century are still written in Roman numerals). We have more ample early evidence of the use of Indian numerals in the Eastern part of the Islamic world. These numerals were written in the 'Eastern forms' and it is these forms that were adopted by certain Greek mathematicians in the twelfth century, and by Latin scholars in Northern Italy (especially Pisa). Again, in all three cultures, the use of Indian numerals is limited; Greek and Arabic mathematicians, as one would expect, preferred to use the letters of their own alphabets with their numerical values.

The translator of Ptolemy's *Almagest* also translated Euclid's *Elements* from Greek, from a Greek manuscript that uses the Eastern forms (or from a closely related one). We find these Eastern forms used in the earliest manuscript of this translation of the *Elements*, and Paris, BNF, la 7373. This manuscript, however, is said by palaeographers to have been written in Tuscany,[15] and this opinion would seem to be confirmed by the fact that Leonardo of Pisa (Fibonacci), writing in Pisa in the early thirteenth century used a compendium of books XIV and XV of the *Elements* that is found uniquely in this manuscript.[16] There is other evidence of these Eastern forms being used in Pisa in the mid- to late-twelfth century, and the forms of the numerals

[10] Ibid., p.73: "Tabule iste directe sunt ad meridiem Sicilie per instrumenta certissima que Wilielmus rex Sicilie felicis memorie in civitate Palormi fieri fecit anno domini Christi .1188. indictione .6. ...quia vero ipse tabule sunt cum motu .8. spere ipsas Latinas placuit appellare sive Constantinopolitanas eo quod magister fabricator instrumentorum venit ab ipsa civitate et facte fuerunt anno domini Christi .1080."

[11] "Hos (the books of the *Almagest*) autem ... quendam ex nunciis regis Sicilie quos ipse Constantinopolim miserat agnomine Aristipum largitione susceptos imperatorio Panormum transvexisse"; preface to the anonymous translation of the *Almagest*, Haskins (note 1 above), p.191, with comments on p.159f.

[12] Richard Lemay, "De la scolastique à l'histoire par le truchement de la philologie", in *La diffusione delle scienze islamiche nel medio evo europeo*, ed. Biancamaria Scarcia Amoretti, Rome 1987, pp.399–535 (see p.466), quoting MS Vat. Pal. lat.1371, fol.67v: "Hoc fuit necesse, ut ait Albategni, tabulas diversitatum apponere."

[13] See Takahashi et al. (as in note 2).

[14] For this history see Karl Menninger, *Zahlwort und Ziffer* (2nd ed 2 vols., Göttingen 1958; Georges Ifrah, *The Universal History of Numbers*, trans. from the French by David Bellos et al., London 1998; Char Burnett, "Indian Numerals in the Mediterranean Basin in the Twelf Century, with Special Reference to the 'Eastern Forms'", in *From Chito Paris: 2000 Years Transmission of Mathematical Ideas*, eds. Yvon Dold-Samplonius et al., Stuttgart 2002, pp.237–88; Paul Kunitzse "The Transmission of Hindu-Arabic Numerals", in *The Enterprise Science in Islam: New Perspectives*, ed. by Jan P.Hogendijk and Abdhamid I. Sabra, Cambridge Ma. and London 2003, pp.3–21.

[15] François Avril et al., *Manuscrits enluminés d'origine italienne*, 2 vo Paris 1980–84, vol.1, p.56, no.95.

[16] Busard (as in note 2), pp.18–20, and Menso Folkerts, "Leonardo bonacci's Knowledge of Euclid's Elements", to be published in the pceedings of the conference *Leonardo Fibonacci: Matematica e soci nel Mediterraneo del secolo XIII*, Pisa and Florence, 20–23 Novem 2002 (Bollettino di Storia delle scienze Matematiche).

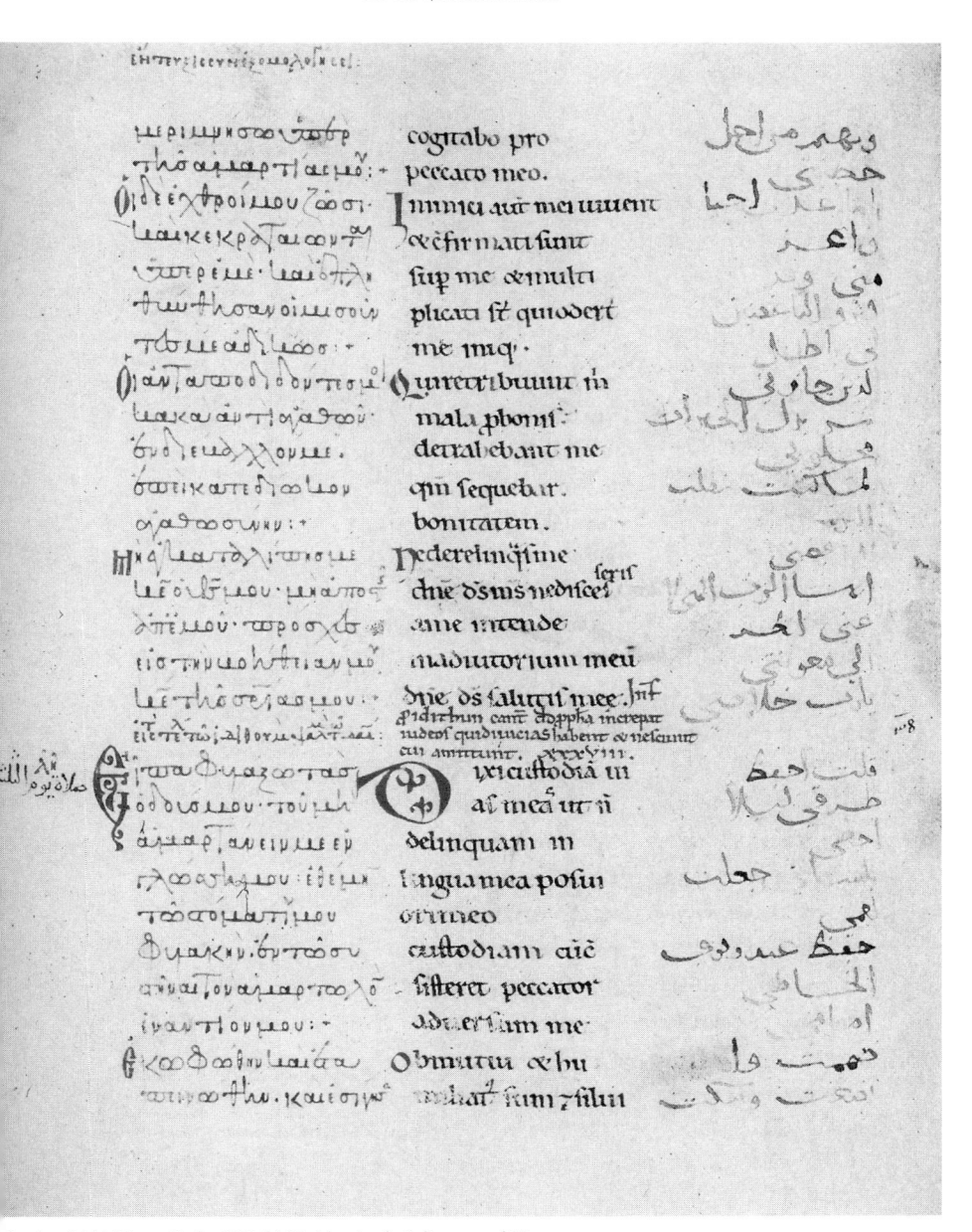

London, British Library, Harley 5786, fol. 50v (showing the Indian numeral 38)

2. London, British Library, Harley 5786, fol. 75r (showing the Indian numeral 57)

The Use of Arabic Numerals

London, British Library, Harley 5786, fol. 77v (showing the Indian numeral 59)

4. London, British Library, Harley 5786, fol. 124v (showing the Indian numerals 99 and 100)

The Use of Arabic Numerals

5. London, British Library, Harley 5786, fol. 162r *(showing the Indian numerals 133 and ‹13›4)*

Cat... et expositio continentium medios progressus Lune

Motus Lune in longitudine						
Anni collecti	Pres	Ꝯīa	Scda	Tcia	Qrta	Qinta	Sexta

Motus lune in diuersitate						
Anni collecti	Pres	Ꝯīa	Scda	Tcia	Qrta	Qinta	Sexta

Motus lune in latitudine						
Anni collecti	Gras	Ꝯīa	Scda	Tercia	Qrta	Qinta	Sexta

Motus lune in loco suo a sole						
Anni collecti	Gras	Ꝯīa	Scda	Tcia	Qrta	Qinta	Sexta

6. *Vatican City, BAV, Pal. lat. 1371, fol. 69v*

The Use of Arabic Numerals

in the Paris manuscript are similar to those that we know from the city.[17] So, the contents of the Paris manuscript may well have been copied in Pisa from a Sicilian manuscript.[18] We do, however, have some good evidence for the forms of numerals used in Palermo itself, and this is what I would like to concentrate on in the rest of my talk.

The first witness is the 'Harleian Psalter' (London, British Library, Harley 5786; figs. 1–5).[19] It consists of a Psalter written in three languages: the Greek text on the left, the Latin text in the centre, and the Arabic text on the right. Each psalm is numbered in the numeral system appropriate to the language. Thus alphabetical numeration is used for the Greek, and Roman numerals for the Latin, and Indian numerals for the Arabic. But here we have neither the Western forms, nor the Eastern forms of the Indian numerals, but a kind of intermediate form. The '6' and '8' are in the Western form, but '2', '3', '5' and '7' have the characteristics associated with the Eastern forms.

The Harleian Psalter was written before 1153, for this date has been written on the last folio: "anno incarnationis dominice .m.c.liii. ind. <i> mensis ianuarii die octavo die mercurii."[20] Since the Psalter accommodates a community in which three languages were used for the liturgy, its most likely place of origin is the Palermo of Roger II. The Arabic script is a mixture of Eastern and Western forms: its general appearance is typical of the Eastern Arabic script used in the Royal Diwan,[21] but the *maghribi* forms of the Arabic letters *fā'* and *qāf* (with a single dot above and below the loop respectively) have been substituted for the Eastern forms with one and two dots under the loop. Thus, for both the letters and the numerals, Western forms have been preferred in cases where the Eastern and Western forms differ most greatly. This may be explained as being due to a scribe who is used to writing Arabic in the Eastern way, but makes

concessions to the local population of Sicily, who would be more familiar with *maghribi* script.

The second witness is the anonymous Greek-Latin translation of Ptolemy's *Almagest*, which, as I have already indicated, was made in Palermo, and includes references to Arabic scholarship. The most accurate, and probably the earliest, of the four known manuscripts of this translation is Vatican, Pal. lat. 1371, and it is this manuscript that contains the numerals that are very similar to those of the Harleian Psalter.[22] Here too we find the Western forms of '8' and '6' alongside the Eastern forms of the other numerals. Moreover, the *ductus* of the individual numerals is similar, when one takes into account the fact that one scribe is writing from right to left, the other from left to right. For example, in writing the letter five, both scribes have extended the ascender, producing a form which is not found in other manuscripts.[23]

The third witness is the Arabic coins of Roger II. Two dated coins survive with the *hijra* date given in Indian numerals: a 'folles' of 533 (=1138–1139 A.D.) and another 'folles', struck in Messina in 543 (=1148–1149) (figs. 7–8). Although only three numerals are represented, their forms agree with those of the Harleian Psalter and the Greek-Latin *Almagest*; and include the '5' with its distinctive *ductus*.

Now, it is extraordinary that Indian numerals should be used on coins at all. Not only is this by far the earliest example of Indian numerals on European coinage but also, we do not find it on Arabic coins until much later. How can we explain this? I have already indicated that, in the twelfth century, Indian numerals, whether of the Western or Eastern form, were used in very restricted contexts, and are not likely to have been familiar to the public at large. Does their presence on the coins of Roger II indicate that they were better known in Sicily? Or is this, rather, a direct reflection of the mathematical prowess of the leading administrators in Roger II's court? After all, both Henricus Aristippus[24] and the Admiral Eugene were keen on mathematics as well as being leading politicians, and Henricus had been royal tutor to Roger II's son, the future William I.[25] One may draw a parallel from Norman England. It is very likely that Adelard of Bath, after his sojourn in Sicily, was involved in some capacity in the English exchequer, whilst also translating mathematical texts from Arabic. An important twelfth-century manuscript, containing texts using both the Western

---

17 See Burnett (as in note 14), p. 251.

18 It is interesting to note that the manuscript ends with computus tables, within which the date 1181 occurs. Could the person responsible be the same astronomer who mentions king William II's ordering of the drawing up of astronomical tables in 1188 (see notes 9–10 above)?

19 For the MS see K. and S. Lake, *Dated Greek Miniscule Manuscripts to the Year 1200*, vol. I–X, Boston Mass., 1934–9, no. 80 (plates 140f.), and the discussion in Peter E. Pormann, "The Parisinus Graecus 2293", *Arabic Sciences and Philosophy*, 13 (2003), pp. 137–61 (at p. 151f.). Nigel Wilson argues that it was produced together with a medical manuscript, Vat. gr. 300, at the Norman court of Palermo: Nigel G. Wilson, "The Madrid Scylitzes", *Scrittura e Civiltà*, 2 (1978), pp. 209–19.

20 This date was read by Haskins (as in note 1), p. 184, but is now almost impossible to decipher, and may not be exactly correct, since 8 January fell on a Tuesday in 1153.

21 Jeremy Johns has suggested that Roger may have invited scribes from Fatimid Egypt, whose chancery he used as a model for his own: see Jeremy Jones, *Arabic Administration in Norman Sicily: The Royal Diwan*, Cambridge 2002.

22 See figure 6.

23 A comprehensive table of the Eastern forms in Latin manuscripts (with a selection of forms from Arabic and Greek manuscripts) can be found in Burnett (as in note 14), pp. 265–67.

24 Nothing is known of Aristippus before he became Archdeacon of Catania in 1156.

25 See Haskins (as in note 1), pp. 160 and 171.

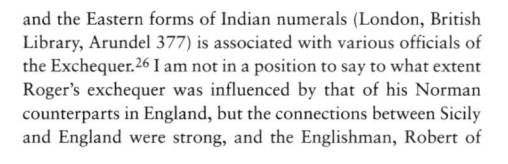

7. A "folles" of 533

8. A "folles" of 543

and the Eastern forms of Indian numerals (London, British Library, Arundel 377) is associated with various officials of the Exchequer.[26] I am not in a position to say to what extent Roger's exchequer was influenced by that of his Norman counterparts in England, but the connections between Sicily and England were strong, and the Englishman, Robert of Selby, was head of Roger's chancery. But, instead of an influence from England, one could simply suggest that it was natural that, in a chancery or exchequer which included some of the best mathematicians of the age, Indian numerals might have been used for calculating tax returns, and indeed, for marking the dates of coins. In any case, the Palermitan examples are a unique testimony, within the Latin, Greek and Islamic worlds of the time, of a common use of Indian numerals shared by the three cultures.

[26] See Charles Burnett, *The Introduction of Arabic Learning into England*, London 1997, p. 58 f.

Photo credits: Courtesy British Library 1–5; Biblioteca Apostolica Vaticana 6; L. Travaini, *La monetazione nell'Italia normanna,* Rome 1995, pl. 12 and table 15: 7–8.

Why We Read Arabic Numerals Backwards

In Arabic script the 'figures of the Indians' were written and read with the lowest values first, just as one might say 'four and twenty' and write it '42'. When the same numerals were taken over in the West (as our 'Arabic numerals'), it became the norm for the higher values to be written first, and this has remained so until the present day. The fixing of the direction of writing is crucial in the case of Arabic numerals, since, unlike Greek and Arabic alphabetical numerals, their value depends entirely on their position. This short note is a contribution to the study of how the present direction of writing Arabic numerals became established in the West.[1]

Since Arabic was written from right to left, but Latin from left to right, this change in the order of the writing resulted in the fact that the actual order of Hindu-Arabic numerals on the page is the same in Arabic and Latin: The higher numerals are on the left and the lower on the right.[2] One may explain the change in the direction of writing by the fact that the Latins simply copied the Hindu-Arabic numerals as they appeared in the Arabic manuscript or on the calculating board, like the geometrical diagrams that they also found there.[3] This is particularly likely considering that the nu-

---

[1] I am sure Wilbur Knorr will have appreciated this question, given his abiding interest in the visual representation of mathematical concepts, especially in the course of their transmission from one language to another. For me, the question has arisen out of the study of different experiments in the representation of numbers in the Latin Middle Ages, some aspects of which are dealt with in Appendix II of Burnett (1999). I am grateful to the advice of Paul Kunitzsch and David d'Avray.

[2] This is, in fact, the order in which the numerals are written by the Indians. The number 1999 is set out with the higher numerals on the left in Sanskrit, Arabic and Latin.

[3] It was usual for geometrical diagrams in Latin translations from Arabic to be copied 'straight', but there is at least one example of them being copied back to front (i.e., as mirror

merals were commonly separated from the text, either appearing as examples of calculation, or, if within the text, being placed in boxes.[4] Just as most of their individual shapes did not substantially change when passing from Arabic into Latin,[5] so their position relative to one another remained the same. This is indicated most obviously by the fact that, in the earliest Latin representations of Arabic numerals, the row of number-forms is always written 'backwards', with '9' coming first. This is the case for the numerals copied into two manuscripts of Isidore's *Etymologies* in the Rioja in the late tenth century, and for the first Latin texts on calculation with Arabic numerals.[6]

Another reason for the Latins' writing the numerals with the higher values on the left was the precedents set by the Roman numerals and Greek alphabetical numerals with which they were familiar. In these, that the higher values were generally placed on the left was perhaps due to the fact that this was the order in which the numbers were spoken and written out in words. Nevertheless, the similar custom in Arabic of writing alphabetical numerals or writing out the numbers in full with the higher values first,[7] did *not* lead to the adoption of the same direction for Hindu-Arabic numerals. Moreover, there are several indications that, when Arabic numerals were introduced into Europe, the direction in which they were to be written was not self-evident.

This is already apparent in the earliest instructions for using these numerals. These are called 'algorisms' after the name of the ninth-century Arabic mathematician, al-Khwārizmī, whose book 'on Indian calculation' described the principles of calculation with Hindu-Arabic numerals having place value. What we have are Latin *remaniements* of one or more lost original translations of al-Khwārizmī's lost Arabic text. The Arabic order of

---

images) to match the change of direction of the writing: MS Dresden, Landesbibliothek, Db. 87, f. 1–71v, a unique copy of a translation of Ptolemy's *Almagest* made from Arabic: see Burnett (1999).

[4]See the Plates in Folkerts and Kunitzsch (1997). This includes a useful, up-to-date account of the transmission of Hindu-Arabic numerals to the West.

[5]It has been suggested that the observable differences between the forms of the numerals in Arabic and in Latin is, in fact, due principally to the fact that they have been written from opposite directions; mirror images of the numerals are rarely found: see Lemay (1977).

[6]E.g., *Dixit Algorizmi* (Folkerts and Kunitzsch (1997), pp. 28–9): 'et he sunt figure, in quibus illa est diversitas: [9 8 7 6] 5 4 3 2 [1]'; and the *Liber Ysagogarum Alchoarismi* (ed. Allard (1992), p. 25): 'in quibus etiam his .viiii. figuris 9 8 7 6 5 4 3 2 1 tam integros quam minutias significantibus utuntur'.

[7]In the case of numbers spoken or written out as words, the Arabs are less consistent than the Latins. The units come before the tens (as in 'four and twenty'), but '1999' can be pronounced and written either as 'one thousand, nine hundred and nine and ninety' or 'nine and ninety, nine hundred and a thousand' (see Wright (1964), I, 259D) and, for a ninth-century example of the reverse direction Yamamoto and Burnett (2000), pp. 130-132.

writing the numerals is betrayed by the phraseology of some of these Latin texts. For example, in the version which is closest to a translation (that known as *Dixit Algorizmi*), we read:

> Et erit dispositio numeri ita: omne unum cum fuerit in *priori* differentia, unum erit, in *posteriori* vero erit .x., et quod fuerit .x. in *posteriori*, erit unum in *priori*[8].

> This is how the numbers will be set out: every '1' which is *in an earlier position* will be 'one', but *in a later position* it will be 'ten', and whatever is 'ten' *in a later position*, will be one *in the earlier position*.

This is probably a literal translation of an Arabic sentence, in which 'earlier' and 'later' refer to the order of writing. That these adjectives were felt to be incongruous, or at least ambiguous, in the context of writing from left to right is indicated both by the fact that the writer of *Dixit Algorizmi* immediately goes on to add 'et erit inicium differentiarum *in dextera scriptoris*, et hec erit prima earum, et ipsa posita est unitatibus' ('the beginning of the positions will be *to the right of the writer*, and this will be the first of them and the position of the units'), whereas, a variant version of the same text substitutes 'superiori' and 'inferiori' for 'priori' and 'posteriori'[9]. The Arabic order is implied again in the description of the role of the zero in *Dixit Algorizmi*:

> Cum autem ponerentur .x. in loco unius et fierent in secunda differentia essetque figura eorum figura unius, necesse fuit eis figure decenorum aliquid *preponere*, ut per hoc scirent quid esset .x. *Preposuerunt* igitur ei loco differentie circulum parvum in similitudine 'o' littere[10].

> When ten were written instead of one and they were put in the second position and the numeral '1' was used for them, it was necessary that something should be placed *in front of* the numeral for the tens, so that they should know through it that it was 'ten'. Therefore *they placed before it* a small circle like the letter 'O' in the <first> position.

A similar statement appears in another early algorism, known as *Liber Alchorismi de practica arismetice*, but this time the Latin writer specifies that the ten is in the 'second position *towards the left of the writer*', whereas

---

[8]Folkerts and Kunitzsch (1997), p. 32. Emphases are my own here and in the following examples.
[9]This is the text printed in a parallel column in Folkerts and Kunitzsch (1997), p. 32.
[10]Ibid., p. 32.

the zero is in the 'first position *towards the right of the writer*'.[11] In an early attempt to explain how to calculate 'the Saracen way', which introduces the principle of place value, but in which the Arabic numerals are replaced by the first nine Roman digits, the author gets quite confused as to whether to describe a digit at one end of a compound number as 'first' or 'last'.[12]

By the early thirteenth century most of the algorisms had fully adapted the Arabic numerals to their Latin context. The row of numerals were written from '1' to '9', and the zero was described as being placed 'after' the numeral indicating a ten.[13] But this was not universally the case. In one instance, for example, where the first nine letters of the alphabet are used as equivalent to the nine Hindu-Arabic digits with the addition of 0 for zero, the compound numbers are written with the lower values on the left; hence 0A = 10, AA = 11, 0B = 20, 0I = 90, 00A = 100, A0A = 101 etc.[14] More interestingly, in an algorismic text itself, copied into a manuscript perhaps shortly before the middle of the thirteenth century, the Arabic order of writing is still found. This is London, British Library, Arundel 206,[15] whose first part (fols 1–65v) consists mainly of sermons and other theological writings, and appears to have been written in Germany.[16] The text of the algorism has been written on fol. 61vb, and runs as follows:

Unitas	binarius	ternarius	quaternarius	Quinarius
.1.	.2.	.3.	.4.	.5.
Senarius	septenarius	octonarius	novenarius	
.6.	.7.	.8.	.9.	

Regula est quod omnis figura preposita significat se tantum, postposita, decies tantum. Verba gratia: .1. preposita unum tantum significat, postposita, decies tantum. .0. autem nichil significat, sed confert tantum aliis sig-

---

[11]Allard (1992), pp. 69–70: 'Ut enim prime differentie 9 numeros representent, primo loco quelibet illarum poni precipiuntur, sed ut numeros secunde, non iam primo loco, sed secundo *versus sinistram scriptoris* ponuntur, preposito circulo in primo loco *versus dexteram scriptoris*, ut per hec prima differentia vacua esse ostendatur'.

[12]Cf. *Helcep Sarracenicum*, sentence 91: 'Duco ergo primum—id est ultimum—superioris ordinis in primum inferioris ordinis' ('I therefore multiply the first <digit>—that is the last—of the higher row by the first and the second of the lower row'); ed. in Burnett (1996), p. 284.

[13]For example, in the algorism in MS Cashel (Tipperary), G. P. A. Bolton Library, Medieval MS 1, p. 41, where we read 'Sunt autem figure .ix. quibus omnes numeri representantur, cum circuli addicione, et sunt littere Indorum et huiusmodi: 1 2 3 4 5 6 7 8 9, cifra vel circulus 0... Cum ergo volueris scribere .x., scribe .1. et *postea* circulum, sic: .10.'

[14]See Gilissen (1976). The manuscript was probably written in the late thirteenth century in Flanders.

[15]I owe my knowledge of this text to David d'Avray, to whom I am very grateful.

[16]There is a German gloss on fol. 18rb.

nificare, quod patet in figuris positis: .12. viginti unum, .21. xi<i>, .13. .31. .14. .41. .15. .51. .16. .61. .17. .71. .18. .81. .19. .91. .09 et sic de aliis omnibus. 001 .c., pone binarium et sunt .cc., ternarium .ccc. et sic per omnes figuras.

This text requires a brief exposition. First of all, the author writes out the Arabic numerals in forms which are used in thirteenth-century England (They are the forms used, for example, by Robert Grosseteste in his autograph notes in MS Oxford, Bodleian Library, Savile 21, and are distinctive for the use of the gnomon shape—⅂—for '2'). Each has a name ending in '-arius', perhaps best rendered in English by prefixing the definite article to the number: e.g., 'the unity', 'the two', 'the three' etc. This terminology was not used in the Golden Age of classical Latin but is regular in Boethius's *De arithmetica* and Calcidius's commentary on the *Timaeus* of Plato, as well as in other algorisms.[17] Then he explains how the value of these numerals depends on their position (the principle of 'place value'): 'The rule is that every numeral, when it is placed before another numeral, signifies itself only, but when it is placed after another numeral, signifies ten times its value'. He is referring here only to units and tens, but in the next phrase appears to suggest that the rule can be extended *mutatis mutandis* to other decimal positions: 'and this is true concerning all the others'. Then, he describes the rule of the zero: '0, however, signifies nothing, but only confers signification to other numerals'. This is a little elliptical, and so the author proceeds to give examples ('as is apparent in the disposition of the figures'). Here it is clear that he understands the lower value to be on the left, the higher on the right: '12' is twenty-one, '21' is twelve, '09' is ninety, '001' is one hundred, '002' is two hundred, and '003' is three hundred, etc.

This little text is internally self-consistent and is an adequate, though rather abbreviated, explanation of how to write Arabic numerals. The difference from what became the normal order in writing Arabic numerals may be an aberration on the part of the author, but it is more likely that it represents a stage in the introduction of the numerals into the West in which the direction had not yet become fixed.   *

---

[17]E.g., *Liber Ysagogarum Alchoarismi*, in Allard (1992), p. 25 and *Liber Alchorismi de practica arismetice*, ibid, p. 64.

* References

Allard, A. (1992). *Mohammad Mūsā al-Khwārizmī, Le calcul indien*. Paris and Namur: A. Blanchard.

Burnett, C. (1996). *Algorismi vel helcep decentior est diligentia*: The arithmetic of Adelard of Bath and his circle. In M. Folkerts (Ed.), *Mathematische Probleme im Mittelalter: Der lateinische und arabische Sprachbereich*. Wiesbaden: Harrassowitz Verlag. Pp. 221–331.

Burnett, C. (In press) The Transmission of Arabic astronomy via Antioch and Pisa in the second quarter of the twelfth century. In A. I. Sabra and J. P. Hogendijk (Eds.) *Perspectives on Science in Medieval Islam*. Cambridge: MIT Press.

Gilissen, L. (1976). Curieux Foliotage d'un manuscrit de droit civil: la somma d'Azzon (Bruxelles 9251 et 9252). In I. Forchielli and A. M. Stickler (Eds.), *Studia Gratiana*, 19, Rome. pp. 303–11.

Lemay, R. The Hispanic origin of our present numeral forms. *Viator*, 8, 435–62.

Folkerts M & Kunitzsch, P. *Die älteste lateinische Schrift über das indiscbe Rechnen nach al-Hwārizmī*. Bayerische Akademie der Wissenschaften, philosophisch-historische Klasse, Abhandlungen, neue Folge, 113. Munich: Bayerische Akademie der Wissenschaften.

Wright, W. (1964), *A Grammar of the Arabic Language*. 3 vols. Third Edition. Cambridge: University Press.

Yamamoto, K. & Burnett, C, Eds. (2000). Abū Ma'sar, *On Historical Astrology*, 2 vols, Leiden: E. J. Brill

VIII

The Toledan *Regule* (*Liber Alchorismi,* part II): A Twelfth-century Arithmetical Miscellany

Charles Burnett   Ji-Wei Zhao   Kurt Lampe

The mid-twelfth century was a critical period for the introduction of Arabic mathematics to the West. The forms of the Hindu-Arabic numerals had become known in the late tenth century when they were introduced onto the counters of a peculiar kind of abacus called the 'Gerbertian abacus'(Ifrah 1998: 579-85). The details of calculation using these numerals with place value were not known to Latin scholars until an Arabic version of al-Khwarizmi's account of Indian arithmetic (possibly the *Kitāb fi'l-jam' wa-l-tafrīq,* 'Book on addition and subtraction') was translated (Folkerts/Kunitzsch 1997: 169). While the Arabic text itself is lost, at least four Latin versions of it are known: a literal, but interpolated, translation ('Dixit Algorizmi' = DA), a text incorporated into an introduction to the four mathematical arts of the quadrivium ('Liber Ysagogarum Alchorismi' = LY), and the *Liber Alchorismi* (LA) and the *Liber pulveris* (LP) which share a common source. All these Latin versions have been critically edited by André Allard (Allard 1992). The *Liber Alchorismi,* however, is followed in all the manuscripts (except Dresden C 80, which is fragmentary) by a miscellany of arithmetical texts which have been called 'the second book' of the *Liber Alchorismi.* This has long been known to scholars, ever since its transcription, from Paris, BNF, Bibliothèque nationale de France, lat. 7359, by Baldassare Boncompagni (Boncompagni 1857). No critical edition, translation or study of this text has been made up to now.[1] We offer here a working edition based on what appears to be the best manuscript, Paris, Bibliothèque nationale de France, lat. 15461, together with an English translation and a mathematical commentary. The work is important not so much because of any originality or ingenuity in its mathematical content, as because it documents the kinds of problem that were being tackled in arithmetic, number theory and algebra, just at the time when Hindu-Arabic numerals, calculation with numerals of place value, and algebra were being adapted to a Latin context. As will be argued, this context would appear to be Toledo in the third quarter of the twelfth century, when the new arithmetic of the Arabs was being brought into the Latin curriculum. A humanistic Latin context is evident from the literary quality of the language and the presence of a philosophical

---

[1] André Allard has promised a critical edition of the text.

essay on the rightness of there being nine digits and three orders of numbers.[2]

Although the work always follows the *Liber Alchorismi de pratica arismetice* of 'magister Iohannes', and can, in some way, be regarded as a continuation of the work, it is never called 'the second book' in the manuscripts. Instead we find the incipit: 'Hic incipiunt regule et primum (prius) de aggregatione' in manuscripts P and M, and 'Incipiunt regule' in E; NU have no title. No manuscripts contain a title at the end, which may confirm the status of this work as a series of appendixes attached to the *Liber Alchorismi*; there is no proper ending to the text. Since 'regule' on its own would be an insufficiently distinctive title for the work, we have decided to call it the 'Toledan *regule*' out of consideration for its provenance. The work occurs in the following manuscripts:[3]

Erfurt, Ea, Q 355, s.14, fols 105r-115r (= **A**)
Florence, BN, Conv. soppr. J.V.18, s.13/14, fols 65v-70r (= **c**)
Oxford, BL, Selden supra 26, s.13, fols 96r-100r (= **E**)
Paris, Mazarine, 3642, s.13, fols 113v-117v (= **M**)
Paris, BNF, lat. 7359, ca.1300, fols 101v-111r (= **N**)
Paris, BNF, lat. 15461, s.13, fols 9v-14v (= **P**)
Paris, BNF, lat. 16202, s.13, fols 73v-81v (= **U**)
Salamanca, BU, 2338, s.14, fols 30r-49v (= **S**)
Vatican, Pal. lat. 1393, s.13, fols 51r-60v (= **L**)

Characteristics of the individual manuscripts

Of the manuscripts that have been fully collated, **P** (Paris, BNF, lat. 15461) has been carefully written and corrected. Its contents and context will be described below.

E (Oxford, Bodleian Library, Selden supra 26) may contain the earliest copy of the text. The manuscript is composite. The relevant codex consists of the folios numbered in a recent hand from 96 to 129. The original order of the quires of which this codex consists appears to have been quires II and III (fols 106-29) and

---

[2] This essay – part F below – has been described in detail in Lampe 2005. A text with the incipit of <35.3> in part F ('Cum enim in omni lingua ⋯') is listed amongst the works lent to Franciscan and Dominican friars by the archbishop of Santiago de Compostela in 1225: see Luis García-Ballester, 'Nature and Science in Thirteenth-Century Castile. The Origins of a Tradition: the Franciscan and Dominican Studia at Santiago de Compostela (1222-1230)', in *Medicine in a Multicultural Society: Christian, Jewish and Muslim Practitioners in the Spanish Kingdoms, 1222-1610*, Aldershot, 2001, article II.

[3] We are very grateful to Menso Folkerts for the loan of microfilms and the relevant pages of a history of mathematics in the Middle Ages being prepared by H. L. L. Busard and himself.

I (fols 96-100). In this rearranged order the codex contains an incomplete copy of the first part of the *Liber Alchorismi* (106-121v),[4] an incomplete set of instructions for the use of the Pisan tables (122-129v) and the Toledan *regule* (96-100).[5] The codex is characterised by the use of the eastern forms of Hindu-Arabic numerals, which bring it into the ambit of mathematicians associated with Abraham Ibn Ezra in Pisa and Lucca in the mid twelfth century (Burnett 2002a). The text appears to be derivative from that represented by PN. E.g., in most cases 'morabotini', a coinage distinctive of the Iberian peninsula, has been replaced by 'aurei', a term of more general use. Only parts A and C (see next section) are fully represented, but the broken off fragment of part B suggests that more of the text was available in the exemplar. **E** shows some evidence of intelligent engagement with the subject matter. Paraphrases are added to some passages, and the word 'res' for the (square of) the unknown, in part C, is replaced by 'quadratus'. The order in **E** is as follows: **A** 1.1-2, 3.1-3, 2.1-3, 3.4-6, alternate version of 3.5-6, 4.1-5.2, 5.3-4, 6.1-7.2, 8.2-3, 8.1, 8.4, 9.1-12.8. 12.10-14, 7.1-2 (=E²), **C** 18.1-7, **B** 13.1-4 (breaks off incomplete).

N (Paris, BNF, lat. 7359) provides a text which is very close to that of P, and reproduces some of its mistakes (such as 'numerum eorum' for 'numerorum' in 17.11; 'multiplitio' for 'multiplicatio' in 23.4; the spelling 'duodenerii' in 27.5) including errors in reading the Hindu-Arabic numerals ('3' for '2' in 17.10) and in calculation (in 20.2, 32.3). Occasionally N gives the correct reading where P is wrong (24.2 ductu; 30.1 ut; 30.2 qui; 34.1 aggregata; 34.3 qui; 44.1 etiam constituta) which suggests that it cannot be a direct descendant of P. Moreover, it provides an extra paragraph at the end of the text. On the whole, however, its readings are less reliable, and scribal error is more frequent. Among the variants and characteristic traits of this text (most of which are not mentioned in the apparatus criticus) are: Roman numerals where P has Hindu-Arabic (4.2: ii, 4, 6, 8, 10), or *vice versa* (11.6 '30' for 'xxx'; in 7.2 'iii' has been expunged and replaced by '3'!), or numerals written in full (9.6: 'septem' for '7') or *vice versa* (12.3: '6' for 'sex'), confusion of Hindu-Arabic numerals with each other (especially '2' and '3'; 27.5 '100' for '900') or with the tironian 'et' (read as '2' in 7.2, as '3' in 21.3 and as '7' in 26.3), ambiguity in the use of Hindu-Arabic numerals (e.g. 6.7: '3. 9. 2. 7. 8. 1. 2. 4. 3' for '3, 9, 27, 81, 243'); misreading of abbreviations in the original (17.12 'modo' for 'ergo'; 37.1 '-ant' for '-atur'; 43.4 'qui' for 'quia'; N could be correct here), misreading of minims (4.1. 'inde' for 'vide'), reading 'quod' for 'quot' (6.1 and 6.3 quodlibet), 'id est' for 'scilicet' (16.3), -ationem' for '-antem' (19.1 'denominationem' for 'denominantem'), and so on. N favours spellings with single consonants ('quatuor', 'milia', 'galina'), but writes 'hiis' for 'his'. Frequently phrases that the scribe had at first omitted are

---

[4]One folio has been cut out between fols 121 and 122.

[5]Fol. 100v is blank, while fols 101-105v contain an acephalous text written in a later hand ('...quanta fuerit differentia partis et cursus mediate...et ad evidentiam directionem quandam subtraximus').

added by him in the margin, or a phrase is written twice (23.5: et fiunt 148 (va-) quos duos productos simul aggrega (-cat) centum 48 —the two halves of 'vacat' added as superscript indicate the redundant text).

Nature of the Text

The repetition and the abrupt changes in subject matter suggest that we are dealing with at least seven distinct elements (indicated by the letters A to G); the multiplication tables between A and B and the numerical magic square at the end of the text, all of which have no accompanying text, may be regarded as further elements. The distinguishable elements may be classified as follows:

A 1.1-12.15. Arithmetical rules, especially on progressions
Multiplication tables for sexagesimal and decimal calculation respectively
B 13.1-17.18. Further arithmetical rules, with some overlap with A but with the addition of fractions, and examples from real life.
C 18.1-7. Algebra
D 19.1-25.2. Further arithmetical rules, on division and multiplication
E 26.1-34.6. Further arithmetical rules, on establishing the 'places'('columns') in multiplication, and on finding a 'hidden' number.
F 35.1-44.7. A philosophical introduction to the principles of Hindu-Arabic numeration
G 45.1-46.2. Brief arithmetical rules
A numerical magic square.

Within each of these elements a certain amount of ordering and planning may be discerned. There are slight changes of terminology between sections. D and E employ the phrase 'multiply a numeral (*figura*) by a numeral (*figura*)', while A and B use 'number' (*numerus*) in the same context. On one occasion (41.4) the author of F seems to be using 'nodus' in place of 'articulus', though he generally uses 'articulus'. But on the whole the terminology and the style are the same from section to section. There are, however, internal disruptions and hesitations in regard to order: at the end of 9.2 P adds a reference mark, while N and E each indicate that 'another method' is about to be presented; phrases have been copied in the wrong order in 11.1; there is break between 15 and 16, where P has placed a reference sign whose referent is not clear. The difference in order in E may also indicate an earlier different arrangement in the texts and within the texts. Perhaps what we are dealing with are 'working copies' of arithmetical texts, dating from a time when Hindu-Arabic numerals were still a novelty, and different ways of teaching and learning their use were being experimented with. Only one source of this material is mentioned: namely, the 'liber qui dicitur gebla mucabala', or 'book on algebra'

(18.1), from which the excerpts in **C** are taken.[6]

There is little that is unique to the *Regule*. Progressions are found widely. The rule for multiplying digits by each other (9.1-9.6) can also be found in LY II (Allard 1992: 27, 22-23) and *Helcep Sarracenicum* 43-45 (Burnett 1996: 272-3). Notable is that the examples in the short section on algebra do not all follow al-Khwarizmi's algebra but include one example which is also in Ibn Turk (18.5).

The context of the text

To what extent can we say that the context is Toledan? The majority of the manuscripts indicate that the diffusion of these texts is within and from North Italy. In E the material is associated with the tables drawn up for Pisa. N includes some mathematical works translated in Italy. P belongs to a group of manuscripts copied in Northeast Italy — probably in the vicinity of Padua. This latter group, however, consists of texts which are distinctive products of Toledo. Two of the group (Paris, BNF, lat. 9335 and Vatican City, BAV, Ross. lat. 579) contain only translations of mathematical texts made by Gerard of Cremona (1114-87). P itself does not include any text by Gerard, but rather a calendar, whose saint days mark it out as belonging to Toledo Cathedral after 1156 (the date of the transfer of the relics of St Eugenius, Toledo's patron saint), and a computus referring to the 'present years' of 1143 and 1159. The other text in the manuscript is the *Liber mahameleth*, which shares passages both with the *De divisione philosophiae* of the archdeacon resident at Toledo cathedral, Dominicus Gundissalinus (d. after 1180) and with the Toledan *regule*.[7] Both the *Liber mahameleth* and the Toledan *regule* refer, in their examples, to 'morabotini/morabatini', the coinage introduced into Arabic Spain by the Almoravids (after which it is named), and used by both the Islamic and the Christian kingdoms within the Iberian peninsula, but not elsewhere.

Characteristics of the text:

1) The term used for x^2 is *res* ('thing'), where other Latin texts use *substantia* or *census* (both suggesting 'wealth'): see 18.1-6. The 'thing' in Arabic (*shai'*) is, rather, the root, of which the square is the *mal* ('wealth'). Nevertheless, in both cases it is the 'unknown' that is the 'thing' and, in that the 'unknown' is what is ultimately sought, it is more properly the square than the root. On the other hand, 'res' can also mean 'property' (i.e., wealth) in legal language. To indicate this special use of *res* 'thing' we have put the 'thing' in italics in the translation.

2) 'cifre' together with the symbol '0' is used to indicate an 'empty place' which is to be filled by the result of the calculation; i.e., it serves as the unknown (19.1).

---

[6]Høyrup 1998: 16-17 places this excerpt on algebra in context.

[7]See Burnett, 2002b: 66-70.

3) 'Limes' is used as a category of number, rather than with the meaning '(decimal) place' which is found, e.g., in *Helcep Sarracenicum* (cf. Burnett 1996: 262-3: 5 *Ordines igitur numerorum sive limites a primis numeris, qui digiti vocantur et sunt novem...6 Sunt autem in unoquoque limite numerorum novem termini...*). For the latter the text uses 'differentia'. The four categories of number are *digitus, articulus, limes*, and *compositus* (26.1). While the other categories are those found generally in works on the algorism, *limes* is an extra one, which is used in the sense of the *first* number in each decimal place: i.e., 10, 100, 1,000, 10,000 etc. (this is implied in 14.3, 23.2). In other words it is, strictly speaking, a subcategory of 'articulus' which is *any* number that is divisible by 10. On the other hand, it could be interpreted as any number which is followed by more than one zero (when 'articulus' is a number followed by a single zero). In section F it is specifically used to define the genus to which the ten, the hundred and the thousand belong (35.5), and acquires the meaning of 'place' as in other arithmetical texts (43.5-6).

4) 'Articulus' sometimes appears to be used for any number divisible by 10, at other times, only those belonging to the tens (10, 20, 30, 40, 50, 60, 70, 80 and 90). As such it belongs to the series *digitus, articulus, centenus, mille* (27.2-29.2). The hesitation between accepting 'articulus' as only referring to tens and as referring to any number divisible by 10 is indicated by the fact that on one occasion (31.1) 'vel centeni' has been added after 'articulus', to restrict it to the tens.

5) The 'multiplicatus' is the smaller number, while the 'multiplicans' is the larger number (30.1-3, 30.5, 32.1-2).

6) 'Sequens' ('following') denotes progression towards the left (the hundreds 'follow' the tens): 22.1, 26.2-3, 26.5, 30.1-2.

7) 'Figura' is used for any numeral ('0' is not regarded as a numeral), clearly indicating written numbers rather than the counters on an abacus.

8) 'Vel', in most cases, does not introduce a different procedure, but expresses the same procedure in different words. It is equivalent to 'i.e.'

9) Numerals always take a plural verb. Hence the apparent incongruity of 15.16: 'cuius sexta pars sunt binarius'.

10) 'Unus' is usually used as the adjective (17.16, 19.4, 34.1), 'unum' as the noun (2.2, 7.2, 13.3, 18.7, 42.9; exceptionally 'unus' is a noun in 30.2 and 36.6).

11) Some confusion remains (perhaps originating in the exemplar) concerning the use of zero. It may be omitted where it is required ('3' for '30' in 19.2) or added where it is not required (32.2, 32.3).

12) Illustrative figures which originally must have been separate from the text (probably 'boxed off' from the text) have erroneously been incorporated, wholly or partially, into the text. E.g., the figure accompanying 17.12 must be reconstructed from two fragments in the margin (20–12 and 34, 2, 7 under each other) and a '7' in the text of P; in N 20, 12 and 7 have all been incorporated into the text and 34, 2, 7 are missing. In 23.2 the figure demonstrating the four numbers in proportion appears

as four number added to the end of the last sentence.

13) The text is written in good Latin. The sentences are joined by conjunctions or relatives ('ergo' is used more than 'igitur', 'vero' and 'autem' are of about the same frequency). There is much subordination, and conditional, final clauses (etc.) are generally correctly constructed.

Editorial Principles

The text is based on P, all readings of which have been reported. Specimens of the readings of M and U have been given for the opening paragraphs. N and E have been fully collated against P, but their readings have been reported only if (1) they have been adopted when P is clearly corrupt, (2) they offer a plausible alternative to the readings of P. Scribal errors in EN and variants which are purely orthographic, have not been reported. Specimens of the readings of M and U have been given for the opening paragraphs.

The orthography of P has been followed. This entails the use of 'e' for Classical Latin 'ae' and 'oe'. However, where P has been inconsistent, the most common spelling has been adopted (e.g. 'millia', 'millies' and 'millium' instead of 'milia', 'milies' and 'milium'). We have, however, not followed P in writing 'vicesima' instead of 'vicesima', though we have kept the variation between 'octuagesimus' and 'octogesimus' that P shares with N.

In most cases in the manuscripts numbers are set off from the text by being surrounded by *puncta*: this applies as much to Hindu-Arabic numerals ('.2.') as to Roman numerals ('.ii.'), and is even found where the numerals are written out in code (34.1: '.dxp.'). The *puncta* have been omitted in this edition. Angle brackets indicate editorial additions; square brackets, deletions. [8]

Bibliography

Abu Kamil: *The Algebra of Abu Kamil*, trans. by Martin Levey, London, 1966.

Allard 1992: André Allard, Muhammad ibn Musa al-Khwarizmi, *Le Calcul Indien (Algorismus)*, Paris/Namur.

Boncompagni 1857: Baldassare Boncompagni, *Trattati d'aritmetica* II. Ioannis Hispalensis *Liber algorismi de pratica arismetrice*, Rome, 1857.

---

[8]This research was made possible by a British Academy Visiting Fellowship which enabled Ji-Wei Zhao to spend four months working with Charles Burnett at the Warburg Institute, London. Burnett is primarily responsible for the introduction, and the edition of the Latin text, parts A-E and G, Zhao is responsible for the mathematical translation and notes, and Lampe for the edition and translation of part F. All three scholars have collaborated on the article as a whole.

148

Burnett 1996: Charles Burnett, '*Algorismi vel helcep decentior est diligentia*: the Arithmetic of Adelard of Bath and his Circle', in *Mathematische Probleme im Mittelalter: Der lateinische und arabische Sprachbereich*, ed. M. Folkerts, Wiesbaden, 1996, pp. 221-331.

Burnett 2002a: id., 'Indian Numerals in the Mediterranean Basin in the Twelfth Century, with Special Reference to the "Eastern Forms"', in *From China to Paris: 2000 Years Transmission of Mathematical Ideas*, eds Y. Dold-Samplonius, J. W. Dauben, M. Folkerts and B. van Dalen, Stuttgart, 2002, pp. 237-88.

Burnett 2002b: id., 'John of Seville and John of Spain: a *mise au point*', *Bulletin de philosophie médiévale*, 44, 2002, pp. 59-78.

Cashel I: Charles Burnett, *Learning Indian Arithmetic in the Early Thirteenth Century*, Boletín de la Asociación Matemática Venezolana, vol. IX, no.1(2002), pp. 15-26.

Folkerts/Kunitzsch 1997: *Die älteste lateinische Schrift über das indische Rechnen nach al-Ḫwārizmī*, ed., trans. and comm. Menso Folkerts and Paul Kunitzsch, Munich, 1997.

Høyrup 1998: Jens Høyrup, 'A New Art in Ancient Clothes: Itineraries chosen between Scholasticism and Baroque in order to make algebra appear legitimate, and their impact on the substance of the discipline', *Physis*, 35,1998, pp. 11-50.

Ibn Turk: Aydin Sayili, *Logical Necessities in Mixed Equations by 'Abd al Hamid ibn Turk and the Algebra of his Time*, Turk Tarih Kurumu Basimevi, Ankara, 1962.

Ifrah 1998: George Ifrah, *The Universal History of Numbers*, English translation by D. Bellos et al., London, 1998, pp. 579-85.

Khwarizmi: al-Khwarizmi, *The Algebra of Mohammed ben Musa*, ed. and trans. by F. Rosen, New York, 1986.

Lampe 2005: Kurt Lampe, 'A Twelfth-Century Text on the Number Nine and Divine Creation: A New Interpretation of Boethian Cosmology?' *Mediaeval Studies*, 67, 2005, pp. 1-26.

Liber mahameleth: Anne-Marie Vlasschaert, 'Le "*Liber mahameleth*". Edition critique, traduction et commentaires', unpublished PhD thesis, Université catholique de Louvain, année académique 2002-2003.

A

Hic[1] incipiunt regule et primum[2] de aggregatione.
<1.1> Omnis numerus naturali dispositione circum se positorum numerorum sibi ipsis aggregatorum usque dum occurrens unitas terminum ponat, medietas est. **<1.2>** Ut quinarius quaternarii et senarii inter se coniunctorum medietas est, similiter ternarii et septenarii coniunctorum, similiter[3] binarii et octonarii, similiter unitatis et novenarii et similiter[4] de omnibus.

<2.1>[5] Item omnis numerus in se multiplicatus tantam summam reddit quanta ex circumpositorum inter se et differentiarum inter se, quas habent circumpositi ad medium, multiplicatione provenerit.[6] **<2.2>** Tantum enim fit[7] ex quinquies 5 quantum ex quater sex adiuncta multiplicatione differentiarum quas habent ad 5, id est unum et unum, que multiplicata in se non faciunt nisi unum. Et quantum provenerit[8] ex ter septem, vel ex bis octo, vel ex uno et novem, set adiecta semper multiplicatione differentiarum quas habent extremitates ad medium. **<2.3>** Item[9] omnis numerus usque ad 10 in se ductus, tantum reddit quantum primi[10] duo circumpositi in se ducti, set unitate dempta de multiplicatione medii.

<3.1> Si vis scire ex aggregatione numerorum ab uno naturaliter se sequentium quanta summa reddatur, ipsum in quo desieris, si par fuerit, multiplica per medietatem sui et adde ipsam medietatem et hec erit summa que ex ipsis efficitur.[11] **<3.2>** Verbi gratia, dispone in ordinem[12] i, ii, iii, iv, 5, 6, 7, 8. Multiplica ipsum in quem[13] desiisti, scilicet octo, per medietatem sui, scilicet per[14] 4, et efficies 32 et adde insuper 4 et hec erit summa que ex aggregatione illorum efficiebatur, scilicet 36. **<3.3>** Vel sequentem imparem per medietatem eiusdem paris multiplica et habebis summam. **<3.4>** Si vero in impari desieris, verbi gratia 1, 2, 3, 4, 5, 6, 7, ipsum,

---

[1] E omits

[2] prius M

[3] N adds 'et'

[4] sic E

[5] In E 2.1-3 follows 3.3

[6] proveniunt MU

[7] MU omit

[8] provenit MU

[9] MU omit

[10] M omits

[11] efficietur M

[12] ordine E

[13] quo E

[14] E omits

scilicet[15] septem, multiplica per maiorem partem sui, scilicet 4, 7 enim constant ex tribus et 4, et quater septem efficiunt 28 et hec est summa predictorum. <**3.5**> Vel per medietatem sequentis paris eundem imparem multiplica. <**3.6**> Si autem non ab unitate incipiens continue numeraveris, acsi ab unitate incepisses operandum est et omnes qui sub eo sunt a quo incepisti, aggreges; illud aggregatum a producto subtrahendum, residuum erit summa.[16]

<**4.1**> Si ex solis paribus a binario se per ordinem naturaliter sequentibus aggregatis scire volueris quanta summa reddatur, ipsum in quem desiisti vide quotus est a primo pari et per ipsum a quo denominatur multiplica sequentem se et efficies summam. <**4.2**> Verbi gratia, 2 , 4, 6, 8, 10. Ultimus est 10 et est quintus a primo pari, id est binario. V[17] sequitur senarius. Per 5 ergo a quo denominatur 10 multiplicetur senarius et quinquies sex vel econverso fient 30 et hec erit summa predictorum. <**4.3**> Vel medietatem paris numeri in quo desinis per medietatem proximi sequentis paris multiplica et habebis summam. <**4.4**> Si vero non a binario incipiens continue[18] numeraveris pretereundo impares, acsi a binario incepisses agendum[19] et omnes pares qui sub eo sunt a quo incepisti coacerves; illud[20] coacervatum a producto subtrahendum et residuum erit summa.

<**5.1**>[21] Si autem ex aggregatis imparibus ab uno naturaliter se sequentibus scire volueris quanta summa reddatur, ipsum a quo denominatur ultimus in quem desinis, multiplica in se ipsum et efficies summam. <**5.2**> Verbi gratia, 1, 3, 5, 7, 9. Ultimus est novem et est quintus in ordine. 5 vero[22] a quo denominatur multiplicetur in /9rb/ se et efficies 25. Quinquies enim quinque 25 fiunt[23] et hec erit[24] summa predictorum. <**5.3**> Vel a proximo pari qui est post imparem in quo desinis, medietatem subtrahe, ablatam per se multiplica et habebis summam. <**5.4**> Si vero non ab unitate inicipiens <continue numeraveris>, acsi ab unitate incepisses agendum est et omnes impares qui sub eo sunt a quo incepisti coacerves; illud coacervatum a producto

---

[15] M omits

[16] E continues with an alternative version of the previous sentences: vel a proximo pare qui est post imparem in quo desinis medietatem subtrahe ab altero per se multiplica et habebis summam. Si vero non ab unitate incipies, acsi ab unitate incepisses agendum est et omnes impares qui sunt sub eo a quo incepisti illum coacervatum a producto residuum erit summa.

[17] E adds 'autem'

[18] continue E, continuo PN

[19] E adds 'est' superscript

[20] illum E

[21] E adds heading 'De imparibus'

[22] ergo E

[23] efficiunt E

[24] erat P

subtrahendum; residuum erit summa.[25]

<**6.1**> Si autem ex aggregatis quotlibet duplicibus a primo per ordinem[26] se naturaliter sequentibus scire volueris quanta summa reddatur, ultimum in quem desinere volueris, multiplica per primum parem, scilicet binarium, et que summa ex eorum multiplicatione provenerit,[27] subtracto primo pari, illa eadem ex predictorum aggregatione colligitur. <**6.2**> Verbi gratia, 2, 4, 8, 16, 32. Ultimum qui est 32 multiplica per binarium et efficies 64 et subtrahe ipsum[28] a quo inceperis et quod fiet hec est[29] summa predictorum, scilicet 62. <**6.3**> Vel aliter: quotlibet duplicium a primo naturaliter se sequentium ultimus duplatus est summa omnium aggregatorum, subtracto primo pari. <**6.4**> Vel aliter generalius: ex quorumlibet duplorum aggregatione si volueris scire quanta summa proveniat, duplica extremum in quem desieris et subtrahe primum a quo inceperis et quod fit ex duplatione ultimi cum subtractione primi, hec est summa quam requiris.[30] <**6.5**> Vel si ab uno incipere volueris, duplicationi ultimi ipsum addere debebis et precedentium summam habebis.

<**6.6**> Si ex quibuslibet aggregatis triplis vis scire quanta summa reddatur, ultimum in quem desieris divide et minorem eius partem triplica et quod ex eius triplicatione provenerit, precedentium summa erit. <**6.7**> Verbi gratia, 3, 9, 27, 81, 243. Ultimi medietas minor 121, hec triplicata reddit 363, quam summam aggregati precedentes tripli reddunt.

<**7.1**> Si volueris[31] scire ex aggregatione omnium quadratorum quorumlibet numerorum ab uno naturaliter se sequentium, quanta summa proveniat,[32] vide prius ex aggregatione ipsorum numerorum quanta summa provenit. Deinde duas tertias numeri, quot ipsi sunt naturaliter se sequentes,[33] addita tertia parte unius, multiplica in predictam summam et quod inde provenerit, summa aggregationis quadratorum ipsorum[34] numerorum erit. <**7.2**> Verbi gratia, sint numeri naturaliter se sequentes isti quattuor, scilicet i, ii, iii, 4, quorum aggregationis summa est 10. Quadrati vero ipsorum sunt 4, scilicet unum et 4 et 9 et 16. Si ergo vis scire ex aggregatione istorum quadratorum quanta summa proveniat, duas tertias partes quaternarii, quot scilicet

---

[25] E omits 5.3-4

[26] E omits 'per ordinem'

[27] efficitur E

[28] illud E

[29] erit E

[30] queris E

[31] vis E²

[32] provenit E²

[33] numeri quot...sequentes] ipsorum numerorum naturaliter se sequentium E²

[34] E² omits

sunt,[35] que sunt 3 minus tertia, multiplica in predictam summam numerorum que erant[36] 10 et cum additione tertie partis unius fiunt[37] 30. Et hec summa provenit si predictos quadratos aggreges, scilicet 1, 4, 9, 16.

<8.1>[38] Omnes numeri equa differentia se vincentes,[39] si fuerint in impari numero, tantum reddit medius in se duplatus quantum extremi sibi aggregati et extremi extremorum, ut 2, 4, 6, 8, 10, 12, 14.[40] <8.2> Omnes numeri equa proportione a se distantes, si fuerint impares numeri, tantum reddit medius in se multiplicatus quantum duo circumpositi et circumpositorum circumpositi in se ducti usque ad unitatem, ut 2, 4, 8, 16, 32. <8.3> Si vero pares fuerint, tantum duo medii in se ducti quantum duo extremi et extremi extremorum in se ducti, ut 2, 4, 8, 16, 32, 64, 128, 256. <8.4> Si vero pares[41] fuerint, tantum efficiunt duo medii sibi aggregati quantum extremi et extremi extremorum usque ad unitatem, ut 2, 4, 6, 8.

De multiplicatione digitorum in se[42]

<9.1> Omnis[43] numerus infra denarium multiplicatus in se ipsum reddit summam sue[44] denominationis decuplate, subtracta inde multiplicatione differentie ipsius ad denarium facta in se ipsum. <9.2> Verbi gratia, sexies sex dicantur fieri 60, que est denominatio a sex decuplata. Differentia autem senarii ad denarium est quaternarius, qui multiplicatus in sex facit 24. /9va/ His ergo 24 de sexaginta subtractis remanent 36 quam summam reddunt sexies sex. Et in omnibus sic ab uno usque ad decem.[45] <9.3> Si autem maiorem per minorem vel econverso multiplicare volueris, differentiam maioris ad denarium multiplica in minorem et ipsam multiplicationem subtrahe a denominatione facta a minore et quod remanserit est summa que provenit ex multiplicatione diversorum numerorum. <9.4> Verbi gratia, cum multiplicaveris quinquies 7, dic fieri quinquaginta, que est denominatio a quinque qui erat ibi minor numerus. Set differentia maioris numeri, scilicet 7, ad denarium sunt tres. Qui tres multiplicati in minorem numerum, scilicet 5, fiunt 15. His ergo subtractis de 50 remanent 35 et hec est summa quam faciunt quinquies 7 vel econverso. <9.5> Vel

---

[35] E² omits

[36] erat P

[37] fient E²

[38] E adds heading: 'De equa proportione' and puts 8.1 after 8.2-3

[39] iungentes E

[40] 12, 14] 20, 40 E

[41] impares *MSS.*

[42] De multiplicatione digitorum in se] Item E

[43] N adds 'namque'

[44] due P, sue M

[45] P adds a triangle of dots. N adds 'vel aliter'. E adds 'Item alio modo'

aliter: denominatio fiat a maiore et differentia minoris ad denarium multiplicetur in maiorem et subtrahatur de summa prime denominationis. <**9.6**> Verbi gratia, quinquies 7 dicatur fieri 70 qui denominatur a 7 et est maior in illa multiplicatione. Set[46] differentia minoris, scilicet quinarii, ad denarium, est 5 qui multiplicatus in maiorem, scilicet 7, facit 35. His subtractis a 70 que erat decuplata denominatio a maiore, scilicet septenario, remanent .35. et hanc summam reddunt multiplicati alter per alterum, quinquies 7 vel econverso septies 5.

<**10.1**> Omnium ergo trium numerorum eiusdem proportionis, si multiplicaveris[47] primum in tertium, tantum provenerit[48] ex multiplicatione eorum quantum ex ductu solius medii in se. Quorum si prepositis[49] primo et medio, tertius tantum fuerit incognitus, multiplica medium in se et quod inde provenerit divide per primum et quod exierit de divisione erit tertius. <**10.2**> Aut si primus tantum fuerit incognitus, multiplica medium in se et divide per tertium et exibit primus. <**10.3**> Aut si medius tantum fuerit incognitus, multiplica primum in tertium et radix eius quod inde provenerit est medius, quoniam medius in se ductus tantum reddit quantum duo extremi, quod in prepositis numeris facile notabis.

<**11.1**> Si ergo aliqui quattuor numeri fuerint proportionales, scilicet ut quomodo se habet primus ad secundum, sic se habet[50] tertius ad quartum, tunc tantum proveniet ex multiplicatione primi in quartum quantum ex multiplicatione secundi in tertium. In his autem 4 terminis socii sunt primus et quartus, secundus et tertius. Unde generaliter: ex omnibus qualiscumque ignoretur unumquemlibet reliquorum duorum per socium ignorati divide et quod exierit per socium dividendi[51] multiplica et proveniet ignoratus terminus. Item si[52] aliquis eorum ignoratur productus ex aliis duobus per ignorati socium dividatur et proveniet incognitus.[53]

<**11.2**> Unde si prepositis[54] tribus quartus fuerit tantum incognitus, multiplica secundum in tertium et quod inde provenerit divide per primum et quod exierit erit quartus. <**11.3**> Aut si primus tantum fuerit incognitus, multiplica secundum in tertium et divide per quartum et exibit primus. <**11.4**> Aut si secundus tantum

---

[46] set ME, scilicet PN

[47] multiplicabis E

[48] provenit E

[49] propositis E

[50] habeat E

[51] dividendi *scripsi*, dividentis PNE

[52] Item si] Item P, Item \si/ N, Vel cum E

[53] In P these two sentences are reversed, but an inserted (a) and (b) indicate that the order should changed. (b) is missing in N. E has the correct order, followed by our edition.

[54] propositis E

154

fuerit incognitus, multiplica primum in quartum et divide per tertium et exibit secundus. <**11.5**> Aut si tertius fuerit incognitus, multiplica[55] primum in quartum et divide per secundum et exibit tertius.

<**11.6**> Verbi gratia, ut si 10 modii vendantur pro xxx aureis, tunc pro duobus modiis sex aurei debentur. Hic quattuor numeri sunt proportionales, scilicet 10 modii, 30 aurei, duo modii, sex aurei.[56] Que enim proportio est decem modiorum ad 30 aureos, quod est pretium eorum, eadem proportio est duorum modiorum ad sex aureos quod est pretium eorum. Cum ergo multiplicaveris primum numerum qui est 10 modii in quartum numerum qui est sex morab<otini> proveniunt inde 60. Tantum similiter proveniet ex multiplicatione secundi numeri qui est 30 morab<otini> in tertium numerum qui est duo modii. /9vb/

<**11.7**> Cum ergo aliquis occultans tibi quartum numerum qui est sex aurei, dicat: 'Cum decem modii vendantur pro 30 aureis, quantum debeatur pro duobus?' Multiplica tunc 30 morabo<tinos>,[57] qui est secundus numerus, in duos modios qui est tertius numerus et quod provenerit divide per decem modios qui est primus numerus et exibit quartus,[58] scilicet[59] sex aurei, qui debentur pro duobus. <**11.8**> Similiter si occultans primum[60] qui est decem modii, dicat: 'Duo modii venduntur pro sex aureis, quot modii habebuntur[61] pro 30?' Multiplica tunc secundum numerum qui est 30 aurei in duos modios qui est tertius numerus et quod inde provenerit divide per sex qui est quartus numerus et exibit primus, qui est scilicet 10[62] quot dantur pro 30. <**11.9**> Similiter si occultans secundum numerum qui est 30 morab<otini>[63] dicat: 'Postquam pro duobus modiis habui sex aureos, quot morab<otinos> habeo pro decem modiis?' Multiplica tunc primum numerum qui est decem modii, in quartum numerum qui est sex aurei et quod inde provenerit divide per duos modios qui est tertius numerus et exibit secundus numerus, scilicet 30 morab<otini>,[64] qui debentur pro decem modiis. <**11.10**> Similiter si occultans tertium qui est duo modii, dicat: 'Postquam decem modii dantur pro 30 aureis, pro sex aureis quot modios

---

[55]multiplicam P

[56]The four values are numbered 1 to 4 superscript in PN, though N puts the '4' on the line of writing.

[57]aureos E

[58]E adds 'numerus'

[59]scilicet] qui est E

[60]E adds 'numerum'

[61]habentur E

[62]E adds 'modii'

[63]aurei E

[64]scilicet 30 morabotini] qui est 30 aurei E

dabunt[65] ?' Multiplica tunc primum numerum qui est decem modii in quartum numerum qui est sex aurei et divide per secundum qui est 30 morab<otini> et exibit tertius numerus qui est duo modii. <**11.11**> In his autem interrogationibus summopere notandum est quid primum dicatur et quid secundum, scilicet si res prius[66] nominatur aut pretium. Quicquid enim prius dicitur, illud tertio loco repetitur et quod secundo nominatur illud est quartum, sive sit dictum sive occultatum.

<**12.1**> Si autem tertius et quartus ignorantur[67] set eorum aggregatio tibi sola proponitur, cum eos invenire volueris, aggrega primum et secundum et eorum aggregatio respectu prime proposite sit tibi secunda. Deinde multiplica primum numerum in aggregationem primam et quod inde provenit[68] divide per aggregationem secundam et quod de divisione exierit[69] erit numerus tertius. <**12.2**> Similiter ad inveniendum quartum, multiplica numerum secundum in aggregationem primam et quod inde provenerit divide per secundam et quod exierit erit quartus.[70] <**12.3**> Similiter etiam econverso: si ignoraveris primum et secundum, [set] eorum aggregatione cognita, invenies eos per tertium et quartum secundum predictam regulam. <**12.4**> Verbi gratia, sint quattuor numeri, primus duo, secundus quattuor,[71] tertius tres, quartus 6. Si ergo tertius et quartus ignorantur, scilicet tres et sex, set eorum aggregatio tibi ostenditur que est novem, cum eos invenire volueris, aggrega primum et secundum, scilicet duo et quattuor et efficies sex. Deinde multiplica primum qui est duo in aggregationem primam que fuit novem et fient decem et octo, quos divide per aggregationem secundam que est sex et exibit tertius qui est tres. Similiter[72] de aliis, ut subiecta figura declarat:

6 3	4 2
9	6

<**12.5**> Si vero tertius et quartus ignorantur, set quod ex diminutione minoris eorum ex maiore remanet tibi proponitur, si eos invenire[73] volueris, diminue primum de secundo vel econverso, semper[74] minorem de maiore, et quod remanserit respectu prime proposite voca diminutionem secundam. Postea multiplica primum numerum in diminutionem primam et quod inde provenerit divide per diminutionem secun-

---

[65]dabunt E, dabuntur PN

[66]primum E

[67]P first wrote 'occultantur' and then underscored it

[68]provenerit E

[69]exibit E

[70]'Similiter...quartus' added in the margin by the same hand in P, with a triangle of dots to connect it to the right position in the text.

[71]quarto PN

[72]E adds 'etiam'

[73]aggregare PNE

[74]scilicet PNE

dam et quod exierit erit numerus tertius. <**12.6**> Similiter ad quartum inveniendum, multiplica numerum secundum in diminutionem primam et quod provenerit divide per denominationem secundam et quod exierit erit numerus quartus. <**12.7**> Similiter etiam econverso: si ignoraveris primum et secundum, ostensa tibi eorum diminutione invenies eos per tertium et quartum secundum predictam regulam, ut in predictis numeris patet. <**12.8**> Si enim ignoraveris tertium et quartum, scilicet tres et sex, set quod remanet ex diminutione alterius ex altero tibi ostenditur, quod est tres, tu minue primum de secundo vel econverso, semper minorem de maiore, et remanent duo que est diminutio secunda. Deinde multiplica primum in diminutionem primam et quod inde provenerit divide per diminutionem secundam et quod inde exierit erit numerus tertius. Similiter etiam in aliis ut subiecta figura declarat.

6 3	4 2
3	2

<**12.9**> Similiter etiam hoc idem invenies, si re/10ra/bus et earum pretiis hoc ipsum adaptes.[75]

<**12.10**> Si vero tertius et quartus tantum fuerint tibi incogniti, set eorum multiplicatio sola tibi proponitur, tu multiplica primum[76] in ipsam multiplicationem et quod inde provenerit divide per secundum[77] et radix eius quod de divisione exierit erit tertius. <**12.11**> Similiter ad inveniendum quartum, multiplica secundum in ipsam multiplicationem et quod inde provenerit divide per primum et radix eius quod de divisione exierit, erit quartus. <**12.12**> Verbi gratia, si predictorum numerorum tertium et quartum, scilicet tres et sex, ignoraveris, set eorum multiplicatio tibi sola proponitur, que est decem et octo, tunc multiplica primum qui est duo in ipsam multiplicationem et provenient 36.[78] Quos 36 divide per secundum qui est quattuor et exibunt de divisione novem, cuius radix, scilicet ternarius, est tertius. <**12.13**> Similiter ad inveniendum quartum, multiplica secundum qui est quattuor in ipsam multiplicationem et provenient 72.[79] Quos divide per primum et exibunt 36 cuius radix est senarius et hic est quartus, ut subiecta figura declarat.

6 3	4 2
18	8

<**12.14**> Si vero duo medii fuerint incogniti, set eorum aggregatio tibi sola proponitur, tu multiplica primum in quartum et quod inde provenerit pone per se. Deinde divide ipsam aggregationem in tales duas partes quarum altera multiplicata in alteram reddat multiplicationem primi in quartum et ipse partes erunt medii incogniti.

---

[75]E omits 12.9

[76]E adds 'numerum'

[77]secundam PNE

[78]36 is placed in a box in PN

[79]72 is placed in a box in PN.

<12.15> Verbi gratia, ut si quattuor et tres ignores, set eorum aggregatio tibi proponitur, scilicet septem, multiplica primum qui est duo in quartum qui est sex et efficies 12. Deinde propositam aggregationem que est septem divide in tales partes que in se multiplicate efficiant[80] 12 et ipse erunt medii numeri incogniti, tres igitur et quattuor, ut subiecta figura declarat.[81]

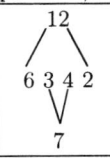

gradus	minuta	secunda	tertia	quarta	quinta	sexta	septima	octava	nona
minuta	secunda	tertia	quarta	quinta	sexta	septima	octava	nona	dec
secunda	tertia	quarta	quinta	sexta	septima	octava	nona	dec	undec
tertia	quarta	quinta	sexta	septima	octava	nona	dec	undec	duodec
quarta	quinta	sexta	septima	octava	nona	dec	undec	duodec	tertiad
quinta	sexta	septima	octava	nona	dec	undec	duodec	tertiad	quartad
sexta	septima	octava	nona	dec	undec	duodec	tertiad	quartad	quintad
septima	octava	nona	dec	undec	duodec	tertiad	quartad	quintad	sextad
octava	nona	dec	undec	duodec	tertiad	quartad	quintad	sextad	septimad
nona	dec	undec	duodec	tertiad	quartad	quintad	sextad	septimad	octavad.

1	2	3	4	5	6	7	8	9	10
2	4	6	8	10	12	14	16	18	20
3	6	9	12	15	18	21	24	27	30
4	8	12	16	20	24	28	32	36	40
5	10	15	20	25	30	35	40	45	50
6	12	18	24	30	36	42	48	54	60
7	14	21	28	35	42	49	56	63	70
8	16	24	32	40	48	56	64	72	80
9	18	27	36	45	54	63	72	81	90
10	20	30	40	50	60	70	80	90	100

B

<13.1> /10rb/ Omnis numerus usque ad decem in se ductus tantum reddit quantum primi duo circumpositi in se ducti, si dempta unitate de multiplicatione medii.
<13.2> Vel aliter generalius: omnis numerus in se ductus tantum reddit quantum duo extremi et extremi extremorum usque ad unitatem, set adiectis differentiis in

---

[80]efficiunt E

[81]E continues with a second copy of 7.1-2.

se ductis, quas habet medius ad extremos.[82] <**13.3**> Tantum enim reddit 5 in se ductus quantum quater sex cum differentiis in se ductis que sunt due unitates, et quantum ter septem cum differentiis in se ductis que sunt duo binarii, et quantum bis 8 cum ductis in se differentiis que sunt duo ternarii, et sic usque ad unum. <**13.4**> Omnis numerus tantum reddit in se ductus quantum duo in se[83] ducti equa proportione ab illa distantes.[84]

<**14.1**> Omnis numerus in se ductus tantum efficit quantum due partes eius, si utraque in se ducatur et altera in alteram bis. <**14.2**> Omnis numerus ductus in alium tantum reddit quantum ductus in omnes partes eius. <**14.3**> Cum aliquis numerus multiplicat alium, tantum provenit quantum si idem multiplicet limitem, subtracto eo de summa quod differentia multiplicati ad limitem, ducta per multiplicantem efficit.

<**15.1**> Omnis numerus per alium dividendus aut est ei equalis aut maior aut minor. Si est equalis, tunc unicuique dividentium singule unitates dividendi proveniunt. <**15.2**> Si vero maior, tunc quotiens dividens in dividendo fiunt (sic!), tot integri unicuique dividentium proveniunt. <**15.3**> Si vero aliud superfuerit, per fractiones dividendum erit. <**15.4**> Ita ut quota vel quote partes minor numerus est maioris, tota pars vel partes unicuique dividentium proveniunt. <**15.5**> Cum fractiones fractionibus aggregas, si idem est numerus fractionum et denominationis earum, tunc ex aggregatione integer surgit. <**15.6**> Ut ex tribus tertiis vel quattuor quartis unum integrum redditur. <**15.7**> Si vero minor est numerus fractionum numero denominationis, tunc qua proportione se habet numerus fractionum ad numerum denominationis earum, eadem proportione habent se fractiones ille ad integrum, <**15.8**> ut sex duodecime sic se habent ad integrum ut senarius ad duodenarium. Sunt ergo eius medietas. <**15.9**> Si vero maior fuerit, tunc quotiens maior fuerit tot integra fractiones aggregate constituent. <**15.10**> Ut sex tertie duo integra restituunt, quoniam numerus fractionis bis continet numerum denominationis. <**15.11**> Si vero continet eum aliquotiens et insuper aliquam[85] vel aliquas eius partes, tunc quotiens eum continet, tot integra constituunt fractiones aggregate et insuper totam vel totas partes unius integri quota vel quote partes est numerus ille qui superest numeri denumerantis fractioness. <**15.12**> Ut est octo tertie, octonarius bis continet ternarium et eius duas tertias.

<**15.13**> Si scire volueris quomodo pars partis cuiuslibet se habet ad integrum, numeros a quibus denominantur fractiones in se multiplica et qualiter unitas se

---

[82]extrema E

[83]E breaks off here, leaving the rest of the page blank.

[84]distante s P

[85]aliqua P

habuerit ad summam illam, sic pars partis habebit se ad integrum. <**15.14**> Ut tertia pars unius quarte, duodecima est unius integri. Nam ter quattuor duodecim fiunt. <**15.15**> Si scire volueris partes partis cuiuslibet quomodo se habent ad integrum, numeros a quibus denominantur fractiones in se multiplica et qualiter numerus coacervans[86] habet se ad numerum iam productum, sic fractiones ille aggregate ad integrum. <**15.16**> Ut due tertie partes unius quarte, sexta pars sunt unius integri. Nam ter 4 12 sunt, cuius sexta pars sunt (*sic*) binarius.

<**15.17**> Si vero a duobus diversis numeris denominantur fractiones, tunc qua proportione maior numerus se habet ad minorem, taliter fractio denominata a minori habet se ad fractionem denominatam a maiore. <**15.18**> Ut tertia pars alicuius continet duas sextas eius. Nam senarius continet duos ternarios. <**15.19**> Si diversorum numerorum vel diversarum quantitatum fractiones denominantur ab eodem numero, sicut integra habent se ad invicem, sic et fractiones et econverso. <**15.20**> Nam sicut duodenarius habet se ad novenarium, sic tertia pars duodenarii ad tertiam novenarii et econverso. <**15.21**> Sin autem fractiones quot/10va/cumque a diversis nominibus denominatas coacervare volueris, numeros a quibus denominantur fractiones coacerva et per summam inde provenientem coacerva fractionem denominatam a numero qui surgit ex multiplicatione numerorum fractiones denominantium. <**15.22**> Nam si scire volueris tertia pars et quarta aggregate quid efficiant, numeros a quibus denominantur fractiones, scilicet ternarium et quaternarium, aggrega et fiunt 7. Quo septenario coacerva[87] fractiones denominatas a numero qui surgit ex multiplicatione numerorum fractiones denominantium, scilicet duodecim. Nam ter 4 12 fiunt. Sunt igitur tertia et quarta pars alicuius septem duodecime partes, que coacervate quid constituant superius ostensum est.[88]

<**16.1**> ...Vel ipsum per se multiplica et multiplicationi inde provenienti ipsum adde et hoc totum in duo equa divide et illa medietas est tota summa illius et omnium infra se ipsum contentorum. <**16.2**> Si ex numeris equa differentia se vincentibus sibi aggregatis vis scire quanta summa reddatur, si in impari numero coacervandi fuerint, vide quot sint et toto numero medium aggregandorum multiplica et tota est summa. <**16.3**> Verbi gratia, sint tres, 4, 5 aggregandi. Ternario igitur quot ipsi sunt medium eorum, scilicet[89] 4, multiplica et tanta erit summa quam reddunt prepositi numeri aggregati. <**16.4**> Vel sint tres, 5, 7, 9, 11 aggregandi. Quinario igitur, tot enim sunt aggregandi, medium eorum, scilicet septem, multiplica et tanta erit summa prepositorum aggregatorum. <**16.5**> Vel sint 2, 5, 8 aggregandi. Ternario igitur medium eorum, scilicet 5, multiplica et tanta erit summa illorum. <**16.6**>

---

[86] coacervans N, 'coacreveras' corrected from 'coacerveris' P

[87] coacreva P

[88] This is followed in P by a siglum (a fancy equals sign), whose significance is not clear.

[89] id est N

Sin autem pari numero sunt aggregandi, sub eadem proportionalitate unum maiorem prioribus adiunge. Deinde vide quot sint et toto numero medium eorum multiplica et ablato illo quem illis aggregandis adiunxeras summa aggregatorum efficitur. <**16.7**> Ut sint 2, 4, 6, 8 aggregandi. Proportionalitatis illius proximum, id est x, adiunge. Quinario itaque, quia tot sunt aggregandi, medium eorum, scilicet senarium, multiplica eruntque 30. Huic itaque summe denarium quem adiunxeras aufer et 20 qui remanent summa sunt aggregatorum. <**16.8**> Si impar quilibet cum omnibus imparibus sibi subpositis et unitate aggregatur summa que excrescit numerus quadratus erit.

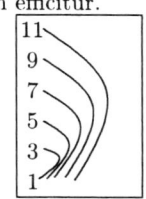

<**17.1**> Si duorum numerorum quadrata pariter accepta fuerint numerus quadratus, necesse est quorumlibet numerorum duorum eadem proportionalitate ad se relativorum quadrata pariter accepta, numerum quadratum esse. <**17.2**> Si ad quantitatem aliquam quantitates quotlibet proportionentur diversis set notis proportionibus, que pariter accepte summam notam faciunt, eadem prima quan-

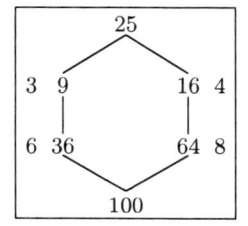

titas quanta sit invenire. <**17.3**> Verbi gratia, sit a quantitas sitque etiam ut b et c et d proportionibus notis ad ipsum a proportionentur. Sit quoque ut b et c et d[90] pariter accepte quantitatem component g, que g quanta sit notum sit. Proponitur itaque ut etiam quanta a fuerit inveniatur. <**17.4**> Ut si proponatur Socrates bis tot nummos habere quot Plato et insuper eorum quos Plato habet duas tertias partes, totumque quod Socrates habet simul acceptum quindecim nummos esse. Proponitur itaque ut quantum Plato habeat inveniatur. Sumo itaque numerum ad quem duo numeri predictis proportionibus proportionantur. Hic[91] autem ternarius est. Senarius enim bis ipso maior est et binarius eiusdem due tertie partes. Hos itaque duos numeros, senarium et binarium, coniungo et fiunt 8. Considerato igitur qua proportione 15 ad 8 se habeat, eadem proportione Platonem ad tres nummos se habere pronuntio. Quindecim enim continet totum octo et eius septem octavas. Similiter quinque nummi et obolus et quarta pars oboli tres nummos et[92] eorum septem octavas continent. Dico itaque Plato 5 nummos et obolum[93] et quartam partem oboli habere. Bis enim tantum et eiusdem due tertie partes 15 nummi sunt. Si enim quinque nummos et obolum[94] et quartam partem oboli bis sumpseris xi nummi et quarta pars unius nummi erunt. Eorumdem etiam quinque nummorum et

---

[90] a PN

[91] his P, hiis N

[92] et *bis* P

[93] obulum P

[94] obulum P

oboli et quarte partis oboli, si duas tertias sumpseris, tres nummos et tres quartas unius facient. <**17.5**> Quod sic accipe: novem obolorum duas tertias[95] partes accipe, id est vi obolos, et sunt tres nummi. Reliquus vero nummus et quarta pars /10vb/ oboli novem octave partes unius sunt nummi. Quoniam due tertie partes, id est sex octave nummi, tres quarte partes unius nummi sunt, quas si cum prioribus undecim nummis et quarta parte nummi et tribus nummis iunxeris, 15 nummi sunt. <**17.6**> Vel aliter facilius: multiplica numerum denominationis partium, scilicet tres, in 15 et proveniunt 45. Deinde eundem ternarium multiplica in <duo et> duas tertias et proveniunt octo. Nam tres in duo fiunt 6 et additis duabus tertiis fiunt octo. Postea divide primum productum, scilicet 45, per ultimum productum, scilicet 8, et exeunt 5 et quinque octave, hoc modo:

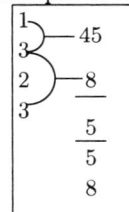

<**17.7**> Cum queritur quot minora sint in aliquot maioribus, per numerum minorum qui sunt in uno maiorum maiora multiplices et numerus qui excrescet quot minora sint in tot maioribus ostendet. <**17.8**> Si vero quot maiora in aliquot minoribus queritur, per numerum minorum qui sunt in uno maiorum minora divides et numerus qui de divisione exibit quot maiora sint in tot minoribus ostendet.[96] <**17.9**> Verbi gratia, solidus minus est quam libra. Ergo si queritur quot solidi sint in c libris, quere quot minora sint in maiori aliquo. Per numerum itaque minorum, id est solidorum, qui sunt in uno maiorum, id est libra una, id est per xx, viginti enim solidi faciunt libram unam, numerum maiorum, id est librarum, videlicet centenarium, multiplices et sunt 2,000. Scito igitur quia tot solidi, videlicet 2,000, sunt in c libris. <**17.10**> Rursus sit questio quot libre sint in 24[97] milibus nummorum. Queritur ergo quot maiora sint in aliquot minoribus, eo quod nummi minus sunt quam libre. Per numerum itaque minorum, id est nummorum, qui sunt in uno maiorum, id est in libra una, scilicet per ccxl, tot enim denarii sunt in una libra, numerum minor<um>, id est denariorum, videlicet 24,000,[98] divides et exibunt inde 100. Scito ergo quia tot libre, id est c, sunt in 24,000[99] nummorum.

<**17.11**> Cognita summa quotarumlibet partium alicuius totius, totum ipsum invenire. Primo numeros partes propositas denominantes aggrega. Deinde alterum per alterum multiplica et sic habebis quattuor proposita; summam scilicet partium propositarum, summam numerorum partes denominantium, et aggregatum ex eis-

---

[95] tertias *bis* P

[96] 'Si vero quot maiora in aliquot minoribus queritur, per numerum minorum qui sunt in uno maiorum minora divides et numerus qui de divisione exibit quot maiora sint in tot minoribus ostendet' repeated in PN.

[97] 34 PN

[98] 34,000 PN

[99] 34,000 PN

dem, quartum est totum quod ignoratur. Nam que proportio est totius ad summam partium propositarum, eadem est producti ex numeris partes denominantibus ad totum aggregatum ex eisdem. Multiplicetur ergo summa partium propositarum per productum ex numeris eas denominantibus et productus inde [7]^{100} dividatur per aggregatum ex easdem denominantibus et exibit totum quod ignoratur, per precedentem regulam quattuor numerorum[101] proportionalium: si primus fuerit[102] incognitus, multiplica secundum in tertium et divide per quartum; exibit primus.

<17.12> Verbi gratia, sint tertia et quarta peccunie mee 20 nummi. Proponitur ergo invenire quota sit summa peccunie. Aggregatis ergo denominantibus, scilicet ternario et quaternario, fiunt septem. Eorumdem alterum per alterum multiplica et productus erit 12. Habes ergo quattuor proposita. Multiplica ergo secundum per tertium primo, id est 20 per 12, et producentur 240. Hoc ergo divide per 7 et exeunt 34 et due septime, qui numerus in eius generis rebus constituendus est, nummis scilicet vel solidis vel libris, in quo partes proposite fuerunt et hec est summa quesita.[103]

20	12
\/	
240	
7	
34	
2	
7	

<17.13> Ex tribus numeris proportionalibus si primus ducatur in tertium exibit quadratus medii.

<17.14> Pluribus hominibus diversas summas peccunie ad lucrandum simul conferentibus, si ex lucro quod ex tota collectione provenerit vis scire quanta sors unumquemque eorum iure contingit, portiones quas apposuerunt aggrega et portionem cuius volueris in summam lucri multiplica. Deinde quod ex multiplicatio/11ra/ne fit, per aggregatum divide et quod ex divisione exierit, hec illius cuius sortem multiplicasti pars erit. <17.15> Vel econverso: divide sortem per aggregatum et quod exierit in summam lucri multiplica et quod inde provenerit, hec ipsius sors erit. Similiter de ceteris singillatim facies. <17.16> Verbi gratia, tres mercatores peccuniam ad lucrandum contulerunt, unus sex solidos, alius 8, alius 12 qui omnes fiunt 26. Ex his lucrati sunt lx. Si vis ergo scire quantum de lucro contingat quemque secundum quantitatem collate peccunie, partes omnium simul aggrega et fiunt 26. Deinde multiplica per se unamquamque partem quam quisque contulit in summam lucri. Deinde divide id quod ex multiplicatione provenit in summam collati capitalis, scilicet 26, et quod de divisione exierit, hoc est quod debetur illi cuius sortem multiplicasti. Sic facies de unoquoque per se. Sunt ergo hi quattuor

---

^{100}inde 7] in.30127. N

^{101}numerorum] numerum eorum PN

^{102}fuit P

^{103}NP add (P *in marg.*): Hic numerus continet totum 20 et insuper eius 5 septimas sicut duodecim septem.

numeri, scilicet sors cuiuslibet et 26, tertius est incognitus, quartus 60, qui sunt proportionales per suprapositam regulam. Multiplica igitur primum in quartum, id est partem cuiuslibet in 60 et productum divide per secundum, scilicet 26, et exibit tertius, scilicet sors que contingit eum cuius partem posuisti primum terminum.

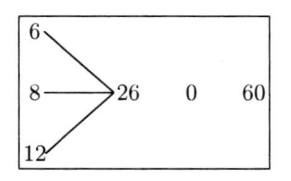

<17.17> Si vis scire de aliqua certa summa multis hominibus debita quantum proveniat aliquibus illorum, numerum ipsorum aliquorum de quibus vis scire in ipsam summam multiplica et quod ex multiplicatione provenerit in totum numerum multorum divide et quod provenerit, hoc est quod debetur illis. <17.18> Verbi gratia, vigintiquattuor nummi debentur octo hominibus et vis scire quantum proveniat tribus illorum. Multiplica ergo ipsos tres in 24 et quod ex multiplicatione provenerit, divide per 8 et videbis quid proveniet illis. Et secundum hanc regulam similiter probabis quantum proveniat aliis quinque, ipsos scilicet quinque multiplicando per 24 et quod ex eorum multiplicatione provenerit dividendo per octo. Sunt igitur hi tres termini 8, 24, 3, quartus ignoratur, scilicet quantum debetur tribus. Per precedentem ergo regulam multiplica secundum in tertium et productum inde divide per primum, scilicet 8, et exibit quartus incognitus, scilicet pretium trium. Sicut enim octo habet se ad tres, sic 24 ad pretium trium. Continet ergo illud bis et insuper eius duas tertias partes.

8	24	3	0

C

<18.1> Exceptiones de libro qui dicitur gebla mucabala. Fit hic quedam trimembris[104] divisio per opposita. Quia queritur[105] aut que res[106] cum totiens[107] radice sua efficiat numerum, aut que res[108] cum tali[109] numero efficiat totiens[110] radicem, aut que totiens radix cum tali numero efficiat rem[111] .

<18.2> Queritur ergo que res cum x radicibus[112] suis, id est decies accepta[113] radice

---

[104]trimenbris P

[105]E adds 'ut'

[106]que res] queres N, quadratus E

[107]E omits

[108]que res] queres N, quadratus E

[109]E omits

[110]E omits

[111]aut...rem] aut radix cum numero efficiat quadratum E

[112]que res...radicibus] quis quadratus est qui cum radicibus decem E

[113]multiplicata E

sua, efficiat 39.[114] Ad hoc inveniendum medietatem radicum prenominatarum multiplica in se et quod inde provenerit adde priori numero et eius quod inde excreverit[115] accipe radicem et de ipsa radice minue[116] medietatem radicum prenominatarum et quod inde remanserit est radix rei,[117] quam si in se multiplicaveris, res provenit quam[118] queris. <**18.3**> Verbi gratia, quoniam superius decem radices fuerant[119] proposite, medietatem earum que est 5 si in se multiplicaveris, 25 efficis.[120] Quos adde predicto numero qui est[121] 39 et efficies[122] 64, cuius radix est octo. De qua radice, scilicet 8, si minueris[123] medietatem radicum[124] que est quinque, remanent tres, qui sunt radix rei,[125] scilicet novenarii, qui cum x radicibus suis, id est x ternariis, efficit[126] 39. Ergo novenarius est res que[127] queritur.

<**18.4**> Item que est res que[128] cum novem sibi additis efficit sex radices[129] sui?[130] Ad hoc inveniendum medietatem[131] radicum multiplica in se et ex eo quod inde provenerit numerum predictum diminue[132] et eius quod[133] remanserit radicem de medietate radicum minue et quod remanserit erit radix rei quam[134] queris. <**18.5**> Verbi gratia, sex radices fuerunt proposite, quarum medietatem que est[135] tres in se multiplica et efficies 9, de quibus 9 predictum numerum, scilicet 9, minue et remanet nichil. Et huius quod remanet, scilicet nichil, /11rb/ radicem que similiter

---

[114]R<espondetur> E

[115]provenerit E

[116]diminue E

[117]quadrati E

[118]res provenit quam] quadratus provenit quem E

[119]fuerunt E

[120]efficies E

[121]predicto numero qui est] prenominate summe scilicet E

[122]efficis E

[123]minuis E

[124]radicis E

[125]quadrati E

[126]efficit E] efficis PN

[127]quadratus qui E

[128]Item...que] Similiter queritur quis est quadratus qui E

[129]sex radices] sexies radicem E

[130]R<espondetur> E

[131]medietatem *bis* PN

[132]minue E

[133]quod E] que PN

[134]quadrati quem E

[135]sunt E

est nichil de medietate radicum que est[136] tres minue et quia de tribus nichil minuisti, remanent tres qui sunt radix rei quam[137] queris, scilicet novenarii, qui[138] cum novem sibi additis fit 18, qui sunt sex radices novenarii, id est sexies tres, qui ternarius est[139] radix novenarii.[140]

<**18.6**> Item[141] que sunt[142] radices que cum 4 sibi additis efficiunt rem suam?[143] Ad hoc inveniendum multiplica medietatem radicum in se et quod inde provenerit adde predicto numero et radicem eius quod inde excreverit,[144] adde medietati radicum et quod inde excreverit[145] est radix rei quam[146] queris. <**18.7**> Verbi gratia, tres radices fuerunt preposite[147] quarum medietas est unum et dimidium, que multiplicata in se efficiunt duo et quartam, que adde priori numero qui est 4 et efficies sex et quartam. Cuius radix est duo et dimidium, quam radicem[148] adde medietati radicum, que est unum et dimidium et fient 4, qui sunt radix rei quam[149] queris, scilicet sedecim.[150] Quos sedecim[151] efficiunt tres radices sui, id est tres quaternarii vel ter quattuor cum[152] additis sibi 4.[153]

D

<**19.1**> Quamcumque datam quantitatem secundum quascumque datas proportiones si dividere volueris vel numerum indivisum ad modum divisi dividere, primo

---

[136]sunt E

[137]quadrati quem E

[138]E adds 'novenarius'

[139]ternarius est] sunt E

[140]E adds 'Similiter in aliis omnibus'

[141]Item] Similiter queritur E

[142]EN adds 'tres'

[143]qudratum suum. R<espondetur> E

[144]radicem...excreverit] eius quod inde provenerit radicem E

[145]provenerit E

[146]quadrati quem E

[147]proposite E

[148]E omits

[149]quadrati quem E

[150]sexdecim P (*passim*)

[151]Quos sedecim] Quem quadratum E

[152]E omits

[153]E continues with: 'Sicut in predictis numeris ita et in omnibus aliis invenire poteris, in rebus et in earum pretiis si diligenter observaveris has predictas regulas. Explicit Liber algvorismorum.' This is followed by the beginning of **B** above, which ends abruptly after a few lines.

proportiones partium propositarum in terminis sunt ordinande, deinde termini aggregandi, aggregatum vero primo ponendum. Numerus vero proportionis cum proportionalis ei quod queritur secundo ponendus. Datam vero quantitatem tertio ponendum. Cifre vero quarto[154] ponendum. Multiplicatur igitur data quantitas per numerum proportionis et productum inde[155] dividatur per aggregatum ex ipsis terminis et exibit quod queritur, per regulam quattuor proportionalium. Si quartus ignoratur, multiplicetur tertius in secundum et productum dividatur per primum et exibit quartus. Quod si per unum inventum cetera habere volueris, multiplicia habebis multiplicando[156] ipsum per numerum denominantem proportiones. Multiplicia autem dicuntur que aliquotiens continent aliquem numerum. Submultiplicia dicuntur que aliquotiens continentur. Si vero aliquis numerus post divisionem remanserit, ille numerus erit fractionum quantitatis in divisione a dividente denominatarum.

<19.2> Verbi gratia, proponitur nobis dividere 40 solidos 4 hominibus ita quod secundus habeat quadruplum ei quod habet primus, tertius vero quincuplum secundo, quartus triplum tertio. Multiplicentur ergo quattuor per unum et exeunt 4. Set sit primus terminus unum, secundus vero quattuor. Item multiplicentur[157] 4 per 5 et fiunt 20. Erunt ergo 20 tertius terminus. Item multiplicentur[158] 20 per 3[159] et fiunt 60. Erunt ergo 60 quartus terminus. Aggregatis autem terminis, fiunt 85. Ponatur ergo primo[160] 85. Si vis autem scire quid accidit secundo, pone secundum terminum secundum. Similiter si vis scire quid accidit tertio, pone tertium terminum secundum. Similiter de singulis.

85	primus	1		40	0
	secundus	4	quadruplus		
	tertius	20	quincuplus		
	quartus	60	triplus		

<19.3> Pone ergo secundum terminum, scilicet 4, secundum, datam vero quantitatem pone tertio, çifre[161] vero quarto. Multiplicetur data quantitas, scilicet 40, per 4 et fiunt 160. Dividatur autem 160 per 85 et exeunt unum et 75 octuagesime quinte. Qui numerus in eius generis rebus constituendus est, nummis vel solidis vel libris in quo data quantitas proposita fuit. Et hoc est quesitum. <19.4> Si vis autem scire per hoc quid accidit primo numerum inventum, id est unum inte-

---

[154]quartum PN

[155]in P

[156]dividendo P

[157]multiplicantur N

[158]multiplicantur N

[159]30 P

[160]primum PN

[161]ziffre N

grum, scilicet 12 denarios qui sunt unus solidus, divide per 4 et exeunt 3 d<enarii.> Item numerum numerantem fractiones, scilicet 75, per eundem 4 divide et exeunt 18 octuagesime quinte et tres quarte unius octogesime quinte et hec est pars primi, scilicet 3 d<enarii.> et 18 octogesime quinte et tres quarte unius octogesime quinte. <**19.5**> Item si per numerum prius inventum cetera habere volueris, scilicet multiplicia eius, ipsum numerum inventum, scilicet 1 et 75 octogesimas quintas, multiplica per numerum denominantem tertiam proportionem, verbi gratia per 5, et fiunt 9^{162} solidi et 35 octogesime quinte et hec est portio tertii. Et ita de singulis multiplicibus. <**19.6**> Si vero submultiplicia eius habere volueris, ipsum numerum prius inventum per numerum denominantem proportionem divide. <**19.7**> In probando /11va/ adde integra integris.

<**20.1**> Si quascumque proportiones datas ad quemcumque terminum minorem continuare volueris, primam datarum per terminum minorem multiplica et productum per sequentem multiplica et ita procede multiplicando productum sequentis per sequentem et productum ex penultimo sequenti multiplicatum per ultimum sequentem, erit primum continuandorum. Hoc facto secundum datarum per eundem terminum minorem multiplica et productum per sequentem multiplica et ad modum predictum procede multiplicando et productum ex penultimo sequenti per ultimum sequens multiplicatum erit secundum continuandum. Item tertiam datarum per eundem terminum minorem multiplica et productum per sequentem multiplica et ad modum predictum procede multiplicando et productum ex penultimo sequenti per ultimum sequens multiplicatum erit tertium continuandorum. Et ita de singulis. <**20.2**> Verbi gratia, sint date proportiones 4, 3, 2, 1 et sit minor terminus 6. Multiplicetur 4 per 6 et erit productum 24. Hoc autem multiplica per tres et productum erit 72.[163] Quod multiplicetur per 2 et fient[164] 144.[165] Et si hoc multiplicetur per 1, exibit[166] idem. Hoc ergo, scilicet 144,[167] est primum continuandorum. <**20.3**> Item multiplicetur 3 per 6 et erit productum 18, hoc autem multiplica per 2 et fit 36 secundum continuandum. <**20.4**> Item multiplica duo per 6 et fit 12. Erit ergo 12 tertium continuandorum. <**20.5**> Item multiplica 1 per 6 et fit 6. Erit ergo quartum continuandorum.

<**21.1**> Si radicem cuiuslibet propositi quadrati invenire volueris propositum quadratum per quemlibet[168] alium quadratum multiplica et producti radicem ex-

---

[162]5 PN

[163]64 PN

[164]fiet P

[165]124 PN

[166]N adds 'unum'

[167]24 P, 124 N

[168]quelibet P

trahe et radicem per radicem quadrati per quem multiplicasti quadratum propositum divide. Quod exibit erit radix quesita. <**21.2**> Verbi gratia, si radicem duorum et quarte invenire volueris, multiplica illud in aliud quadratum, qui scilicet sit 4, et proveniunt 9. Duo enim in quattuor fiunt 8 et quarta in quattuor fit unum; unde fiunt 9. Horum autem 9 radix sunt 3. Quos divide per radicem quaternarii que est

duo. Exibit unum et dimidium, hoc modo: $\begin{array}{|c|} \hline 1 \\ 3 \\ 2 \\ \hline \end{array}$

<**21.3**> Si radicem propinquioris quadrati invenire volueris, 3^{169} numerum propositum per quemlibet quadratum multiplica et predicto modo operare.

<**22.1**> Si articulum in articulum vel digitum in digitum vel compositum in compositum multiplicare volueris, figuram per figuram multiplica. Deinde numeros denominantes differentias aggrega, ab aggregato unum subtrahe et principium sequentis differentie ab eo quod remanet denominate totiens excrescit quot unitates excreverint ex multiplicatione figurarum et quot excreverint denarii, totiens principium sequentis differentie excrescit.

<**23.1**> Cum multiplicaveris unum numerum in alium, vide quilibet eorum quota pars sit alicuius articuli sive limitis et accipe tantam partem de altero eorum, quam multiplica in illum articulum vel limitem et productum inde est id quod ex ductu unius in alterum provenit. <**23.2**> Verbi gratia, si multiplicaveris 32 in 25, quota pars est 25 de centum limite, scilicet quarta, tanta accipe de 32, scilicet quartam que est octo. Quos octo multiplica in centum et productum inde est id quod ex ductu 32orum in 25 provenit. Probatio. Sicut enim habet se 8 ad 32 sic 25 ad centum. Tantum ergo fit ex ductu duorum mediorum in se quantum ex ductu duorum extremorum per predictam regulam 4 numerorum proportionalium et sic in omnibus. <**23.3**> Similiter etiam de articulo. Verbi gratia, 25 medietas est de quinquaginta. Set medietas de 32 sunt 16. Quos multiplica in 50. Idem provenit quod ex ductu 32 in 25. $\boxed{8 \ \ 32 \ 25100}$

<**23.4**> Duorum numerorum compositorum ex diversis vel eisdem digitis set eodem articulo vel limite /11vb/ cum multiplicaveris unum in alium, ut sedecim in 18 et huiusmodi, multiplica digitum in digitum et articulum in articulum et productos inde aggrega. Deinde digitum digito aggrega et aggregatum in articulum vel limitem multiplica et productum inde priori aggregationi adde et illud totum aggregatum est summa que unius compositi in alium multiplic<at>io^{170} efficit. <**23.5**> Verbi gratia, proponantur multiplicandi 16cim in 18. Multiplicetur ergo digitus in digitum, id est sex in octo, et fiunt 48. Deinde articulum in articulum, id est decem in decem,

---

^{169}et N

^{170}multiplitio PN

et fiunt 100. Quos duos productos simul aggrega et fiunt centum 48. Deinde digitum digito aggrega et fiunt 14. Quos in articulum, scilicet decem, multiplica et fiunt 140. Quos aggrega priori aggregationi que fuit 148 et fiunt 288. Et hec est summa que ex ductu 16 in 18 provenit.

<**24.1**> Cum volueris multiplicare radices aliquorum numerorum, ipsos numeros in se multiplica et producti radix est productus ex ductu unius radicis in aliam. <**24.2**> Verbi gratia, si volueris multiplicare radices denarii et quadraginta, multiplica 10 in 40 et proveniunt 400. Horum autem 400orum radix est 20. Qui 20 sunt numerus productus ex multiplicatione radicis denarii in radicem 40 per regulam trium numerorum proportionaliter se habentium, qui cum sicut se habet primus ad secundum, sic secundus ad tertium, tunc quantum fit ex ductu medii in se, tantum ex ductu[171] extremorum hoc modo: $\boxed{10\ 20\ 40}$

<**25.1**> Si vis scire quantum vixerit qui vivens tantum quantum vixit et iterum tantum et dimidium tanti et dimidium dimidii c annos complet, aggrega que proponuntur et per aggregatum divide summam que completur et quod exierit hoc est quod vixit. <**25.2**> Verbi gratia, cum proponitur tantum quantum vixit et iterum tantum et dimi<dium> tanti et dimi<dium> dimidii, simul[172] aggregata quattuor minus quarta fiunt, que sunt 15 quarte. Per quas 15 quartas si dividis centum conversum prius in quartas, exeunt 26 et due tertie. Que quater <minus quarta> simul accepta, centum complent et hoc est quod vixit.

E

<**26.1**>...Vel aliter: quoniam omnis numerus vel est digitus vel articulus vel limes vel compositus, ideo quotiens multiplicatur numerus in numerum, aut multiplicatur digitus in digitum vel in articulum vel limitem vel compositum vel econverso, aut compositus in compositum vel articulum vel limitem vel digitum et econverso. <**26.2**> Cum autem articulum in articulum multiplicare volueris, figuram per figuram multiplica. Deinde quarum differentiarum sint ipsi articuli considera et numeros a quibus denominantur eorum differentie aggrega. Ab aggregato autem unum subtrahe et in differentia denominata a numero remanenti productum ex multiplicatione figurarum si fuerit tantum digitus pone. Si autem articulus tantum in sequenti eam. Si vero digitus et articulus, digitus in differentia denominata a numero remanenti, articulum vero ponatur in differentia sequenti. Et quod ibi significaverit est summa que ex ductu unius articuli in alium provenit. <**26.3**> Verbi gratia, si multiplicare volueris 20 per 70, figuras quibus representantur, scilicet 2 et 7, in se multiplica et fiunt 14. Set quia secunde differentie est utraque que est decenorum, ideo numeros denominantes differentiam utriusque, scilicet

---

[171]ducto P

[172]similiter PN

duo et duo, secunda enim a duobus denominatur, aggrega et fiunt quattuor. A quibus quattuor subtracto uno remanent tres. A quibus denominatur tertia differentia, que est centenorum. Et quia ex ductu figurarum in se provenerat[173] 14 qui est digitus et articulus, ideo digitum, scilicet 4, in eadem differentia, scilicet tertia, pone et articulum, scilicet decem, in sequenti que est quarta et habebis 1,400. Et hec est summa que ex ductu unius articuli in alium provenit, scilicet 20 per 70. Similiter etiam faciendum est si digitus in articulum vel limitem vel compositum multiplicetur et econverso. <**26.4**> Cum autem /12ra/ compositum in compositum multiplicare volueris, predictam regulam observabis, hoc adiecto, ut unusquisque superiorum multiplicetur in unumquemque inferiorum, videlicet digitus in digitum et articulus et articulum in digitum et articulum, quotquot fuerint, singuli superiorum in omnes inferiores. <**26.5**> Verbi gratia, 23 in 64 cum multiplicare volueris, digitum superiorem, scilicet tres, in digitum inferiorem, scilicet 4, multiplica et fiunt 12. Per priorem ergo regulam, digitus erit in prima differentia, articulus in secunda. Deinde eundem digitum, scilicet 3, multiplica in figuram inferioris articuli, qui est sex et fiunt 18. Aggregatis autem numeris denominantibus differentias, scilicet uno et duobus, articulus enim est secunde differentie et digitus prime, fiunt tres. De quibus subtracto uno remanent duo, a quibus denominatur secunda differentia. Pone ergo in secunda digitum, scilicet 8, et in sequenti, scilicet tertia, 1. Deinde figuram articuli superioris que est 2 multiplicabis in inferiorem digitum, qui est 4, et fiunt 8. Aggregatis ergo numeris denominantibus differentias, scilicet uno et duobus, fiunt tres. A quibus subtracto uno remanent[174] duo, a quibus denominatur secunda. Ideo digitus, scilicet 8, ponatur in secunda. Postea figuram suprapositi articuli multiplicabis in figuram subpositi, scilicet 6, et fiunt 12. Et quoniam uterque est secunde differentie, ideo aggregatis

numeris denominantibus differentias, scilicet duobus et duobus, fiunt 4. A quibus subtracto uno remanent tres, a quibus denominatur tertia differentia. Pone ergo digitum in tertia, scilicet duo, et articulum in sequenti, scilicet quarta, et fiunt hoc modo:

			1	2
		1	8	
			8	
	1	2		

<**26.6**> Que sic posita aggrega et fiunt 1472. Et hec est summa que provenit ex multiplicatione 23 in 64. <**26.7**> In hac eadem regula docetur qualiter etiam compositus in articulum vel limitem vel digitum multiplicetur.

<**27.1**> Cum multiplicaveris digitum in digitum aut proveniet tantum digitus aut tantum denarius aut digitus cum denario semel vel aliquotiens aut denarius multotiens. <**27.2**> Cum multiplicaveris digitum aliquem in aliquem articulorum qui sunt usque ad centum, multiplica figuram in figuram et quot unitates fuerint in

---

[173]provenerit N

[174]remanet P

digito qui provenerit, tot denarii erunt. Quot autem denarii fuerint in articulo qui provenerit, tot centenarii erunt. <**27.3**> Verbi gratia, cum multiplicare volueris septem in 70 multiplica figuram in figuram et proveniunt 49. Novem autem unitates sunt in digito. Tot ergo denarii erunt, qui sunt 90. Quater autem denarius[175] est in articulo, tot ergo centenarii erunt, qui sunt quadringenti. Ergo 490 est summa que ex ductu illorum in se provenit. Similiter fit in omnibus aliis. <**27.4**> Cum multiplicaveris digitum in aliquem centenorum qui sunt usque ad mille, multiplica figuram in figuram et quot unitates fuerint in digito si provenerit, tot centenarii erunt. Quot autem denarii in articulo si provenerit, tot millenarii erunt. <**27.5**> Verbi gratia, si multiplicaveris 3 in 900 multiplica figuram in figuram et fiunt 27. Septem sunt unitates in digito et duo denarii[176] in articulo. Ergo 2,700 est summa que ex ductu illorum in se provenit. Similiter fit in omnibus aliis.

<**28.1**> Cum multiplicaveris unum articulorum in alium de his qui sunt usque ad centum, multiplica figuram in figuram et quot unitates fuerint in digito qui provenerit, tot erunt centenarii. Quot autem denarii in articulo, tot erunt millenarii. <**28.2**> Verbi gratia, si multiplicaveris 30 in 70, multiplica figuram in figuram et fiunt 21. Duo denarii sunt in articulo et unitas semel in digito. Ex ductu igitur priorum proveniunt duo millia et centum.[177] Similiter in omnibus aliis. <**28.3**> Cum multiplicaveris aliquem articulorum qui sunt usque ad centum in aliquem centenorum qui sunt usque ad mille, figuram in figuram multiplica /12rb/ et quot unitates fuerint in digito, tot erunt millenarii. Quot autem denarii in articulo, totiens decem millia. <**28.4**> Verbi gratia, si multiplicaveris 30 in 500, multiplica figuram in figuram et proveniunt 15. Et quia quinque sunt unitates in digito, erunt quinque millia. Denarius autem est semel in articulo. Ex ductu igitur priorum proveniunt quindecim millia hoc modo: 15,000. Similiter fit in omnibus.

<**29.1**> Cum multiplicaveris aliquem centenorum qui sunt usque ad mille in alium ex his multiplica figuram in figuram et quot unitates fuerint in digito qui provenerit, totiens erunt decem millia. Quot autem denarii in articulo, totiens erunt centum millia. <**29.2**> Verbi gratia, cum multiplicaveris 300 in 500, multiplica figuram in figuram et proveniunt 15. Quinque autem unitates sunt in digito et denarius semel in articulo. Ex ductu igitur suprapositorum proveniunt 150,000, que sunt centum quinquaginta millia.

<**30.1**> Cum multiplicaveris aliquem digitorum in aliquem articulum millenorum, ut[178] decies vel vigies millia et huiusmodi, aut iteratorum millium, ut decies millies

---

[175]denerius P

[176]duodenerii PN

[177]cetum P

[178]vel P

millium et quantum iterare mille volueris, multiplica figuram in figuram et digitum si provenerit pone in differentia multiplicantis et articulum in sequenti. $<$**30.2**$>$ Verbi gratia, sex si multiplicaveris in triginta millia, multiplica figuram et figuram et proveniunt 18. Digitus ergo, qui est octo, ponatur in eadem differentia multiplicantis qui[179] est tres et articulus qui est unus in sequenti hoc modo: 180,000, et proveniunt centum octoginta millia. $<$**30.3**$>$ Cum multiplicaveris aliquem articulorum in aliquem sepe repetitorum millium, ut decies vel vigies millies millies millium et quotiens mille iterare volueris, multiplica figuram in figuram et digitum si provenerit pone in differentia secunda a multiplicante, articulum vero in tertia ab ipso. $<$**30.4**$>$ Verbi gratia, cum multiplicaveris 30 in quater millies m. m. millium et quotiens iterare volueris, figuram que est 3 multiplica in figuram que est 4 et proveniunt 12. Digitum ergo qui est duo pone in differentia secunda post 4^{180} et articulum in tertia post 4^{181} hoc modo: 4,000,000,000,000, qui fiunt centies vigies millies .m.m. millium quater: 120,000,000,000,000. $<$**30.5**$>$ Cum multiplicaveris aliquem centenorum in aliquem millenorum sepe iteratorum multiplica figuram in figuram et digitum si provenerit pone in differentia tertia a multiplicante, articulum vero in quarta ab eo. $<$**30.6**$>$ Verbi gratia, cum multiplicaveris ducenta in quinquies millies .m.m.m. quater, multiplica duo in quinque et, quoniam articulus provenit, ponatur in quarta differentia a quinque hoc modo: 1,005,000,000,000,000.

$<$**31.1**$>$ Cum autem volueris scire de qua differentia sint digiti vel articuli \vel centeni/[182] sepe iteratorum millium, vide quotiens iteratur mille et in tres multiplica[183] numerum quo iterantur et productum inde retine. Si autem voluisti scire de differentia digitorum iteratorum millium ut bis vel ter vel quater usque ad novies millies .m. millium, quotiens repetere volueris mille, de qua differentia sint semper adde unum primo producto retento et a numero qui inde excrescit denominatur differentia de qua sunt digiti repetitorum millium. $<$**31.2**$>$ Verbi gratia, si volueris scire de qua differentia sunt ter millies .m. m. millium, numerum quo iteratur mille, sicut hic est quattuor, multiplica in tres et fiunt 12. Quibus adde unum et fiunt 13. Tredecima ergo est differentia predictorum. $<$**31.3**$>$ Si autem voluisti scire de qua differentia sint articuli sepe iteratorum millium, ut decies vel vigies millies millium et quotiens mille repetere volueris, priori producto retento semper adde \duo/[184] et a numero qui inde excrescit denominatur differentia de qua sunt articuli sepe iteratorum mil/12va/lium. $<$**31.4**$>$ Verbi gratia, si volueris scire de qua differentia

---

[179]que P

[180]3 PN

[181]3 PN

[182]P adds 'vel centeni' in the margin; N gives 'vel centum' in the text.

[183]PN reverse the phrases: 'in tres multiplica' and 'vide quotiens iteratur mille et'

[184]P adds 'duo' in the margin

sint suprapositi articuli sepe iteratorum millium, ut quinquies millies .m.m. mille quater, numerum quo iteratur mille, sicut hic est quattuor, multiplica in tres et fiunt 12. Quibus adde duo et fiunt 14. De quartadecima ergo differentia sunt articuli suprapositorum millium iteratorum. <**31.5**> Si autem voluisti scire de qua differentia sint centeni millium iteratorum, priori producto semper adde tres et a numero qui inde excrescit denominatur differentia centenorum iteratorum millium. <**31.6**> Verbi gratia, si volueris scire de qua differentia sint centies vel ducenties et huiusmodi millies mille et quotiens iterare volueris mille, numerum quo hic iteratur mille, scilicet duo, multiplica in 3 et fiunt 6. Quibus adde tres et fiunt novem. Nona est ergo differentia de qua sunt predicti centeni iteratorum millium.

<**32.1**> Cum volueris multiplicare quodlibet millies vel decies vel centies millies millium sepe iteratorum in aliud[185] quodlibet ex illis, reiecta iteratione de millies a multiplicato et multiplicante, ea que de utroque remanent multiplica inter se et productum inde retine. Deinde aggrega numeros iterationis utriusque et summam inde excrescentem pone sub prius producto et quod inde fit est numerus qui ex ductu unius in alterum provenit. <**32.2**> Verbi gratia, cum volueris multiplicare digitos millenorum inter se, ut ter millies mille in septies millies .m. m. mille, pretermissis numeris iterationis utriusque, scilicet duo et quattuor, in multiplicato etenim bis numeratur mille et in multiplicante quater, remanent tantum figure utriusque scilicet 3 et septem. Quarum altera multiplicata in alteram fiunt 21. Deinde aggrega ipsos numeros iterationis utriusque scilicet duo et quattuor, et fiunt sex. Quos pone sub producto prius, scilicet 21, hoc modo: $\begin{array}{|c|} \hline 21 \\ 6 \\ \hline \end{array}$

Et dices quia 20 et una vicibus millies m. m. m. m. mille sexies proveniunt ex ductu unius predictorum in alterum. Per suprapositam igitur regulam si numerum iterationis multiplicaveris in tres, fient 18. Quibus addito uno fiunt 19. Digitus ergo suprapositus qui est un[i]us erit in nonadecima differentia et articuli 20 in vicesima.

<**32.3**> Cum autem centenos iteratorum millium inter se multiplicare volueris, ut quingenta[186] millies millia in tre[s]centa millies .m.m. millium quater, reiecto numero iterationis utriusque qui est duo et quattuor, in multiplicato etenim bis iterabatur mille et in multiplicante quater, remanent de multiplicato 50<0> et de multiplicante 300. Que in se ducta fiunt centum quinquaginta millia. Quos retine. Deinde aggrega numeros iterationis utriusque, scilicet duo et quattuor, et fiunt sex. Quos pone sub prioribus et significabitur summa que ex ductu unius suprapositorum in alterum fit, scilicet centies quinquagies millies .m.m.m.m.m.m. septies, hoc modo: 15<0,000,000,000,000,000,000,000>. Per priorem ergo regulam quinquaginta millia

---

[185]alium N

[186]quinquaginta PN

erunt in vicesima tertia[187] erunt differentia et centum in vicesima quarta. <**32.4**> Similiter etiam fit cum multiplicaveris digitos iteratorum millium in decenos iteratorum millium et centenos et econverso. Similiter etiam fit cum multiplicaveris decenos iteratorum millium inter se, vel in centenos.

<**33.1**> Cum autem habueris aliquam differentiam et volueris scire quis numerus est illa, numerum a quo denominatur differentia divide per 3. Si autem de divisione nichil remanserit, illa differentia centenorum millium totiens vel totiens iteratorum erit. <**33.2**> Si autem volueris scire numerum iterationis, id est quotiens iteratur, ab eo quod de divisione exit, extrahe unum et quod remanserit numerus iterationis erit illorum centenorum millium qui illius differentie sunt. <**33.3**> Verbi gratia, si habueris duodecimam differentiam et volueris scire quis numerus est illa, divide duodecim a quo denominatur dif/12vb/ferentia per tres et exibunt 4. A quibus extracto uno remane<n>t 3 et quoniam de divisione nichil remansit, duodecima differentia centenorum millies millies m. ter iteratorum erit. <**33.4**> Si autem de divisione remanserit duo, illa differentia erit decenorum millium totiens iteratorum quotus fuerit numerus qui de divisione exit. <**33.5**> Verbi gratia, cum habueris undecimam differentiam et volueris scire quis numerus est illa, divide undecim per tres et exibunt 3 integri et remanebunt duo. Undecima ergo differentia est decenorum millium ter iteratorum. <**33.6**> Si autem de divisione remanserit unum, illa differentia erit digitorum millium totiens iteratorum quotus est numerus qui de divisione exit. <**33.7**> Verbi gratia, si habueris decimam differentiam et volueris scire quis numerus est [in] illa, divide decem per tres et exibunt de divisione 3 remanente uno. Decima ergo differentia est millium ter iteratorum.

<**34.1**> Numerum quem quis occultatum[188] in corde suo tenet ipso non significante sic invenies. Primum precipe ut ipsum numerum triplicet. Deinde triplicationis summa<m> in duo dividat. Postea an pares sint partes interroga. Si autem impares fuerint, unum retine et ut maiorem partem triplicet iterum precipe et summam in duo dividat. Que si interrogatus responderit esse imparia, tu duo retine. Que cum priore uno aggrega<ta> tria fiunt. Deinde ut de ipsa parte maiore novem reiciat precipe et alios iterum novem et sic donec non remaneat unde novem reiciat. De unoquoque autem novem tu quattuor accipe. Qui aggregati cum primis sunt numerus quem occultavit. Aut si de parte maiore novem eicere non potuerit, primi tres erunt numerus occultus. Aut si utraque divisio fuerit per paria, tu nichil accipies, set quaternarii unus vel plures sumpti de novenario uno vel pluribus, numerus occultus erunt.[189] Quotiens autem divisio fuerit per imparia, de prima accipe xnxm[190] de se-

---

[187]PN add a tironian et with 'm' suprascript.

[188]occultum N

[189]erit N

[190]*i.e. unum*

cunda dxp.[191] <**34.2**> Verbi gratia, sit binarius numerus quem occultat. Triplicatus autem efficit 6. Senarius vero in duo equa dividitur. Cuius iterum altera<utra> pars triplicata[192] efficit novem, qui dividitur in duo inequalia. Unde quia secunda divisio est, duo retineo. Et quia de parte maiore novem non possunt reici dxp quos retknxk sxnt n;.;s[193] pccxlt;.;s.[194] Et hpc (hoc) f[195] (est) quod proppsxkm;.;s.[196]

<**34.3**> Item 1l3c53 4cc5lt1nt3[197] quot solidos habeant, dic ut de tuis <nummis> tot singulos vel tot binos vel ternos et huiusmodi quot volueris accipiat. Et de omnibus tuis nummis unum aliquid ut pote gallinam unam vel aliquid huius modi[198] emat. Deinde de omnibus suis solidis secundum idem pretium quot potuerit gallinas emat. Tu ergo tunc divide solidum in numerum quem sibi dedisti, videlicet in unum vel duo vel tres. Et exeunti de divisione numero adde unum et quod exit addito uno est numerus eorum que[199] emuntur. <**34.4**> Verbi gratia, ponamus quod occultet quinque solidos. Acceptis de meis totidem nummis quot sunt solidi, scilicet 5, gallina una ematur[200] et secundum idem pretium de suis solidis duodecim galine emuntur. Divide ergo solidum in 12 nummos in numerum quem sibi dedisti, scilicet unum, exibunt 12 qui est numerus galinarum emptarum. Si vero binos dedisses galline sex essent, aut si ternos, essent quattuor et sic de ceteris.

<**34.5**> Si aliqua duo equalia occultantur et de uno eorum duo accepta alteri addantur et de augmentato equale residuo addatur, necessario quattuor remanebunt, aut si tres necessario sex et ita semper remanet duplum eius quod primo accipitur. <**34.6**> Verbi gratia, si occultantur 5 solidi in una manu et quinque in alia, duos acceptos de una appone aliis quinque et fiunt 7 in una, remanentibus tribus in alia. Cum ergo de septem acceperis equale residuo, id est tribus, necessario quattuor remanent.

F

<**35.1**> Queritur cur non omnes[201] vel plurimos numeros propriis nominibus designamus /13ra/, vel cur non semper per adiectionem novorum set post decem per

---

[191] *i.e. duo*

[192] triplica P, triplicata N

[193] n;.;s has an abbreviation mark above it in PN, indicating that it is short for 'numerus'

[194] pccxt;.;s P, pccxle;.;s N

[195] 'f' has a line over it in P; N reads it as 's'

[196] *i.e. duo quos retinui sunt numerus occultus. Et hoc est quod proposuimus.*

[197] *i.e. alicui occultanti*

[198] alquid hoc modo PN

[199] quem P, qui N

[200] emat P

[201] omnis PN

repetitionem priorum[202] semper numeramus. <**35.2**> Ad quod dicitur quia non fuit possibile ut omnes numeri propria nomina haberent, idcirco quod numerorum in infinitum crescit multitudo, nominum autem in qualibet lingua infinita non potest esse inventio. <**35.3**> Cum enim in omni lingua certa et terminata sint[203] instrumenta et eorum definite naturaliter modulationes, quibus vox articulata formatur et unde[204] litterarum figure apud omnes gentes et earum varie set diffinite sunt secundum ordinem preponendi et postponendi ad representanda rerum omnium nomina compositiones, necessario omnes numeri cum[205] cum sint infiniti, nomina non potuerunt nec debuerunt habere singuli, precipue cum et homines in omni pene re numeris utentes nimis impedirentur, si in numerationibus suis infinitam numeralium nominum multitudinem in promptu semper habere numerandi necessitate cogerentur. <**35.4**> Idcirco[206] necesse fuit infinitam numerorum progressionem certis limitibus terminare, paucis nominibus illos[207] designare, ne cogeretur homo in numerando per novas additiones tam numerorum quam nominum semper procedere, set per repetitionem priorum brevem quantamlibet summam paucis nominibus possit comprehendere.[208] <**35.5**> Unde cum omnes numeros habere nomina fuerit[209] impossibile et aliquos necesse, ratio exegit natura predicante ut ex omnibus numeris soli 12 nomina haberent: tres limites, videlicet denarius centenarius et millenarius et novem primi numeri ab uno usque ad novem infra decem constituti.

<**36.1**> Quam rationem novenarius pre aliis omnibus numeris proprio privilegio merito vindicavit, ut pote continens in se omnes pene species numerorum et numeralium proportionum. <**36.2**> In ternario etenim quamvis deo dicato predicta ratio consistere non debuit, quia sibi deerat primus perfectus, qui est senarius; set nec propter hoc in senario, quia deerat ei primus cubus qui est octonarius; set nec ideo in octonario, quia deerat ei prima vera superficies, que est in novenario. <**36.3**> Ex hac ergo plenitudine virtutum novenarius promeruit ut in se ratio numerandi et numeros appellandi consisteret, ultra quam nisi tres tantum limites, nullus numerus proprium nomen haberet. <**36.4**> Nimirum cum ad instar novenarii tam celestia quam terrestria, tam corpora quam species formata et ordinata esse videantur; novem enim sunt spere celestium corporum, novem etiam sunt ordines celestium spirituum, novem etiam complexiones omnium corporum. <**36.5**> Novem igitur

---

[202]primorum P

[203]*Liber mahameleth* adds 'loquendi'

[204]*Liber mahameleth* adds 'et'

[205]omnes numeri cum] idcirco cum numeri *Liber mahameleth*

[206]unde *Liber mahameleth*

[207]*Liber mahameleth* omits 'illos'

[208]*Liber mahameleth* omits 'set...comprehendere'

[209]Unde cum...fuerit] quoniam et...fuit *Liber mahameleth*

debuerunt esse compositiones numerorum in quibus solis tota consisteret infinitas numerorum, sicut ex complexionibus novem universitas corporum. <**36.6**> Sicut enim in complexionibus una est equalis et altera inequalis, una vero tantum temperata, sic et in numeris unus est par, alius impar et inter omnes sola est unitas ex nulla parte sibi dissimilis, semper eadem, semper equalis.

<**37.1**>Sic creature a similitudine sui creatoris qualicumque modo non recederent[210] dum intra illum numerum se continerent,[211] quia primo impari in se multiplicato generatur qui post unitatem deo solus consecratus est, quia numero deus impare gaudet. <**37.2**> Unde et soli tres limites preter novenarium inter alios nomina sortiti sunt, ut per hoc videlicet trinitatis que[212] verus limes est omnium, A et Ω principium et finis, qualemcumque similitudinem teneant et a radice novenarii numquam recedant. <**37.3**> Idcirco igitur ratio postulavit ut, quia universitas rerum intra novenarium continetur, similiter et numerorum infinitas intra novenarium coartaretur novem et nominibus designaretur et novem figuris representaretur. <**37.4**> Omne enim exemplum similitudinem sui retinet exemplaris; alioquin non esset alterum alterius exemplar vel exemplum, et quia ut predictum est pene omnia condita sunt ad instar novenarii, ipsa quoque numerorum infinitas rationabiliter debuit sub novenario coartari,[213] ut numerus etiam ab ea forma non discederet ad quam creator cuncta componeret et quam a numero rerum universitas mutuaret.

<**38.1**> Unde et homines primevam naturam imitantes non nisi solis novem numeris nomina imposuerunt et ad omnes representandos[214] non nisi novem figuras adinvenerunt. <**38.2**> Set quia quedam species numeri adhuc /13rb/ deerat quam novenarius intra se non continebat, scilicet numerus superfluus, qui primus est duodenarius, ideo post novenarium tribus tantum limitibus nomina sunt imposita,[215] ut novenarius cum radice sua, scilicet ternario, omnes dignitates et proprietates numeri intra se contineret et nichil proprietatis nichil misterii in numeris possit inveniri quod in toto novenario cum radice sua non videtur contineri. <**38.3**> Cum igitur non omnes nec plurimi set pauci numeri propriis nominibus necessario fuerant designandi,[216] propter predictas causas novem tantum numeris et tribus limitibus nomina sunt indita, ut per commoditatem paucitatis humanis usibus melius deservirent et rerum occulta misteria quibuscumque signis exprimerent et a nature rationibus non discederent. <**38.4**> De his <h>actenus.

---

[210]recederet PN

[211]continent PN

[212]qui PN

[213]coaritari P

[214]representandas P

[215]impositam P

[216]P adds 'et'

<**39.1**> Unitas est origo et <prima> pars numeri; omnis enim numerus naturaliter ex unitatibus constat et ipsa omnem numerum natura precedit quoniam simplex est. <**39.2**> Et quia simplex est, ideo per multiplicationem sui nichil nisi id per quod multiplicatur generare potest, quod non fit in aliis qui simplices non sunt; ex cuiuslibet enim numeri multiplicatione in se vel in alium necesse est alium provenire diversum. <**39.3**> Unitas autem per se multiplicata non generat nisi se; semel enim unum unum est; per quemcumque enim numerum multiplicaveris non nisi ipsum per quem multiplicas efficis, et quia nullus ex ea generatur nisi ille in quem prius ipsa multiplicatur, idcirco in principio cum nichil esset cui ipsa adiungi posset ad generationem primi numeri, necesse fuit ipsam in se congeminari et a se quodam[217] modo alterari, ut ex se ipsa et ex se altera quasi ex diversis posset aliquid generari. <**39.4**> Et hec est prima numeri generatio que apparet in binario; unde et principium alteritatis dicitur, quoniam ex unitate alterata genitus est. Ideo etiam sibi soli et nulli alii contingit quod ex sui in se multiplicatione itidem quod ex aggregatione provenit; non enim constat ex numero. <**39.5**> Et quoniam preter binarium adhuc non erat nisi unitas, ideo ipsa[218] binario tamquam vir femine iungitur, ex quorum copula ternarius nascitur, qui post unitatem primus impar et masculus vocatur.

<**40.1**> Numerus etenim par femina dicitur quasi mollis eo quod facile solvitur, set masculus impar quasi fortis indivisibilis. <**40.2**> Unitas autem nec par nec impar est actu; unde unitas in se nec femina nec masculus est actu, set potestate utrumque. <**40.3**> Unde quando cum femina iungitur inde masculus, scilicet impar, generatur; quando vero cum masculo coit, feminam quia parem gignit. <**40.4**> Unde ex prima generatione unitatis non nisi femina nascitur, scilicet par, quia binarius. <**40.5**> Decebat enim ut unitas in procreatione prime sobolis non nisi vice viri, scilicet dignioris uteretur et ex ea quasi viro femina nasceretur; prima etenim femina ex viro non primus vir ex femina. <**40.6**> Unde in secundo gradu quoniam unitas femine, scilicet binario, iungitur, ternarius qui est masculus generatur; in tertio vero gradu unitas coniungitur masculo et femina procedit, scilicet quaternarius. Similiter in ceteris usque in infinitum. <**40.7**> Unde unitas nec debuit esse par nec impar, quia si par tantum esset, quando paribus iungeretur, sicut ex coniunctione duarum feminarum, nichil procrearetur; si vero impar tantum esset imparibus iuncta, tamquam masculus cum masculo nichil procrearet. <**40.8**> Unde necesse fuit ut neutrum esset actu, set potestate utrumque, ut cum secundum utriusque sexus potestatem[219] omnibus nascentibus vicissim iungeretur, fecunda numerorum soboles in infinitum propagaretur.

<**41.1**> Set quia numerorum prima et naturalis generatio secundum predictum

---

[217]quoddam P

[218]ipsa in P, ipsam N

[219]potestate P

modum videbatur sine fine multiplicari, placuit postmodum diligentie quorumdam hominum eam ad instar humane generationis quibusdam certis gradibus et limitibus terminari. <**41.2**> Hominum etenim sicut et numerorum generatio ab uno secundum sexum geminato per masculum et feminam descendens in infinitum progreditur. <**41.3**> Set humana cura postmodum gradus et limites adinvenit, quibus cognationes inter homines designavit, ut licet ab uno se omnes eque descendisse cognoscerent, /13va/ tamen propter assignatos gradus alii ad alios potius pertinere cognationis gratia non dubitarent et de uno genere esse dicerentur quicumque sub eisdem cognationis gradibus invenirentur. <**41.4**> Similiter et in numeris post naturalem eorum compositionem et essentiam, humana industria radices, nodos et limites, sicut in hominibus truncos et gradus adinvenit et numerorum generationes per novenarios distinxit, ut numeri qui ex eodem limite nascerentur, usque ad nonum gradum, omnes uno cognationis nomine communi ad aliorum differentiam vocarentur; qui autem aliquem novenarium excederent, ad aliam omnino cognationem pertinere se se cognoscerent. <**41.5**> Unde ad distinguendas huiusmodi numerorum cognationes humana adinventio quosdam appellavit digitos, quosdam articulos, quosdam vero compositos, illos autem ex quibus omnes isti nascuntur vocavit limites, quasi singularum generationum primos parentes. <**41.6**> Illos enim quos in prima creatione per aggregationem sui unitas genuerat, usque ad novem, digitos, quia ab unitate primogenitos vocari instituit, ut[220] hic primus novenarius digitorum sive unitatum novenarius diceretur, cuius novenarii primi unitas limes et primus esset, ut pote quos primum ex se unitas genuisset. <**41.7**> Post hunc autem sequitur secundus novenarius qui est decenorum sive articulorum, et huius novenarii sicut et primi limes unitas est, set decupla primi. <**41.8**> Post hunc vero novenarium decenorum sequitur tertius novenarius centenorum, cuius quoque limes unitas est, set decupla secundi. <**41.9**> Post hunc autem tertium sequitur quartus novenarius millenorum, cuius quoque limes unitas est, set decupla tertii. Et sic usque ad infinitum.

<**42.1**> Et quia omnes numeri ab unitate sunt geniti, merito ipsa etiam constituta[221] est limes omnium novenariorum pro varietate positionum, videlicet ut que ex se species omnium genuerat numerorum, eadem etiam limes esset limitum pro diversitate locorum. <**42.2**> Unde in principio omnium generationum prima et limes ponitur, ut ex hoc cunctorum mater esse comprobetur. <**42.3**> Unde fit ut unitas sicut in prima creatione natura primus limes per aggregationem sui cum ipsis genuerat digitos, sic etiam in secunda institutione placuit ut ipsa eadem omnis limes aggregata primis generet compositos, multiplicata per primos procreet articulos. <**42.4**> Digiti ergo sunt dicti numeri qui ab unitate usque ad novem naturaliter sunt geniti; articuli vero qui per multiplicationem primorum a ceteris limitibus gen-

---

[220]et PN

[221]etiam constituta] inconstituta P

erantur; compositi vero numeri dicuntur qui ex digitis et limitibus sive articulis simul iunctis nascuntur, dicti compositi tamquam ex diversis generibus procreati. <**42.5**> Unde et a quibus substantiam sortiuntur, eorum etiam[222] proprietatem secuntur. <**42.6**> Cum enim dicitur 12 vel 23 vel centum viginti, ex digito et limite vel et[223] ex articulo compositi sunt. <**42.7**> Set quod est in eis de limite vel articulo in vi limitis vel articuli sumitur, scilicet pro decem vel pro 20 vel pro c; quod autem de novenario digitorum est, pro tot unitatibus sumitur quot in ipso contineri videntur. <**42.8**> Omnes itaque novenarii ad instar prioris ordinati sunt, unde singuli habent unitates limites, habent binarios suos, habent ternarios suos et sic usque ad novem consequenter singulos, sicut subiecta dispositio declarat.

Differentia centies millies millenorum	Differentia decies millies milenorum	Differentia millies milenorum	Differentia centies milenorum	Differentia decies millenorum	Differentia millenorum	Differentia centenorum	Differentia decenorum	Differentia unitatum sive digitorum
1	1	1	1	1	1	1	1	1
2	2	2	2	2	2	2	2	2
3	3	3	3	3	3	3	3	3
4	4	4	4	4	4	4	4	4
5	5	5	5	5	5	5	5	5
6	6	6	6	6	6	6	6	6
7	7	7	7	7	7	7	7	7
8	8	8	8	8	8	8	8	8
9	9	9	9	9	9	9	9	9

<**42.9**> /13vb/ Sicut enim in primo limite bis unum[224] faciebat binarium unitatum, ita in secundo limite bis decem efficit binarium decenorum qui est 20 et in tertio limite bis centum binarium centenorum qui est ducenti; et sic in singulis per singulos, usque ad novem. <**42.10**> Et quia ex numeris nichil nisi per aggregationem aut multiplicationem primi novenarii nascitur, idcirco in omnibus in se iteratur et omnibus prior esse comprobatur, quia ante omnes genitus naturalem institutionem adhuc servare videtur. <**42.11**> Unde etiam ipsa unitas, que mater est omnium, in quocumque limite fuerit sive per aggregationem sive per multiplicationem iuxta numerum primogenitorum non nisi novem tantum numeros gignit.

<**43.1**> Set quia post novem naturali ordine decem sequitur et ipsa semper post novem nisi in primo limite humana institutione posita invenitur, idcirco necesse est ut per unitatem post primum novenarium positam decem significentur et sic ipsa ex natura loci in denarium genita secundus limes fiat decenorum, sicut prius simpliciter limes fuerat unitatum, ut eadem esset mater articulorum sive compositorum quam constabat matrem etiam fuisse digitorum. <**43.2**> Et quia post novem semper decem naturaliter sequitur in quo loco semper unitas ponitur, ideo post novenarium decenorum sequitur iterum unitas, tertius limes qui est centenorum;[225] et sic sem-

---

[222]N adds 'et'

[223]N omits

[224]unus N

[225]decenorum PN

per post quemlibet novenarium unitas sequitur limes sequentium. <**43.3**> Quoniam autem omnis limes excepto primo post precedentem novenarium sequitur, ideo ipse factus denarius precedentis limitis semper decuplus invenitur, quia ipse post quemcumque novenarium fuerit ex decuplatione precedentis limitis nascitur. <**43.4**> Et quia omnes articuli ex multiplicatione sui limitis per primos nascuntur, necesse est, ne de genere videantur, ut suorum limitum regulam sequantur, videlicet ut sicut limites decupli sunt precedentium limitum, ita et qui²²⁶ ex eorum multiplicatione numeri nascuntur, precedentium numerorum decupli similiter inveniantur. <**43.5**> Sicut enim secundus limes decuplus est primi, ita et articuli decenorum decupli sunt digitorum; et sicut limes tertius decuplus est²²⁷ secundi, ita et articuli centenorum decupli sunt ad articulos decenorum. <**43.6**> Sic semper sequentes limites, articuli, compositi qui interiacent, decupli sunt precedentium limitum, articulorum, compositorum singuli singulorum.

<**44.1**> Omnes itaque limites et articuli et compositi sicut et digiti sub novenario sunt constituti, ita ut primus novenarius sit digitorum, secundus articulorum, tertius compositorum et sic ceteri huiusmodi. <**44.2**> Sic ergo placuit ut omnis numerus in novenarium quasi in ultimum gradum sui generis terminaretur et post novenarium unitas precedentis limitis decupla, quia post novem decima, omnium novenariorum limes constitueretur. <**44.3**> Et sic per generationes suas a limitibus tamquam a progenitoribus suis descendens numerorum fecunda progenies tota per novenos gradus distincta in infinitum extenditur. <**44.4**> Sic novenarius principatum tenet in omnibus infinita restringens, restricta distinguens, qui tamen a limite incipit et limite terminatur, ut non ipse auctor rerum, set in animo auctoris rerum exemplar fuisse ostendatur; unde²²⁸ ipse a ternario in se multiplicato generatur. <**44.5**> Qui enim cuncta condidit ipsum quoque fecit ad cuius exemplar cetera formavit; omnia enim deus fecit in numero, pondere et mensura. <**44.6**> Unde et ipsum numerum si factus est, ad numerum fecit, ut numerus leges numeri non excederet, ad cuius formam cetera componi deberent. <**44.7**> Set numerus ad quem numerus creatus est, sic quidem increatus est.

G

<**45.1**> Multiplicationis sunt octo species et totidem divisionis. Aut enim multiplicamus integros per integros, aut fractiones per fractiones, aut fractiones per integros, aut integros per fractiones, aut fractiones et integros per integros, aut fractiones et integros per fractiones, aut integros per integros et fractiones, aut fractiones per integros et fracti/14ra/ones. <**45.2**> Quotiens per vel impar numerus parem vel par imparem multiplicat, par provenit. Si vero impar, imparem, impar exit.

---

²²⁶quia P

²²⁷P omits

²²⁸N adds 'et'

<**46.1**> Divide minuta per minuta vel secunda per secunda vel tertia per tertia vel quarta per quarta vel quinta per quinta vel sexta per sexta; quicquid provenerit erunt gradus, quandoque[229] unumquodque istorum numerorum multiplicatum <est> in gradus quicquid provenerit erit de genere eiusdem fractionis. <**46.2**> Et si minuta diviserint secunda vel secunda tertia vel quarta quinta vel quinta sexta vel tertia sexta, quicquid exierit de divisionibus erunt denominata a fractionibus maioribus.[230]

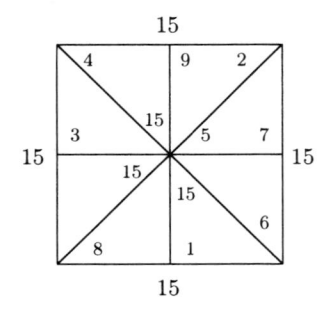

---

[229]quorumque] quare quia (?) N

[230]P omits paragraph 46

The Toledan *Regule*. Translation

A

Here begin the rules and first, concerning addition

<1.1> When the numbers are placed in their natural order, every number is half the sum of the two numbers surrounding it on either side,[1] as far as 1. <1.2> E.g., 5 is half the sum of 4 and 6; similarly, <it is half> the sum of 3 and 7; similarly, of 2 and 8; similarly, of 1 and 9. The same applies in all cases.

<2.1> Likewise, every number multiplied by itself produces the same total as results from the multiplication of the surrounding numbers by each other plus the multiplication by each other of the differences between the surrounding numbers and the middle number. <2.2> For so much results from 5 times 5 as from 4 times 6 plus the multiplication of the differences between them and 5, i.e., 1 in each case. When these are multiplied by each other, they only make 1. Also as much as what results from 3 times 7 or 2 times 8 or 1 times 9, always plus the multiplication of the differences between the end points and the middle. <2.3> Likewise, every number up to 10,[2] when multiplied by itself, produces as much as the first two surrounding numbers multiplied by each other when 1 is taken from the multiplication of the middle number.

<3.1> If you wish to know how great a sum results from adding the numbers following each other naturally from 1, multiply that number at which you ended, if it is even, by its half and add the half itself and this will be the sum which results from them. <3.2> E.g., put in order 1, 2, 3, 4, 5, 6, 7, 8. Multiply that number at which you ended, i.e., 8, by its half, i.e., by 4, and you will get 32. Add to this 4 and this will be the sum which resulted from their addition, i.e., 36. <3.3> Or multiply the following odd number by half the same even number and you will have the sum. <3.4> But if you end at an odd number, e.g., 1, 2, 3, 4, 5, 6, 7, multiply that number (i.e., 7) by its larger part (i.e., 4; for 7 consists of 3 and 4) and 4 times 7 makes 28, which is the sum of the above-mentioned numbers. <3.5> Or multiply the same odd number by half the following even number. <3.6> But if you do not start from 1, but count continuously, you should proceed as if you had started from 1 and you should add all the numbers which are below that from which you started; that sum should be subtracted from the product; the remainder will be the total.

<4.1> If you wish to know what sum will result from even numbers alone added together, when they follow each other in a natural order from 2, consider how many numbers away from the first even number is the number at which you have ended

---

[1]I.e., with the same differences from the given number.

[2]'up to 10' is not necessary.

and multiply the following number by what this is denoted by and you will get the sum. <**4.2**> E.g., 2, 4, 6, 8, 10. The last is 10 and it is fifth from the first even number (i.e., 2). 6 follows 5. 6 then should be multiplied by 5 (by which 10 is denoted) and 5 times 6 (or the reverse) will result in 30 and this will be the sum of the above-mentioned numbers. <**4.3**> Or multiply half the even number at which you ended by half the next following even number and you will have the sum. <**4.4**> But if you count continuously, missing out the odd numbers, not starting from 2, you should proceed as if you had started from 2 and you should add all the even numbers which are below that from which you started; that sum should be subtracted from the product and the remainder will be the total.

<**5.1**> If you wish to know what sum will result from odd numbers added together, when they follow each other in a natural order from 1, multiply by itself the number by which is denoted the last number at which you ended and you will get the sum. <**5.2**> E.g., 1, 3, 5, 7, 9. The last is 9 and it is the fifth in order. The 5, therefore, by which it is denoted, should be multiplied by itself and you will get 25. For 5 times 5 makes 25 and this was the sum of the above-mentioned numbers. <**5.3**> Or from the closest even number which is after the odd number at which you ended, subtract <its> half; multiply the subtracted part by itself and you will have the sum. <**5.4**> If, however, you do not start from 1, you should proceed as if you had started from 1 and you should add all the odd numbers which are below that from which you started; that sum should be subtracted from the result; the remainder will be the total.

<**6.1**> If you wish to know what sum will result from any number of doubles[3] added together, when they follow each other in a natural order from the first, multiply the last number at which you wish to end by the first even number, i.e., 2, and the total that results from their multiplication, when the first even number is subtracted, is gathered from adding the above-mentioned numbers. <**6.2**> E.g., 2, 4, 8, 16, 32. Multiply the last number, which is 32, by 2 and you will get 64. Subtract <the number> from which you started and what will result is the sum of the above-mentioned numbers, i.e., 62. <**6.3**> Or in another way: the last of however many doubles following each other naturally from the first double, when doubled, is the sum of all of them added together, if the first even number is subtracted.[4] <**6.4**> Or in another way, more generally: if you wish to know what sum will result from any doubles added together, double the last number at which you ended and subtract the first number from which you started and what results from the doubling of the last with the subtraction of the first is the sum that you require. <**6.5**> Or if you wish

---

[3]'doubles' here is a technical word which represents 2^n.

[4]It would make more sense if this were 'added', so that this procedure balances the previous procedure, rather than reproducing it.

to start from 1, you should add it to the double of the last <with the 2 subtracted>
and you will have the sum of the preceding numbers.

<**6.6**> If you wish to know what sum will result from any triples[5] added together,
divide the last number at which you ended and triple its smaller part and what
results from its tripling is the sum of the preceding numbers.[6] <**6.7**> E.g., 3, 9,
27, 81, 243. The smaller half[7] of the last number, when tripled, makes 363. The
preceding triples added together make this total.

<**7.1**> If you wish to know what sum will result from adding all the squares of
any numbers when they follow each other naturally from 1, consider first what
sum results from adding the numbers themselves. Then multiply two thirds of the
number, however many are following each other naturally by the above-mentioned
sum, with the addition of one third of the sum and what results will be the sum of
the squares of the numbers themselves. <**7.2**> E.g., let the <number of> numbers
following each other naturally be four, i.e., 1, 2, 3, 4, which, when added together,
make 10. <The number of> their squares are four, i.e., 1, 4, 9 and 16. If you wish
to know what sum will result from adding these squares, multiply two thirds of 4,
however many that is, <namely> 3 minus a third, by the above-mentioned sum of
numbers, which was 10 and with the addition of one third <of ten> this becomes
30. And this total results if you should add the above-mentioned squares, i.e., 1, 4,
9, 16.

<**8.1**> In every case of numbers exceeding each other by an equal difference, if they
consist of an odd number <of numbers>, the middle number when doubled produces
as much as the extremes (and the extremes of the extremes) added together. E.g.,
2, 4, 6, 8, 10, 12, 14. <**8.2**> In every case of numbers distant from each other
by an equal ratio, if they are an odd number <of numbers>, the middle number
when multiplied by itself produces as much as the two surrounding numbers (and
the numbers surrounding the surrounding numbers) multiplied by each other, as far
as 1.[8] E.g., 2, 4, 8, 16, 32. <**8.3**> But if they are an even number <of numbers>,
the two middle numbers, multiplied by each other, <produce> as much as the two
extremes (and the extremes of the extremes) multiplied by each other. E.g., 2, 4,
8, 16, 32, 64, 128, 256. <**8.4**> But if they are an even[9] number <of numbers>, the
two middle numbers, added to each other, produce as much as the two extremes
(and the extremes of the extremes as far as 1).[10]

---

[5]'triples' here is a technical word which represents 3^n.

[6]The text argues the case when the triples follow each other naturally from the first triple.

[7]The 'smaller half' is the same as 'the smaller part'.

[8]'As far as the first term' would be expected.

[9]The text has 'odd'.

[10]The proper place of <8.4> should be immediately after <8.1>, because what is discussed in <8.4>

On the multiplication of digits by each other

<**9.1**> Every number below 10, when multiplied by itself, produces the total of its denomination times 10, when the multiplication of its difference from 10 by itself is subtracted. <**9.2**> E.g., 6 times 6 should be said to be 60, which is the denomination from 6 multiplied by 10. The difference between 6 and 10 is 4, which, when multiplied by 6, makes 24. When this 24 is subtracted from 60, 36 remains, which is the total which 6 times 6 produces. And this applies for all numbers from one to ten.[11] <**9.3**> But if you wish to multiply a larger number by a smaller one or vice versa, multiply the difference between the larger number and ten by the smaller number and subtract <the result of> this multiplication from the denomination made from the smaller number and what remains is the total which results from the multiplication of different numbers. <**9.4**> E.g., when you multiply 5 by 7, say that it becomes 50, which is the denomination from 5 which was the smaller number there. But the difference between the larger number, i.e., 7, and 10 is 3. When 3 is multiplied by the smaller number, i.e., 5, 15 results. When this is subtracted from 50, 35 remains and this is the total produced from 5 times 7 or vice versa. <**9.5**> Or in another way: let the denomination be made from the larger number and the difference between the smaller number and 10 be multiplied by the larger and <the result> subtracted from the total of the first denomination <is the multiplication of different numbers>. <**9.6**> E.g., 5 times 7 should be said to become 70, which is denominated from 7 and this is the larger number in that multiplication. But the difference between the smaller number, i.e., 5, and 10 is 5 which, when multiplied by the larger, i.e., 7, makes 35. When this is subtracted from 70, which was the denomination from the larger number, i.e., 7, multiplied by 10, there remains 35 and this is the total produced by the multiplication of the one number by the other, <i.e.,> 5 times 7 or vice versa: 7 times 5.

<**10.1**> In every case of three numbers in the same ratio, if you multiply the first by the third, as much results from their multiplication as from the multiplication of the middle number by itself. If the first and middle number are known, but the third alone is unknown, multiply the middle number by itself and divide what results by the first number and what results from the division will be the third. <**10.2**> Or if the first number alone is unknown, multiply the middle number by itself and divide by the third and the first will result. <**10.3**> Or if the middle number alone is unknown, multiply the first by the third and the root of what results is the middle

---

and <8.1> is about the arithmetical progression, while <8.2> and <8.3> discuss the geometrical progression. <8.1> discusses the case when the number of the terms is odd, while <8.4> discusses the case when the number of the terms is even (assuming our emendation of the manuscripts' 'odd' to 'even' is correct).

[11] 10 can only be retained instead of 9, if 0 is considered as a number. A reference mark of three dots in a triangular shape appears here in P.

number, since the middle number multiplied by itself produces as much as the two extremes, which you can observe easily in the above-mentioned numbers.

<11.1> If, then, any four numbers are in proportion, i.e., so that the ratio of the first to the second is the same as that of the third to the fourth, then as much will result from the multiplication of the first by the fourth as from the multiplication of the second by the third. For in these four terms the first and fourth are companions, as are the second and third. Hence, generally: in all <numbers>, whichever is unknown, divide either of the other two by the companion of the unknown and multiply what results by the companion of the dividend[12] and the unknown term will result. Or when any of them are unknown, the product from the other two should be divided by the companion of the unknown and the unknown will result. <11.2> Hence, if three numbers are known and only the fourth is unknown, multiply the second by the third and divide what results from this by the first and the result will be the fourth. <11.3> Or if the first alone is unknown, multiply the second by the third and divide <the result> by the fourth and the first will result. <11.4> Or if the second alone is unknown, multiply the first by the fourth and divide <the result> by the third and the second will result. <11.5> Or if the third is unknown, multiply the first by the fourth and divide <the result> by the second and the third will result.

<11.6> E.g., if 10 measures are sold for 30 crowns, then 6 crowns are owed for 2 measures. Here four numbers are in proportion, i.e., (1st) 10 measures, (2nd) 30 crowns, (3rd) 2 measures, (4th) 6 crowns. For the ratio of 10 measures to 30 crowns, which is their price, is the same as the ratio of 2 measures to 6 crowns, which is their price. When, then, you multiply the first number, which is 10 measures, by the fourth number, which is 6 crowns, the result is 60. Similarly so much will result from the multiplication of the second number, which is 30 crowns, by the third number which is 2 measures.

<11.7> When, therefore, anyone, concealing from you the fourth number, which is 6 crowns, should say: "When 10 measures are sold for 30 crowns, how much is owed for 2 measures?" Multiply, then, 30 crowns, which is the second number, by 2 measures, which is the third number and divide what results by 10 measures, which is the first number and the fourth will result, i.e., 6 crowns, which is owed for 2 measures. <11.8> Similarly, if, when concealing the first, which is 10 measures, he should say: "2 measures are sold for 6 crowns, how many measures will be had for 30 crowns?" Multiply, then, the second number, which is 30 crowns, by the 2 measures, which is the third number and divide what results from this by 6 <crowns>, which is the fourth number, and the first will result, which is, namely, 10 <measures> — the

---

[12]MS has 'divisor'.

amount given for 30 <crowns>. <11.9> Similarly, if, when concealing the second number, which is 30 crowns, he should say: "Since I paid 6 crowns for 2 measures, how many crowns should I pay for 10 measures?" Multiply, then, the first number, which is 10 measures, by the fourth number, which is 6 crowns, and divide what results by 2 measures, which is the third number, and the second number will result, i.e., 30 crowns, which is owed for 10 measures. <11.10> Similarly, if, when concealing the third <number>, which is 2 measures, he should say: "Since 10 measures are given for 30 crowns, how many measures will they give for 6 crowns?" Multiply, then, the first number, which is 10 measures, by the fourth number, which is 6 crowns and divide <the result> by the second, which is 30 crowns, and the third number will result, which is 2 measures.

<11.11> In these questions one should note very carefully what is called 'first' and what is called 'second', i.e., whether the thing[13] or the price is named first. Whatever is named first is repeated in the third place and whatever is named second is <placed> fourth, whether it is revealed or concealed.

<12.1> If you do not know the third or the fourth number, but only their sum is given to you, when you wish to find them, add the first and the second and let their addition be the second number for you, in respect to the first known number. Then multiply the first number by the first sum and divide what results from this by the second sum and what results from the division will be the third number. <12.2> Similarly, for finding the fourth, multiply the second number by the first sum and divide what results by the second sum and what results will be the fourth number. <12.3> Similarly also vice versa: if you do not know the first and the second, knowing their sum, you will find them through the third and fourth according to the above-mentioned rule. <12.4> E.g., let there be four numbers, the first 2, the second 4, the third 3, the fourth 6. If, then, the third and fourth are unknown (i.e., 3 and 6), but their sum is shown to you (which is 9), when you wish to find them, add the first and second (i.e., 2 and 4) and you will get 6. Then multiply the first (which is 2) by the first sum (which is 9) and 18 will result. Divide this by the second sum (which is 6) and the third number will result (which is 3). A similar procedure can be used for others, as the figure below makes clear.

6 3	4 2
9	6

<12.5> But if the third and fourth are unknown, but what remains from the subtraction of the smaller of them from the larger is known to you, if you wish to find them, subtract the first from the second or vice versa (always the smaller from the larger) and call what remains the 'second subtraction', in respect to the first that was given. Then multiply the first number by the first subtraction and divide what results from this by the second subtraction and what results from this will be the

---

[13]The 'thing' (*res*) here is the 'measure' of whatever item is being sold.

third number. <**12.6**> Similarly, to find the fourth number, multiply the second number by the first subtraction and divide what results by the second subtraction and what results will be the fourth number. <**12.7**> Similarly also vice versa: if you do not known the first and second, when their subtraction has been shown to you, you will find them through the third and fourth according to the above-mentioned rule, as is clear in the same numerical example. <**12.8**> For if you do not know the third and fourth (i.e., 3 and 6) but what remains from the subtraction of the one from the other is shown to you (which is 3), subtract the first from the second or vice versa (always the smaller from the larger) and 2 remains, which is the second subtraction. Then multiply the first number by the first subtraction and divide what results by the second subtraction and what results from this will be the third number. A similar procedure can be used for others, as the figure below makes clear.

6 3	4 2
3	2

<**12.9**> Similarly also you will find this same thing if you adapt this to things and their prices.

<**12.10**> But if the third and fourth alone are unknown to you, but their multiplication only is given to you, you multiply the first by that multiplication and divide what results from that by the second <number> and the root of what results from the division will be the third number. <**12.11**> Similarly, to find the fourth, multiply the second by that multiplication and divide what results from that by the first number and the root of what results from the division will be the fourth number. <**12.12**> E.g., if you do not know the third and fourth of the above-mentioned numbers (i.e., 3 and 6), but their multiplication alone is given to you (which is 18), then multiply the first (which is 2) by that multiplication and 36 will result. Divide this 36 by the second number (which is 4) and 9 will result from the division, whose root (i.e., 3) is the third number. <**12.13**> Similarly, to find the fourth, multiply the second (i.e., 4) by that multiplication and 72 will result. Divide this by the first number and 36 will result, whose root is 6 and this is the fourth number, as the figure below makes clear.

6 3	4 2
18	8

<**12.14**> But if the two middle numbers are unknown, but their sum alone is given to you, you multiply the first by the fourth and put aside what results from this. Then divide the sum into two parts such that the one multiplied by the other produces the multiplication of the first by the fourth and these parts will be the unknown middle numbers. <**12.15**> E.g., if you do not know 4 and 3, but their sum is given to you (i.e., 7), multiply the first (which is 2) by the fourth (which is 6) and you will get 12. Then divide the given sum (which is 7) into parts such that when multiplied by each other they produce 12 and these parts will be the unknown

middle numbers (i.e., 3 and 4), as the figure below makes clear.

```
        12
       /  \
    6  3  4  2
       \  /
        V
        7
```

degree	minute	second	3rd	4th	5th	6th	7th	8th	9th
minute	second	3rd	4th	5th	6th	7th	8th	9th	10th
second	3rd	4th	5th	6th	7th	8th	9th	10th	11th
3rd	4th	5th	6th	7th	8th	9th	10th	11th	12th
4th	5th	6th	7th	8th	9th	10th	11th	12th	13th
5th	6th	7th	8th	9th	10th	11th	12th	13th	14th
6th	7th	8th	9th	10th	11th	12th	13th	14th	15th
7th	8th	9th	10th	11th	12th	13th	14th	15th	16th
8th	9th	10th	11th	12th	13th	14th	15th	16th	17th
9th	10th	11th	12th	13th	14th	15th	16th	17th	18th.

1	2	3	4	5	6	7	8	9	10
2	4	6	8	10	12	14	16	18	20
3	6	9	12	15	18	21	24	27	30
4	8	12	16	20	24	28	32	36	40
5	10	15	20	25	30	35	40	45	50
6	12	18	24	30	36	42	48	54	60
7	14	21	28	35	42	49	56	63	70
8	16	24	32	40	48	56	64	72	80
9	18	27	36	45	54	63	72	81	90
10	20	30	40	50	60	70	80	90	100

B

<13.1> Every number up to 10, when multiplied by itself, produces as much as the first two numbers on each side multiplied by each other, if 1 is taken from the multiplication of the middle number.[14] <13.2> Or in another way, more generally: every number multiplied by itself, produces as much as the two extremes (and the extremes of the extremes as far as 1), but with the multiplication of the differences between the number and the extremes added.[15] <13.3> For 5 multiplied by itself produces as much as 4 times 6 with the multiplication of the differences by each other, which are two 1s, and as much as 3 times 7 with the multiplication of the differences by each other, which are two twos, and as much as 2 times 8 with the

---

[14]This rule is the same as <2.3>.

[15]This rule is the same as <2.1>.

multiplication of the differences by each other, which are two threes, and so on as far as 1.[16] <**13.4**> Every number produces as much when multiplied by itself as the two numbers multiplied by themselves which are distant from it by an equal ratio.[17]

<**14.1**> Every number multiplied by itself produces as much as its two parts, if each of them is multiplied by itself and one of them is multiplied by the other twice. <**14.2**> Every number multiplied by another produces as much as when it is multiplied by all the parts of that number. <**14.3**> When any number multiplies another, it produces as much as if the same number multiplied the limit, when the product of the difference between the multiplied and the limit and the multiplier is subtracted.

<**15.1**> Every number to be divided by another is either equal or larger or smaller than the other number. If it is equal, then single units in the dividend result for each of the units of the divisors. <**15.2**> If it is larger, then however many times the divisor comes in the dividend, so many whole units result for each of (the units of) the divisors. <**15.3**> But if anything is left over, it will be divided into fractions. <**15.4**> Thus, as the smaller number is such a part or such a number of parts of the larger number, such a part or such parts should be given to each (of the units) of the divisors. <**15.5**> When you add fractions to fractions, if the number of the fractions is the same as that of their denomination, then a whole unit arises from the addition. <**15.6**> E.g., from three thirds or four fourths one whole unit is produced. <**15.7**> But if the number of the fractions is less than that of their denomination, then, whatever ratio the number of fractions has to the number of their denomination, the fractions have the same ratio to the whole unit. <**15.8**> E.g., six twelfths have the same relation to the whole unit as six to twelve. They are its half. <**15.9**> But if it is larger, then however many times it is larger, so many whole units result from the fractions added together. <**15.10**> E.g., six thirds restore two whole units, since the number of the fraction contains the number of the denomination twice. <**15.11**> But if it contains it a certain number of times plus a part or some parts of it, then however many times it contains it, so many whole units result from the fractions added together plus such a part or as many parts of one whole unit as the number remaining is part or parts of the number denumerating the fractions. <**15.12**> E.g., eight thirds: 8 contains 3 twice plus two thirds of it.

<**15.13**> But if you wish to know what relation a part of any part has to the whole unit, multiply by themselves the numbers by which the fractions are denominated and the part of the part will be related to the whole unit in the way that 1 is related to that product. <**15.14**> E.g., a third part of one quarter is a twelfth part of

---

[16]This rule is the same as <2.2>.

[17]This rule is equivalent to <10.1>.

one whole unit. For 3 times 4 becomes 12. <**15.15**> If you wish to know how the parts of any part are related to the whole unit, multiply by themselves the numbers by which the fractions are denominated and those added fractions <denominated by the product> will be related to the whole unit in the way that the number <of fractions> you added is related to the number already resulting. <**15.16**> E.g., two third parts of one quarter are a sixth part of one whole unit. For 3 times 4 are 12, whose sixth part is 2.

<**15.17**> But if <two>fractions are denominated by two different numbers, then by whatever ratio the greater number is related to the smaller, in such a way the fraction denominated by the smaller number is related to the fraction denominated by the larger. <**15.18**> E.g., a third part of anything contains two sixths of it. For 6 contains two threes. <**15.19**> If fractions of different numbers or different quantities are denominated by the same number, as the whole units are related to each other, so the fractions too and *vice versa*. <**15.20**> For just as 12 is related to 9, so a third part of 12 is related to a third of 9 and *vice versa*. <**15.21**> But if you wish to add any number of fractions denominated by different names, add the numbers by which the fractions are denominated and by the total <times> resulting from this add the fraction denominated by the number which arises from the multiplication of the numbers denominating the fractions. <**15.22**> For if you wish to know what the third and the fourth part make when added together, add the numbers by which the fractions are denominated, i.e., 3 and 4, and they become 7. By this 7 <times> add the fractions denominated by the number which arises from the multiplication of the numbers denominating the fractions, i.e., 12 (for 3 times 4 is 12). The third part and the fourth part of something, therefore, are seven twelfth-parts. What they constitute when added has been shown above.

<**16.1**> ...Or multiply it by itself and add it to the multiplication arising from it and divide this total into two equal <parts> and that half is the whole sum of that and of all the numbers below it.[18] <**16.2**> If you wish to know what sum results when numbers exceeding each other by an equal distance are added together, if those to be added consist of an odd number <of numbers>, consider how many they are and multiply the middle of those to be added by that number and the total is the sum. <**16.3**> E.g., let 3, 4, 5 be added together. Multiply the middle number (i.e., 4) by 3, which is how many they are and the result will be the sum the given numbers produce when added together. <**16.4**> Or let 3, 5, 7, 9, 11 be added together. Multiply the middle number (i.e., 7) by 5, for so many numbers are to be added, and the result will be the sum of the given numbers. <**16.5**> Or let 2, 5, 8 be added. Multiply the middle number (i.e., 5) by 3 and the result will be their sum. <**16.6**>

---

[18]In discussing the arithmetical progression, this rule is more general than those in <3.1> to <3.5> because the common difference here is not limited to 1, as is shown by the following examples.

But if the numbers to be added consist of an even number <of numbers>, join a larger number belonging to the same series to the earlier numbers. Then consider how many they are and multiply the middle of them by the whole number and when you have taken away what you had joined to the numbers to be added, the sum of the added numbers is produced. <**16.7**> E.g., 2, 4, 6, 8 should be added. Join the next number in the series (i.e., 10). Multiply its middle number (i.e., 6) by 5, since that is the number of the numbers to be added. The result will be 30. Take away from this total the 10 which you had joined and the 20 that remains is the sum of the added numbers. <**16.8**> If any odd number is added to all the odd numbers below it plus 1, the sum which arises will be a square number.[19]

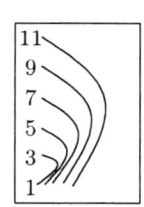 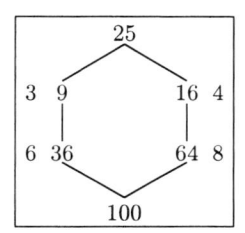

<**17.1**> If the squares of two numbers taken together are a square number, it is necessary that the squares of any two numbers related to each other by the same proportion, when taken together, are a square number.[20] <**17.2**> If any number of quantities are proportioned to a quantity by different, but known, ratios and, when taken together, make a known sum, <the problem is> to find the value of the same first quantity. <**17.3**> E.g., let a be a quantity and let it also be the case that b, c and d are proportioned to a by known ratios. Let it also be the case that b, c and d taken together make up the quantity g and it is known how much g is. It is, then, proposed that one should also find how much a is. <**17.4**> E.g., Socrates is understood to have twice as many pennies as Plato plus two third parts more than those which Plato has and the total that Socrates has, when taken together, is 15 pennies. The problem is to find how much Plato has. I take, then, the number to which the two numbers are proportioned by the above-mentioned ratios. This is 3. For 6 is twice as large as it and 2 is two third parts of it. I add these two numbers, 6 and 2 and 8 results. Having considered, therefore, what ratio 15 has to 8, I pronounce that Plato has the same ratio to the three pennies. For 15 contains

---

[19] A figure is added in the margin of PN without any explanation. But we can easily see that this figure is showing the examples of the rule. It can be expressed as: $1 + 3 = 2^2$; $1 + 3 + 5 = 3^2$; $1 + 3 + 5 + 7 = 4^2$; $1 + 3 + 5 + 7 + 9 = 5^2$; $1 + 3 + 5 + 7 + 9 + 11 = 6^2$.

[20] The figure is in the margin and there is no corresponding explanation about it. It is easy to see that the figure shows an example of the rule. The figure can be explained as follows: since $3^2 + 4^2 = 5^2$ and $6 : 3 = 8 : 4 = 2$, $6^2 + 8^2 = 10^2 = 100$.

the whole of 8 plus seven eighths of it. Similarly 5 pennies and an obol and a fourth part of an obol contain 3 pennies and seven eighths of them. I say, then, that Plato has 5 pennies, an obol and a fourth part of an obol. For twice this amount and two third parts of this are 15 pennies. For if you take 5 pennies, an obol and the fourth part of an obol twice, there will be 11 pennies and the fourth part of one penny. If you take two thirds of the same 5 pennies, one obol and a fourth part of an obol, they will make 3 pennies and three quarters of one penny. <**17.5**> Understand this in the following way: take two third parts of 9 obols (i.e., 6 obols) and they are 3 pennies. But the remaining penny and fourth part of an obol are nine eighth parts of a penny. Since two third parts (i.e., six eighths of a penny) are three quarter parts of one penny, if you add these to the earlier 11 pennies and a fourth part of a penny and 3 pennies, they are 15 pennies. <**17.6**> Or in another way, more simply: multiply the number of the denomination of the parts (i.e., 3) by 15 and the result is 45. Then multiply the same 3 by <two and> two thirds and the result is 8. For 3 <multiplied> by 2 becomes 6 and when two thirds <of three> have been added, it becomes 8. Then divide the first product (i.e., 45) by the last product (i.e., 8) and the result is 5 and five eighths, in this way:

<**17.7**> When the question is: "How many smaller things are in a certain number of larger things?" you should multiply the larger things by the number of smaller things which are in one of the larger ones and the number which arises will show how many smaller things are in so many larger things. <**17.8**> But if the question is: "How many larger things <are> in a certain number of smaller things?" you will divide the smaller things by the number of smaller things which are in one of the larger things and the number which results from the division will show how many larger things are in so many smaller things. <**17.9**> E.g., a shilling is less than a pound. Therefore, if the question is: "How many shillings are in 100 pounds?" find out how many smaller things are in one larger thing. You should, therefore, multiply the number of the larger things (i.e., pounds), i.e., 100, by the number of smaller things (i.e., shillings) which are in one of the larger things (i.e., one pound), i.e., by 20 (for 20 shillings make one pound) and 2,000 results. Know, then, that so many shillings, i.e., 2,000, are in 100 pounds. <**17.10**> Again, let the question be: "How many pounds are in 24,000 pennies?" The question is asked, then: "How many greater numbers are in a number of smaller things?" because pennies are less than pounds. You will divide, then, the number of the smaller things (i.e., pennies), namely 24,000, by the number of the smaller things, i.e., pennies, which are in one of the larger (i.e., in one pound), i.e., by 240 (for so many pennies are in one pound) and 100 will result. Know, then, that so many pounds (i.e., 100) are in 24,000 pennies.

<**17.11**> When the sum of parts of any denomination of a whole is known, <the problem is> to know the whole itself. First, add the numbers denominating the

given parts. Then, multiply one by the other and thus you will have four terms: namely the sum of the given parts, the sum of the numbers denominating the parts and the product of these and the fourth is the whole which is unknown. For the ratio of the whole to the sum of the proposed parts is the same as that of the product of the numbers denominating the parts to the sum arising from their addition. The sum of the proposed parts, therefore, should be multiplied by the product of the numbers denominating the same parts and the product from this[21] is divided by the sum of the numbers denominating them and the total, which is unknown, will result, by the above-mentioned rule of four numbers in proportion: <namely> if the first was unknown, multiply the second by the third and divide <the result> by the fourth and the first will result.

<17.12> E.g., let the third and a quarter of my money be 20 pennies. The problem, therefore, is to find how much is the total amount of money. When the denominators (i.e., 3 and 4) are added, they become 7. Multiply one of them by the other and the product will be 12. You have, then, four terms. First multiply the second by the third, i.e., 20 by 12, and 240 will result. Divide this by 7 and 34 and two sevenths will result. This number should be put into the terms of its genus, i.e.,

20	12
240	
7	
34	
2	
7	

in the pennies, shillings or pounds, in which the proposed parts had been. And this is the total sought.[22] <17.13> When three numbers are in proportion, if the first is multiplied by the third the square of the middle number will result.[23] <17.14> When several men contribute different sums of money for making a profit, if you wish to known how great a part of the profit which arises from the whole amount comes to each one of them by right, add the portions which they had put down and multiply the contribution of whomever you wish by the total profit. Then divide the product of the multiplication by the sum and what results from the division will be the portion of him whose contribution you have multiplied. <17.15> Or the reverse: divide the contribution by the sum and multiply what results by the total profit and what results from this will be his portion. You will follow a similar procedure for each of the others in turn. <17.16> E.g., three merchants have contributed their money for making a profit, one 6 shillings, another 8, another 12, which altogether

---

[21]P has added a '7' here, which has probably strayed into the text from the figure in the margin (see next note). In N the three numbers from the figure —20 (written '30') 12 and 7—have been inserted here.

[22]In P there a defective figure has been added in the margin. Although only 20 and 12 of the first line and the last result $34\frac{2}{7}$ are given in the margin, we can see from the figure of <17.6> that this figure is to show the relevant calculation. Below $34\frac{2}{7}$ is an explanation in the margin: 'This number contains the whole of 20 plus five sevenths of it, just as $\frac{12}{7}$ <of it>', which means $34\frac{2}{7} = 20 + 20 \cdot \frac{5}{7} = 20 \cdot \frac{12}{7}$.

[23]This rule is a repetition of the first sentence of rule <10.1>.

make 26. From this they have made a profit of 60. If you wish, then, to know how much of the profit comes to each one of them according to the quantity of the money contributed, add together the contributions of all of them and they become 26. Then multiply individually the contribution that each has made by the total profit. Then divide what results from the multiplication by the total of the contributed capital, i.e., 26, and what results from the division is what is owed to him whose contribution you multiplied. You will proceed in this way for each person individually. There are, then, these four terms–i.e., the contribution of each one of them, 26, the third is unknown, the fourth is 60– which are proportional according to the above-mentioned

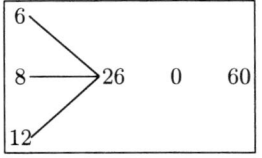

rule. Multiply the first by the fourth, i.e., the contribution of each one of them by 60 and divide the product by the second, i.e., 26, and the third will result, i.e., the portion that comes to the man whose contribution you made the first term.[24]

<**17.17**> If you wish to know how much of any definite sum that is owed to many men comes to some of them, multiply the number of those men about whom you want information, by the sum itself and divide what results from the multiplication by the whole number of the many men and what results is what is due to them.

<**17.18**> E.g., 24 pennies are owed to 8 men and you wish to know how much comes to 3 of them. Multiply those 3 by 24 and divide what results from the multiplication by 8 and you will see what will come to them. And according to this rule likewise you will prove how much will come to the other 5, i.e., by multiplying those 5 by 24 and by dividing what results from their multiplication by 8. There are, therefore, these three terms, 8, 24, 3 and the fourth is unknown, i.e., how much is owed to 3. By the above-mentioned rule, then, multiply the second by the third and divide the product of that by the first, i.e., 8 and the fourth (the unknown) will result, i.e., the amount owed to three men. For as 8 is related to 3, so is 24 related to the amount owed to three men. Therefore it contains it twice plus two thirds of it.

8	24	3	0

C

<**18.1**> Excerpts from the book which is called 'Jabr Muqabala'. A certain three-part division by opposites is made. For the question is either <1> "Which *thing*[25] with so many times its root makes a number?" or <2> "Which *thing* with such a

---

[24]In the figures of <17.16>, <17.17> and <19.2>, '0' is adopted to represent the symbol of the unknown. Robert Kaplan analyses the tradition of this expression in ancient Greek and Indian mathematics (Robert Kaplan, *The Nothing That Is*, London, 1999, pp. 57-67).

[25]In translating 'res' (the unknown) literally as 'thing', we have printed 'thing' in italics. In MS E 'res' is replaced by 'quadratus' ('square').

number makes the root so many times?" or <3> "Which root taken so many times with such a number makes the *thing*?"

<**18.2**> <1> The question is: "Which *thing* with ten of its roots (i.e., with its root taken 10 times), makes 39?" To find this multiply half <the number of> the above-mentioned roots by themselves, add what results from this to the earlier number, take the root of what arises from this and subtract from this root half <the number of> the above-mentioned roots; what remains from this is the root of the *thing*. If you multiply the root by itself, the *thing* which you seek results. <**18.3**> E.g., since 10 roots had been proposed above, if you multiply half them (which is 5) by themselves, you make 25. Add these to the above-mentioned number (which is 39) and you will make 64, whose root is 8. If you subtract from this root (i.e., 8) half <the number of> the roots (which is 5), there remains 3, which is the root of the *thing*, it being 9. This, with 10 of its roots (i.e., 10 threes) makes 39. Therefore 9 is the *thing* which is sought.

<**18.4**> (2) Likewise, what is the *thing* which, with 9 added to it, makes 6 of its roots? To find this multiply half <the number of> the roots by themselves, take away the above-mentioned number from what results and subtract the root of what remains from half <the number of> the roots. What remains will be the root of the *thing* which you seek. <**18.5**> E.g., 6 roots have been proposed. Multiply its half (which is 3) by itself and you will make 9. From this 9 take the above-mentioned number (i.e., 9) and nothing remains. Subtract the root of this which remains (i.e., nothing), which similarly is nothing, from half <the number of> the roots (which is 3). Because you have subtracted nothing from 3, 3 remains, which is the root of the *thing* which you seek, i.e., of 9, which with 9 added to it becomes 18, which is 6 roots of 9, i.e., 6 times 3, this 3 being the root of 9.[26]

<**18.6**> (3) Likewise, what are the roots which with 4 added to them make their *thing*? To find this multiply half <the number of> the roots by themselves, add what results from this to the above-mentioned number and add the root of what arises from it to half <the number of> the roots and what arises from this is the root of the *thing* which you seek. <**18.7**> E.g., 3 roots have been proposed. Its half is 1 and a half, which when multiplied by itself makes 2 and a quarter. Add these to the earlier number (which is 4) and you will make 6 and a quarter. The root of this is 2 and a half. Add this root to half <the number of> the roots, which is 1 and a half and they will become 4. This is the root of the *thing* which you seek, i.e., of 16. Three of its roots (i.e., 3 fours or four three times) with 4 added to them make this 16.

---

[26] This also occurs in Ibn Turk, English trans. p. 100-101. (Aydin Sayili gives no source).

D

<19.1> If you wish to divide any given quantity into any given ratios, or to divide an undivided number in the same way as a divided one, first the ratios of the proposed divisions should be arranged in their terms, then the terms should be added up and the sum should be placed first. When the number of the ratio is in a ratio to what is sought, it should be placed second. The given quantity should be placed third. An empty space should be placed fourth. The given quantity is multiplied by the number of the ratio and the product from that should be divided by the sum of the terms themselves and what is sought will emerge through the rule of four <numbers> in proportion. If the fourth is unknown, the third should be multiplied by the second, the result should be divided by the first and the fourth will emerge. If you wish to know the others through one known number, you will have the multiples by multiplying it by the number denominating the ratios. One calls 'multiples' those which contain a number a certain number of times. One calls 'submultiples' those which are contained <by another number> a certain number of times. If any number remains after division, it will be the number of fractions of the quantity, denominated in the division by the divisor. <19.2> E.g., we are asked to divide 40 shillings between 4 men in such a way that the second has 4 times what the first has, but the third has 5 times the second, the fourth 3 times the third.

4, therefore, should be multiplied by 1 and 4 results. Let the first term be 1, the second 4. Likewise, 4 should be multiplied by 5 and 20 results. The third term, therefore, will be 20. Likewise, 20 should be multiplied by 3 and

	first	1			
85	second	4	four	40	0
	third	20	five		
	fourth	60	three		

the result is 60. The fourth term, therefore, will be 60. When the terms are added, 85 results. Let 85 be put first. Then, if you want to know what comes to the second <man>, place the second term second.[27] Likewise, if you want to know what comes to the third, place the third term second. Likewise, for each one of them.

<19.3> So, place the second term (i.e., 4) second. Place the given quantity third, the empty space fourth. Let the given quantity (i.e., 40) be multiplied by 4 and the result is 160. Let 160 be divided by 85 and there emerge 1 and 75 85ths. This number should be arranged in the denominations of its genus–pennies, shillings or pounds–in which the given quantity was proposed. This is the answer. <19.4> If you wish to know through this what comes to the first (man), divide the given number, i.e., one whole (i.e., 12 pennies which are 1 shilling) by 4 and the result is 3 pennies. Likewise, divide the numerator of the fractions (i.e., 75) by the same 4 and the result is 18 85ths and three quarters of an 85th and this is the portion of the

---

[27]I.e., in the second place in the rule of four.

first (man), i.e., 3 pennies and 18 85ths plus three quarters of an 85th (of a shilling). <**19.5**> Likewise, if you wish to know the others, i.e., its multiples, through the number already known, multiply the known number, i.e., 1 and 75 85ths by the number denominating the third ratio, e.g., by 5 and the result is 9 and 35 85ths shillings and this is the portion of the third (man). (One does) the same with each multiple. <**19.6**> But if you wish to know the submultiples, divide the already known number by the number denominating the ratio. <**19.7**> For proving this add whole numbers to whole numbers.

<**20.1**> If you wish to continue any given ratios to any least term, multiply the first of the given ratios by the least term, multiply the product by the next (ratio) and proceed in this way, multiplying the result of the following (product) by the following (ratio) and the result from multiplying the penultimate following (product) by the last following (ratio) will be the first of the numbers to be continued. Having done this, multiply the second of the given ratios by the same term and multiply the product by the following (ratio) and proceed in multiplying in the above-mentioned way and the result from multiplying the penultimate following (product) by the last following (ratio) will be the second of the numbers to be continued. Likewise multiply the third of the given ratios by the same least term and multiply the result by the following ratio and proceed in multiplying in the above-mentioned way and the result from multiplying the penultimate following (product) by the last following (ratio) will be the third of the numbers to be continued. Likewise concerning each one. <**20.2**> E.g., let the given ratios be 4, 3, 2, 1 and let the least term be 6. Let 4 be multiplied by 6 and the product will be 24. Multiply this by 3 and the product will be 72. Let this be multiplied by 2 and the result will be 144. If this is multiplied by 1, the same number will result. This, therefore (i.e., 144), will be the first of the numbers to be continued. <**20.3**> Likewise let 3 be multiplied by 6 and the result will be 18. Multiply this by 2 and 36 becomes the second number to be continued. <**20.4**> Likewise multiply 2 by 6 and 12 results. 12, therefore, will be the third of the numbers to be continued. <**20.5**> Likewise multiply 1 by 6 and 6 results. This, then, will be the fourth of the numbers to be continued.

<**21.1**> If you wish to find the root of any given square, multiply the square by any other square and take the root of the product and divide the root by the root of the square by which you multiplied the given square. What results will be the root sought. <**21.2**> E.g., if you wish to find the root of 2 and a quarter, multiply that by another square, which could, for instance, be 4 and 9 results. For twice 4 becomes 8 and a quarter times 4 becomes one — hence 9. The root of this 9 is 3. Divide this by the root of 4 (which is 2); 1 and a half will result. In this way: [28]

$$\begin{array}{|c|} \hline 1 \\ 3 \\ 2 \\ \hline \end{array}$$

---

[28] In this figure, 1 is the result of a quarter times 4, 3 is the square root of 9, and 2 is the square

<**21.3**> If you wish to find the root of the nearest square, multiply the third given number by any square and proceed in the above-mentioned way.

<**22.1**> If you wish to multiply an article by an article or a digit by a digit or a composite number by a composite number, multiply the numeral by the numeral. Then add the numbers denominating the columns (decimal places), subtract 1 from the sum and the following column, denominated by what remains, grows so many times as the number of units grow from the multiplication of the numerals, and the following column[29] will grow as many times as the number of tens grow.

<**23.1**> When you multiply one number by another, consider what part one of them is of another article or limit and take the same part from the other number. Multiply that part by that article or limit and the product from that is what results from the multiplication of the one by the other. <**23.2**> E.g., if you multiply 32 by 25, whatever part 25 is of the limit 100 (i.e., a quarter), take the same part of 32, i.e., a quarter, which is 8. Multiply this 8 by 100 and the product from this is what comes from the multiplication of 32 by 25. The proof: just as 8 is to 32, so 25 is to 100. As much arises, then, from the multiplication of the two middle numbers by each other as from the multiplication of the two extremes, according to the above-mentioned rule of four numbers in proportion, and the same is true in all cases. <**23.3**> One can do a similar thing with the article: e.g., 25 is half 50. But half 32 is 16. Multiply this by 50 and the same result is reached as from the multiplication of 32 by 25.

| 8 | 32 | 25 | 100 |

<**23.4**> When you multiply together two composite numbers consisting of the same or different digits but the same article or limit, such as 16 by 18 and the like, multiply the digit by the digit and the article by the article and add the products from this. Then add the digit to the digit and multiply the sum by the article or limit and add the product from this to the first sum and the total sum is the result which the multiplication of one composite number by another produces. <**23.5**> E.g., let us suppose that 16 is to be multiplied by 18. The digit, then, should be multiplied by the digit (i.e., 6 by 8) and the result is 48. Then the article by the article (i.e., 10 by 10) and the result is 100. Add these two products together and they become 148. Then add the digit to the digit and the result is 14. Multiply this by the article (i.e., 10) and the result is 140. Add this to the first sum (which was 148) and the result is 288. This is the total which arises from the multiplication of 16 by 18.

<**24.1**> When you wish to multiply the roots of any numbers, multiply the numbers by each other and the root of the product is the product resulting from the multiplication of one root by another. <**24.2**> E.g., if you wish to multiply the root

---

root of 4.

[29]The Latin text gives 'the beginning of the following column' in both cases.

of 10 and 40, multiply 10 by 40 and 400 results. The root of this 400 is 20. This 20 is the number produced from the multiplication of the root of 10 by the root of 40 through the rule of three numbers related to each other proportionally, of which, when the second is related to the third in the same way as the first is to the second, then whatever results from the multiplication of the middle one by itself, so much results from the multiplication of the two extremes, in this way: $\boxed{10\ 20\ 40}$

<**25.1**> If you wish to know how old a man is who, if he lives as long as he has already lived and the same amount again and half that amount and half of half that amount, completes 100 years, add what is given and divide the total which is completed by the sum and the result is his age. <**25.2**> E.g., when it is proposed that <he lives> as long as he has lived and the same amount again and half that amount and half of half that amount, all these added together make 4 less a quarter, which are 15 quarters. If you divide 100 (turned first into quarters) by these 15 quarters, the result is 26 and two thirds. When these are taken 4 <minus a quarter> times, they complete 100 and this is the (length of time) he has lived.

E

<**26.1**> ...Or another way: since every number is either a digit or an article or a limit or composite, then however many times a number is multiplied by a number, either a digit is multiplied by a digit or by an article or by a limit or by a composite number or *vice versa*, or a composite number (is multiplied) by a composite number or an article or a limit or a digit or *vice versa*. <**26.2**> When you wish to multiply an article by an article, multiply the numeral by the numeral. Then consider which places the articles themselves belong to and add the numbers by which their places are denoted. Take 1 from the sum and in the place named by the remaining number put the product from the multiplication of the numerals if it is only a digit. If, however, it is only an article, <put> it in the <place> following it. But if it is a digit and an article, the digit should be put in the place denoted by the remaining number, but the article should be put in the following place. What is indicated there will be the total that results from the multiplication of one article by another. <**26.3**> E.g., if you wish to multiply 20 by 70, multiply together the numerals by which they are represented (i.e., 2 by 7) and the result is 14. But because both belong to the second place (which is of the tens), add the numbers denoting the place of each of them, i.e., 2 and 2–for the second <place> is denoted by '2'–and the result is 4. Subtract 1 from this 4 and 3 remains. By this 3 is denoted the third place, which is of the hundreds. Because the result of the numerals multiplied by each other had been 14, which is a digit and an article, put the digit (i.e., 4) in the same place (i.e., the third place) and the article (i.e., 10)[30] in the following place

---

[30]'1' would be more correct.

which is the fourth and you will have 1400. This is the total that results from the multiplication of one article by another (i.e., 20 by 70). One should proceed in a similar way if a digit is multiplied by an article or a limit or a composite number and *vice versa.* <**26.4**> When you wish to multiply a composite number by a composite number, you will observe the above-mentioned rule, with this addition: that each of the higher numerals should be multiplied by each of the lower numerals, i.e., the digit by the digit and the article and the article by the digit and the article, <and> however many they are, each of the higher numerals <should be multiplied> by all the lower numerals. <**26.5**> E.g., when you wish to multiply 23 by 64, multiply the higher digit (i.e., 3) by the lower digit (i.e., 4) and the result is 12. According to the earlier rule, then the digit will be in the first place, the article in the second. Then multiply the same digit (i.e., 3) by the numeral of the lower article (which is 6) and the result is 18. When the numbers denoting the places are added (i.e., 1 and 2, for the article is in the second place, the digit in the first), the result is 3. When 1 is subtracted from this, 2 remains; this denotes the second place. So put the digit (i.e., 8) in the second place and 1 in the following place (i.e., the third). Then you will multiply the numeral of the higher article (which is 2) by the lower digit (which is 4) and the result is 8. When the numbers denoting the places (i.e., 1 and 2) are added the result is 3. When 1 is subtracted from this, 2 remains; this denotes the second place. Therefore, the digit (i.e., 8) should be put in the second place. Then you will multiply the numeral of the higher article <(i.e., 2)> by the numeral of the lower article (i.e., 6) and the result is 12. Since each belongs to the second place, when the numerals denoting the places are added together (i.e., 2 and 2) the result is 4. When 1 is subtracted from this, 3 remains; this denotes the third place. So place the digit (i.e., 2) in the third and the article < (i.e., 1)> in the following (i.e., fourth) place and the result looks like this:

			1	2
		1	8	
			8	
1	2			

<**26.6**> When these are arranged like this, add them up and the result is 1,472. This is the total that results from the multiplication of 23 by 64. <**26.7**> In this same rule one also learns how a composite number should be multiplied by an article, a limit or a digit.

<**27.1**> When you multiply a digit by a digit the result will be either a digit only or a ten only or a digit with a ten once or several times or a ten many times. <**27.2**> When you multiply any digit by any article which is below 100, multiply the numeral by the numeral and however many units there are in the digit which arises, there will be so many tens. However many tens there are in the article which arises, there will be so many hundreds. <**27.3**> E.g., when you wish to multiply 7 by 70, multiply the numeral by the numeral and 49 results. There are 9 units in the digit. So, there will be this many tens: i.e., 90. In the article there are four times ten; so there will be this many hundreds, i.e., 400. 490, therefore, is the total that results from

the multiplication of these numbers by each other. A similar procedure is used in all other cases. <**27.4**> When you multiply a digit by any hundred below 1,000, multiply the numeral by the numeral and however many units there are in the digit (if it results), so many hundreds there will be. But however many tens there are in the article (if it results), so many thousands there will be. <**27.5**> E.g., if you multiply 3 by 900, multiply the numeral by the numeral and the result is 27. There are 7 units in the digit and 2 tens in the article. Therefore 2,700 is the sum which results from their multiplication by each other. A similar procedure is used in all other cases.

<**28.1**> When you multiply any article by another which is below 100, multiply the numeral by the numeral and however many units there are in the digit which results, so many hundreds there will be. But however many tens there are in the article, so many thousands there will be. <**28.2**> E.g., if you multiply 30 by 70, multiply the numeral by the numeral and the result is 21. There are 2 tens in the article and one unit in the digit. From the multiplication, therefore, of the earlier numbers there results 2,100. A similar procedure is used in all other cases. <**28.3**> When you multiply any article which is below 100 by any hundred which is below 1,000, multiply the numeral by the numeral and however many units there are in the digit, so many thousands there will be. But however many tens there are in the article, so many ten thousands there will be. <**28.4**> E.g., if you multiply 30 by 500, multiply the numeral by the numeral and the result is 15. Since there are 5 units in the digit, there will be 5 thousands. But there is only one ten in the article, <so there will be one ten thousand>. From the multiplication, therefore, of the earlier numbers there results fifteen thousand, like this: 15,000. A similar procedure is used in all other cases.

<**29.1**> When you multiply any hundred which is below 1,000 by another of them, multiply the numeral by the numeral and however many units there are in the digit which results, so many ten thousands there will be. But however many tens there are in the article, so many hundred thousands there will be. <**29.2**> E.g., when you multiply 300 by 500, multiply the numeral by the numeral and the result is 15. There are 5 units in the digit and one ten in the article. From the multiplication, therefore, of the above-mentioned numbers there results 150,000, which is one hundred and fifty thousand.

<**30.1**> When you multiply any digit by any article belonging to the thousands, such as ten or twenty thousand and so on, or repeated thousands, such as ten thousand thousand and however much you wish to repeat the thousand, multiply the numeral by the numeral and put the digit (if it results) in the place of the multiplier and the article in the following place. <**30.2**> E.g., if you multiply 6 by 30,000, multiply the numeral by the numeral and the result is 18. The digit (which is 8) should

204

be put in the same place as the multiplier (which is 3) and the article (which is 1) should be put in the following place, in this way: 180,000 and there results one hundred and eighty thousand. <**30.3**> But when you multiply any article by any oft repeated thousands, like ten or twenty thousand thousand thousand (however many times you wish to repeat the thousand), multiply the numeral by the numeral and put the digit (if it results) in the second place away from the multiplier, but the article in the third place from it. <**30.4**> E.g., when you multiply 30 by four thousand thousand thousand thousand (and however many times you wish to repeat it), multiply the numeral (which is 3) by the numeral (which is 4) and the result is 12. Therefore place the digit (which is 2) in the second place from the 4, and the article in the third place from the 4, in this way: 4,000,000,000,000 becomes one hundred and twenty thousand thousand thousand thousand (i.e., 4 thousands): 120,000,000,000,000. <**30.5**> When you multiply any hundred by any oft repeated thousands, multiply the numeral by the numeral and place the digit (if it results) in the third place from the multiplier, but the article in the fourth place from it. <**30.6**> E.g., when you multiply 200 by five thousand thousand thousand thousand (i.e., 4 thousands), multiply 2 by 5 and, since an article results, it should be put in the fourth place from the 5, in this way: 1,005,000,000,000,000.[31]

<**31.1**> When you wish to know in which place are the digits, articles, or hundreds of oft repeated thousands, consider how many times the thousand is repeated, multiply by 3[32] the number of times it is repeated and keep in mind the product of this. If you wished to know in which place are the digits of the repeated thousands, e.g., two, three or four and up to nine thousand thousand thousand (however many times you wish to repeat the thousand), always add 1 to the first result kept in mind and the place belonging to the digits of the repeated thousands is denoted by the number that arises from this. <**31.2**> E.g., if you wish to know in which place is three thousand thousand thousand thousand, multiply by 3 the number of times the thousand is repeated (i.e., 4 in this case) and the result is 12. Add 1 to this and the result is 13. The place of the above-mentioned is therefore the thirteenth. <**31.3**> If you wished to know in which place are the articles of oft repeated thousands (e.g., ten or twenty thousand thousand, however many times you wish to repeat the thousand), always add 2 to the earlier result kept in mind and the place belonging to the articles of the oft repeated thousands is denoted by the number that arises from this. <**31.4**> E.g., if you wish to know in which place are the above-mentioned articles of oft repeated thousands, such as fifty thousand thousand thousand thousand (i.e., 4 thousands), multiply by 3 the number of times the thousand is repeated (i.e., 4 in this case) and the result is 12. Add 2 to this and the result is 14. The place of the articles

---

[31] The scribe has shown an interim calculation. In the final result the '5' should be replaced by a '0'.

[32] In the Latin 'multiply by 3' precedes 'consider how many times the thousand is repeated'.

of the above-mentioned repeated thousands is therefore the fourteenth. <**31.5**> If you wished to know in which place are the hundreds of repeated thousands, always add 3 to the earlier result and the place belonging to the hundreds of the repeated thousands is denoted by the number that arises from this. <**31.6**> E.g., if you wish to know in which place is a hundred or two hundred or some other hundred thousand thousand (however many times you wish to repeat the thousand), multiply by 3 the number of times the thousand is repeated (i.e., twice) and the result is 6. Add 3 to this and the result is 9. The place of abovementioned hundreds of repeated thousands is therefore the ninth.

<**32.1**> When you wish to multiply any thousand or ten or hundred thousand of oft repeated thousands by any other thousand, putting aside the repetition of thousands in the multiplied and the multiplier, multiply by each other what remains in each of them and keep in mind what results from this. Then add the number of repeats of each and place the total arising from this under what resulted earlier and what arises from this is the number which results from the multiplication of one by the other. <**32.2**> E.g., when you wish to multiply digits of thousands[33] by each other, such as three thousand thousand by seven thousand thousand thousand thousand, having put aside the number of repeats of <the thousands of> each (i.e., 2 and 4, for in the multiplied the thousand is counted twice, in the multiplier 4 times) and there remain only the numerals of each of them (i.e., 3 and 7). When one of these is multiplied by the other the result is 21. Then add the number of repeats of each of them (i.e., 2 and 4) and the result is 6. Put this under what resulted earlier (i.e., 21) in this way:

21
6

You will say that 21 times a thousand thousand thousand thousand thousand thousand (i.e., 6 thousands) results from the multiplication of one of the above-mentioned numbers by the other. According to the above-mentioned rule, therefore, if you multiply the number of repeats by 3, 18 will result. When 1 is added, 19 results. The above-mentioned digit therefore (which is 1) was in the nineteenth place and the article, 2,[34] in the twentieth. <**32.3**> When you wish to multiply hundreds of repeated thousands by each other, such as five hundred[35] thousand thousand by three hundred thousand thousand thousand thousand (i.e., 4 thousands), having put aside the number of repeats of <the thousands of> each (i.e., 2 and 4, for in the multiplied the thousand was repeated twice, in the multiplier 4 times) there remain 50<0> in the multiplied and 300 in the multiplier. When these are multiplied by each other the result is 150 thousand. Keep this in mind. Then add the number of repeats of each of them (i.e., 2 and 4) and the result is 6. Put this under the earlier numbers and the total will be indicated, which results from the multiplication of one

---

[33] A different word is used for 'thousands' here: 'millena' instead of 'milia'.

[34] Instead of writing '2' the scribes, thinking of the value of the numerals, have written '20'.

[35] The MSS gives 'fifty'.

of the above-mentioned numbers by the other, i.e., one hundred and fifty thousand thousand thousand thousand thousand thousand thousand (i.e., 7 thousands), in this way: 150,000,000,000,000,000,000,000. According to the previous rule, therefore, 5 will be in the twenty-third place and 1^{36} in the twenty-fourth. <**32.4**> The procedure is similar also when you multiply digits of repeated thousands by tens or hundreds of repeated thousands and *vice versa*. The procedure is similar also when you multiply tens of repeated thousands by each other or by hundreds.

<**33.1**> When you have a place and want to know what number it is, divide the number by which the place is denoted by 3. If there is no remainder from the division, that place will belong to a hundred thousand repeated a certain number of times. <**33.2**> If you wish to know the number of repeats (i.e., how many times <the thousand> is repeated) take 1 from what results from the division and what remains will be the number of repeats of that hundred thousand which belong to that place. <**33.3**> E.g., if you have the twelfth place and wish to know what number that is, divide that 12 by which the place is denoted by 3 and 4 will result. When 1 is subtracted from this, 3 remains and since nothing has remained from the division, the twelfth place will belong to a hundred thousand thousand thousand (<i.e.,> repeated 3 times). <**33.4**> But if the remainder from the division is 2, that place will belong to a ten thousand repeated as many times as the number which results from the division. <**33.5**> E.g., when you have the eleventh place and wish to know what number that is, divide 11 by 3 and 3 whole numbers will result and the remainder will be 2. The eleventh place, therefore, belongs to a ten thousand repeated 3 times. <**33.6**> But if the remainder from the division is one, that place will belong to a digit thousand repeated as many times as the number which results from the division. <**33.7**> E.g., if you have the tenth place and wish to know what number that is, divide 10 by 3 and 3 will result from the division with 1 as the remainder. The tenth place, therefore, belongs to a thousand repeated three times.

<**34.1**> You will find in this way the number which someone holds concealed in his heart, without him indicating it. First, tell him to triple the number. Then, let him divide the result of the tripling into two parts. After this ask whether the parts are equal. If they are unequal, keep 1 in mind and tell him again to triple the larger part and divide the result into two parts.[37] If, when asked, he replies that the parts are unequal, keep 2 in mind. When this is added to the earlier 1, 3 results. Then, tell him to subtract 9 from the larger part and again another 9 and continue in this

---

[36] Instead of writing '5' and '1' the scribes, thinking of the value of the numerals, have written '50,000' and '100'.

[37] The two parts are either equal or their difference is 1. This belongs to the field of number theory such as that later developed by Fermat.

way until there does not remain anything from which 9 can be subtracted. For each 9 take 4. The sum of these fours, together with the first <numbers kept in mind> is the number which he concealed. Or if he is not able to subtract 9 from the larger part, the first 3 will be the concealed number. Or if each division results in equal numbers, you will take nothing, but the single or many fours taken from the single or many nines will be the concealed number. However many times the division results in unequal numbers, take 1 from the first (unequal) division and 2 from the second (unequal) division. <**34.2**> E.g., let 2 be the number that he conceals. When tripled it makes 6. But 6 is divided into two equal parts. When either part is again tripled it makes 9, which is divided into two unequal parts. Hence, because it is the second division, I keep 2 in mind. And because 9 cannot be subtracted from the larger part, the 2 which I kept in mind is the concealed number. And this is what we proposed to do.[38]

<**34.3**> Likewise, tell someone who is concealing[39] the number of shillings he has, that he should take from your <pennies> for each <shilling> a single <penny> or 2 <pennies> or 3 <pennies> or any such amount that you wish. With all your pennies he should buy something, such as a chicken or something else like this. Then with all his shillings he should buy as many chickens as he can for the same price. Divide, then, the shilling by the number which you gave him (i.e., by 1 or 2 or 3 <or such like>). Add 1 to the number resulting from the division and what results when 1 is added is the number of those things that are bought. <**34.4**> E.g., let us suppose that he is concealing 5 shillings. Having taken from my pennies as many pennies as there are shillings (i.e., 5), let him buy one chicken and at the same price 12 chickens are bought with his shillings. Divide, then, the shilling into the 12 <pennies> <and divide this> by the number that you gave him (i.e., 1 <each>); 12 will result, which is the number of chickens that were bought. If you had given him 2 for each <shilling>, there would be 6 chickens, if 3 for each <shilling>, there would have been 4 and similarly for the others.

<**34.5**> If any two equal amounts are concealed and 2 taken from one of them is added to the other and from the augmented one something equal to the remainder is subtracted,[40] necessarily 4 will remain, or if 3, necessarily 6 and thus there will always remain the double of what is taken first. <**34.6**> E.g., if 5 shillings are concealed in one hand and 5 in the other, place 2 taken from one hand with the other 5 and they become 7 in one hand, while 3 remain in the other. When then you

---

[38]In these last two sentences the scribe has used a code whereby each vowel is represented by the letter of the alphabet that follows (here I = k, o = p and u = x).

[39]Here a different code is used, whereby the five vowels are represented by the first five numerals (a = 1, e = 2, i = 3, o = 4 and u = 5).

[40]MS gives 'added'.

208

take from the 7 what is equal to the remainder (i.e., 3), necessarily 4 <shillings> remain.

F

<**35.1**> It is asked why we do not designate all or most numbers with their own names, or why we always count, not by adding new names, but, after ten, by the repetition of the first ones. <**35.2**>The response to this is that it was not possible for all numbers to have their own names, because the mass of numbers grows to infinity, but the invention of names cannot in any language be infinite.[41] <**35.3**> For, since in every language the instruments <of speech[42] > are certain and limited and so are their naturally determined variations (by which the articulated spoken word is formed and from which arise the shapes of letters among all people and the compositions of these letters, which are diverse but defined by placement before or after <each other> for representing the names of all things), necessarily, although all numbers <taken together> are infinite, each neither could nor ought to have had its own name, especially since men as well, using numbers in almost every action, would be excessively hindered if in their counting they were always forced by the necessity of counting to have ready to hand an infinite multitude of names for numbers. <**35.4**> For this reason it was necessary to limit the infinite progression of numbers by means of definite limits <and> to designate them with <only a> few names, so that a person would not always be forced to proceed in counting by new additions of both numbers and names, but could embrace any sum with a few names through the brief repetition of earlier names.<**35.5**> Hence, since it was impossible for all numbers to have names, but necessary for some,[43] reason demanded – by the prescription of nature – that only 12 from all numbers have names: three limits, i.e., the ten, the hundred and the thousand and the nine first numbers established under ten from one to nine.

<**36.1**> The nine more than all the other numbers justified this rationale by a rightful privilege, inasmuch as it contained in itself almost all species of numbers and of numerical ratios. <**36.2**> For this rationale could not consist in the three, even though it is dedicated to God, because it lacked the first perfect number, which

---

[41]Cf. *Liber mahameleth* 3.10-11: numerus crescit in infinitum. Unde singuli numeri non potuerunt propriis nominibus designari. From this point until 35.5 (et aliquos necesse) the text follows the *Liber mahameleth* (ed. Vlasschaert, 3.11-24) almost *verbatim*.

[42]Reading <*loquendi*> instrumenta with the *Liber mahamaleth*.

[43]Here the agreement between the *Liber mahameleth* and this text ends. The *Liber mahameleth* continues with 'et quoniam necesse erat eos inter se multiplicari, idcirco dispositi sunt per ordines sive differentias' ('and since it was necessary that they be multiplied by each other, they were arranged in orders or differences, <i.e., places>').

is the six; but neither for this reason could it consist in the six, because it lacked the first cube, which is the eight; but neither for that reason in the eight, because it lacked the first true plane, which is in the nine. <**36.3**> Thus it was from this plenitude of virtues that the nine deserved to have the rationale of counting and of naming numbers in itself, beyond which no number except the three limits would have its own name. <**36.4**> And of course both celestial and earthly things, both bodies and spirits seem to have been formed and put in order according to the model of the nine; for nine are the spheres of celestial bodies, nine are the orders of celestial spirits and nine are the temperaments of all bodies. <**36.5**> It was, therefore, incumbent that there be nine compositions of number in which alone the whole infinity of numbers would consist, just as the universe of bodies from nine temperaments. <**36.6**> For just as in temperaments one is equal and another unequal, but one alone is in the middle (tempered), likewise in numbers also one is even, another odd and among them all only the unit is not unlike itself in any part, is always the same, always equal.

<**37.1**> In this manner creatures will only avoid departing from whatever likeness <they have> to their Creator while they contain themselves within that number, because, when the first odd number is multiplied by itself, that number comes about which, after the unit, alone is consecrated to God (because "God rejoices in the odd number[44] "). <**37.2**> Hence also, beside the nine, among the other <numbers> only the three limits received names, so that by this of course they might maintain some likeness to the Trinity, which is the true limit of all things, "the alpha and the omega, the beginning and the end";[45] and so that they might never depart from the root of the nine. <**37.3**> Because of this, then, reason demanded that, because the universe of things is contained within the nine, in the same manner the infinity of numbers should be restrained within the nine and be designated by nine names and represented by nine figures. <**37.4**> For every copy retains the likeness of its paradigm; otherwise the one would not be the copy or paradigm of the other, and because, as was said, almost all things were founded on the model of the nine, it was necessary that the very infinity of numbers itself be restrained under the nine, so that number also would not depart from that shape according to which the Creator composed all things and which the totality of things borrowed from number.

<**38.1**> Hence men as well, imitating primeval nature, gave names to only nine numbers and invented only nine shapes to represent all numbers. <**38.2**> But because there was still a species of number lacking, which the nine did not contain

---

[44] Virgil, *Eclogues* 8.73-5: "I draw these triple threads with their three different colours around you and thrice I lead this effigy around these altars; god rejoices in the odd number."

[45] *Revelation* 1:8: "I am Alpha and Omega, the beginning and the ending, saith the Lord, which is and which was and which is to come, the Almighty."

within itself—i.e., the "superabundant" number, the first <of> which is the twelve—for that reason, after the nine, names were given only to the three limits, so that the nine with its root, i.e., the three, contained all the dignities and properties of number within itself and no property, no mystery could be found in numbers which does not appear to be contained in the whole nine with its root. <**38.3**> Since, therefore, neither all nor most numbers, but only a few were necessarily to be designated by their own names, for the reasons already stated names were granted only to the nine numbers and the three limits: that they might better serve human needs through their conveniently small number, express the hidden secrets of things through some sort of symbols and not depart from the rules of nature. <**38.4**> So much for these things.

<**39.1**> The unit is <both> the origin and the first part of number;[46] for every number is naturally made up of units and the unit, since it is simple, precedes every number by nature. <**39.2**> Furthermore, because it is simple, it can generate[47] nothing by multiplication of itself except that by which it is multiplied, something which do not occur in the others which are not simple; for from the multiplication of any number by itself, just as <from its multiplication> by any other, it is necessary for a different number to emerge. <**39.3**> When the unit is multiplied by itself, however, it generates only itself; for one times one is one. For by whatever number you multiply it, you make nothing except that number by which you multiply it and because no number is generated from it except that by which it is previously multiplied, for that reason, since in the beginning there was nothing to which it could be joined for the generation of the first number, it was necessary that it itself be doubled in itself and made different from itself in some fashion, so that, taking it both as itself and as something different from itself, as if from different things, it was possible for something to be generated. <**39.4**> This is the first generation of number, which appears in the two; for this reason the two is also called the principle of difference, because it was born from the unit when the latter was made different. For this reason as well, it is only the case with the two and with no other that from its multiplication by itself the same arises as from its addition <to itself>; for it does not consist of number. <**39.5**> And since, beside the two, there was not yet anything except the unit, for this reason the unit is joined with the two like a man to a female; from the joining of these is born the three, which, after the unit, first receives the names "odd" and "male".

---

[46]It would seem necessary to add "prima" (the "first" part) because of the parallel passage in the *Liber Alchorismi* (Allard 1992: 63): unitas est origo et prima pars numeri. Omnis enim numerus ex ea componitur...; see Lampe 2005: 14.

[47]Compounds of *genero* are translated as "generate (generation)" throughout, although at times "produce" would be closer to modern mathematical usage.

<**40.1**> For even numbers are called female as if soft because they are easily divided;[48] but odd numbers <are called> male, as if strong and indivisible. <**40.2**> The unit, however, is neither even nor odd in actuality; hence, the unit per se is neither female nor male in actuality, but potentially both. <**40.3**> Hence when it is joined with the female, then a male, i.e., an odd number, is generated; when it comes together with a male, however, it begets a female, because <it begets> an equal number. <**40.4**> Thus from the first generation of the unit only the female is born (i.e., an equal number), because <it is> the two. <**40.5**> For it was fitting that, in the procreation of its first offspring, the unit only play the role of the man, i.e., of the worthier <gender> and that from the unit, as if from the male, the female be born. For the first female <was born> from a man and not the first man from a woman.[49] <**40.6**> Thus at the second level, inasmuch as the unit is joined to the female (i.e., the two), the three, which is male, is generated. Then, at the third level, the unit is conjoined with the male and a female—i.e., the four—comes forth; similarly in the rest *ad infinitum*. <**40.7**> Thus it was necessary that the unit be neither even nor odd. For, if it were only even, when it was joined to even numbers, nothing would be engendered, just as in the conjunction of two females. If, on the other hand, it were only odd, when it was joined to odd numbers, it would engender nothing, just as a male with a male. <**40.8**> Thus it was necessary that it be neither in actuality, but both in potentiality, so that when it was joined in turn with all <numbers>, being born according to the power of each sex, the fecund offspring of numbers would be propagated *ad infinitum*.

<**41.1**> But because the first and natural generation of numbers, in the fashion previously stated, seemed to be multiplied without end, it pleased the diligence of certain men afterwards that it should be limited by certain definite levels and limits according to the model of human generation. <**41.2**> For the generation of men, like that of numbers, progresses from one individual being doubled in gender and descending by male and female *ad infinitum*. <**41.3**> Later, however, humans invented levels and limits through their labor, by which they designated relationships among people, so that, although they knew that they were all equally descended from one man, yet they did not doubt, because of the assigned levels, that some belong more to others due to <these> relationship<s>; and all those found to be included in the same levels of relation were said to be from one family. <**41.4**> Similarly in the case of numbers too, following their natural composition and essence,

---

[48]This idea depends on the similarity of *mulier* ("woman") to *mollis* ("soft") and the etymology on this basis in Isidore (*Etym.* XI.2.17).

[49]cp. *Gen.* 2:23: "And Adam said, This is now bone of my bones and flesh of my flesh: she shall be called Woman, because she was taken out of Man."

humans invented through their industry roots, nodes[50] and limits, just as it invented trunks (?) and levels in humans and they distinguished the generations of number by the nines, so that all numbers up to the ninth level which are born from the same limit are called by a single common name <indicating> their relationship and distinguishing them from others; those which exceed any given nine, on the other hand, recognize that they belong to an entirely different relationship. <**41.5**> Thus, to distinguish among this sort of relationships among numbers, humans, by their inventiveness, called some 'digits', others 'articles',[51] and others 'composites', but they called those from which all these are born the limits, like the first parents of each generation. <**41.6**> For they established that those which the unit had engendered in the first creation by the aggregation of itself–those up to nine–should be called 'digits', inasmuch as they are the first-born of the unit, so that this first nine was called the nine of digits or of units, since the first limit of this first nine was the unit, given that the unit had first engendered these from itself. <**41.7**> After this follows the second nine, which is that of the tens or of articles and the limit of this nine, as for the first, is a unit, but ten times greater than the first <nine>. <**41.8**> After this nine of tens, there follows the third nine, that of hundreds; the limit of this is also a unit, but ten times greater than that in the second <nine>. <**41.9**> After this third nine follows the fourth nine, that of thousands; its limit is also a unit, but ten times that of the third. And so on *ad infinitum*.

<**42.1**> Because all numbers were engendered by the unit, this was rightly established as the limit of all the nines, according to the variety of their positions, namely so that it came about that what had engendered all numerical species from itself would also be the limit of limits according to the diversity of their places. <**42.2**> Hence the unit is set first at the beginning of all generations as their limit, so that from this it is shown to be the mother of all. <**42.3**> Hence arises the fact that, just as the unit, which by nature is the first limit, had engendered the digits in the first creation by aggregation of itself with them, so in the second establishment it seemed good that this very same unit, as each limit, when added to the first < establishment> should generate the composites and when multiplied by the first < establishment > should produce the articles. <**42.4**> Those numbers have, there- fore, been called digits, which were naturally engendered from the unit up to nine; those <have been called> articles, which are generated from the other limits by the multiplication of the first < establishment >; those numbers are called composites, which are born from digits and limits or articles linked together (for they are called "composite" as if created from different kinds). <**42.5**> Hence they follow the prop- erty of those very things from which they are allotted their substance. <**42.6**> For when we say 12 or 23 or 120, these are composites <created out of> digit<s> and

---

[50]The nodes are presumably the "articuli" referred to in earlier parts of this text and <41.5> below.

[51]Literally articulus, or "joint" (as in the finger joints).

boundar<ies> or else also from article<s>. <**42.7**> But whatever limit or article is in them is taken according to the power of limit or article, i.e., for ten, 20, or 100; whatever is from the nine of digits, however, is taken for as many units as are seen to be contained in it. <**42.8**> Thus, all the nines are organized according to the model of the one preceding, so that each has a unit as its limit, has a two, a three and thus obtains each up to nine, as the table below indicates.

Difference of hundred-millions	Difference of ten-millions	Difference of millions	Difference of hundred-thousands	Difference of ten-thousands	Difference of thousands	Difference of hundreds	Difference of tens	Difference of units or digits
1	1	1	1	1	1	1	1	1
2	2	2	2	2	2	2	2	2
3	3	3	3	3	3	3	3	3
4	4	4	4	4	4	4	4	4
5	5	5	5	5	5	5	5	5
6	6	6	6	6	6	6	6	6
7	7	7	7	7	7	7	7	7
8	8	8	8	8	8	8	8	8
9	9	9	9	9	9	9	9	9

<**42.9**> For just as in the first limit twice one made the two of units, so in the second limit twice ten makes the two of tens, which is 20 and in the third limit, twice one hundred <makes> the two of hundreds, which is two hundred and so on in each by each up to nine. <**42.10**> And because nothing is born from numbers except by either aggregation or multiplication of the first nine, <the nine> is repeated in itself in all numbers; and it is shown to be prior to all because, having been born before all others, it appears still to preserve its natural establishment. <**42.11**> For this reason the unit itself, which is the mother of all <numbers>, only engenders nine numbers in whatever limit it appears, whether through aggregation or multiplication according to the number of the first-born <numbers>.

<**43.1**> But because ten follows after nine in the natural order and the unit is always, except in the first limit, found set after nine by human establishment, for this reason it is necessary that ten be signified by a unit placed after the first nine, and thus the unit, having become a ten by the nature of its location, becomes the limit of the tens, just as earlier it had simply been the limit of the units, so that the mother of the articles or composites is the same as what was clearly the mother of the digits. <**43.2**> And because ten naturally always follows nine and a unit is always placed in that position, for this reason another unit follows after the nine of tens, which is the third limit — that of the hundreds, and thus a unit follows any nine whatsoever as the limit of those which follow. <**43.3**> Since moreover every limit—except the first—follows after the preceding nine, for this reason, when it becomes a ten, it is always found to be ten times the preceding limit, because, no matter what nine it comes after, it is born from the preceding limit multiplied by ten. <**43.4**> And because all articles are born through the multiplication of their limit by the first <nine>, it is necessary, in order that they not appear irregular, for them to follow the rule of their limits, namely, inasmuch as limits are ten times the

limits preceding them and those numbers that are born from the multiplication of these <limits>, those numbers must be found to be ten times the numbers preceding them. <**43.5**> For just as the second limit is ten times the first, so the articles of the tens are ten times the articles of the digits and just as the third limit is ten times the second, so the articles of the hundreds as well are multiplied by ten by comparison to the articles of the tens. <**43.6**> Thus the limits, articles and composites (which lie between them) are each ten times those preceding limits, articles and composites that they follow.

<**44.1**> And thus all limits and articles and composites, just like digits, are established in subordination to the nine, such that the first nine is that of digits, the second, of articles, the third, of composites and so on for the rest of this sort. <**44.2**> For thus it seemed good that every number be terminated in a nine as if in the ultimate level of its kind and that a unit ten times the preceding limit—because <it is> the tenth <element> after nine—should be established after the nine, as the limit for every nine. <**44.3**> Thus the entire fecund progeny of numbers is extended infinitely, separated into sets of nine and descending through its generations from its limits as if from its forebears. <**44.4**> Thus the nine holds the position of leadership in all things, restricting what is infinite, marking distinctions in what is restricted, but in such a way that it begins from and finds its terminus in a limit, so that it is shown to have been, not itself the maker of the world, but rather the paradigm for the world in the mind of the Maker; hence it is generated from the three multiplied by itself. <**44.5**> For he who made all things made that as well according to which, as a paradigm, he formed the rest; for God made all things in number, weight and measure.[52] <**44.6**> Hence if number too was made, He made it according to number, so that number did not exceed the laws of number, since other things were to be composed according to its form. <**44.7**> But the number according to which number was created is in this manner indeed uncreated.

G

<**45.1**> There are eight species of multiplication and the same number of division. For either we multiply whole units by whole units, or fractions by fractions, or fractions by whole units, or whole units by fractions, or fractions and whole units by whole units, or fractions and whole units by fractions, or whole units by whole units and fractions, or fractions by whole units and fractions. <**45.2**> Whenever an even or odd number multiplies an even number, or an even number multiplies an odd number, an even number results. But if an odd number multiplies an odd number, an odd number arises.

<**46.1**> Divide minutes by minutes or seconds by seconds or thirds by thirds or

---

[52]cf. *Wisdom of Solomon* 11:21: "You have disposed all things by measure, number and weight."

fourths by fourths or fifths by fifths or sixths by sixths, and whatever results will be degrees. And when each of these numbers is multiplied by degrees, whatever results will be of the same kind as the fraction. <**46.2**> If minutes divide seconds or seconds thirds or fourths fifths or fifths sixths or thirds sixths, whatever results from the divisions will be denominated by the larger fractions.

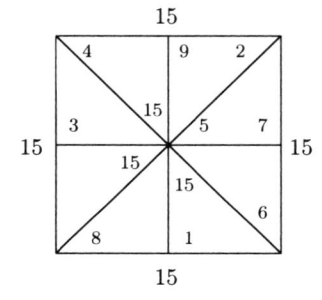

216

Mathematical translation and notes

A

<1> Rule of the mean value

<1.1> $n = \frac{1}{2}[(n - m) + (n + m)]$, $1 \leq m < n$.

<1.2> $5 = \frac{1}{2}(4 + 6) = \frac{1}{2}(3 + 7) = \frac{1}{2}(2 + 8) = \frac{1}{2}(1 + 9)$.

<2> The difference of squares

<2.1> $n^2 = (n - m)(n + m) + m^2$.

<2.2> $5^2 = 4 \times 6 + 1^2 = 3 \times 7 + 2^2 = 2 \times 8 + 3^2 = 1 \times 9 + 4^2$.

<2.3> $n^2 - 1 = (n - 1)(n + 1)$.

<3> The sum of arithmetical progressions of continuous numbers

<3.1> If n is even, $\sum_{i=1}^{n} i = \frac{n}{2} \cdot n + \frac{n}{2}$.

<3.2> $\sum_{i=1}^{8} i = \frac{8}{2} \times 8 + \frac{8}{2} = 36$.

<3.3> If n is even, $\sum_{i=1}^{n} i = \frac{n}{2}(n + 1)$.

<3.4> If n is odd, $\sum_{i=1}^{n} i = $ (the larger part[1] of n) $\cdot n$;

$$\sum_{i=1}^{7} i = 4 \times 7 = 28.$$

<3.5> If n is odd, $\sum_{i=1}^{n} i = \frac{n+1}{2} \cdot n$.

<3.6> $\sum_{i=m}^{n} i = \sum_{i=1}^{n} i - \sum_{i=1}^{m-1} i$.[2]

<4> The sum of arithmetical progressions of continuous even numbers

<4.1> $\sum_{i=1}^{n} 2i = n(n + 1)$.

<4.2> $\sum_{i=1}^{5} 2i = 5 \times 6 = 30$.

---

[1] 'The larger part' and 'the smaller part' of an odd number are two technical words which frequently occur in the text. Sometimes we find in their place 'the larger half' and 'the smaller half' respectively. Let an odd number be denoted by n, then the larger part of it is $\frac{n+1}{2}$, as the rule of <3.5> shows; the the smaller part of it is $\frac{n-1}{2}$.

[2] In discussing the sum of the arithmetical progression of continuous numbers, the cases when the number of the terms is even and odd are considered respectively, The general case is discussed in the rule <16.1>.

<4.3> $\sum_{i=1}^{n} 2i = \frac{2n+2}{2} \cdot \frac{2n}{2}.$

<4.4> $\sum_{i=m}^{n} 2i = \sum_{i=1}^{n} 2i - \sum_{i=1}^{m-1} 2i.$

<5> **The sum of arithmetical progressions of continuous odd numbers**

<5.1> $\sum_{i=1}^{n} (2i - 1) = n^2.$

<5.2> $\sum_{i=1}^{5} (2i - 1) = 5^2 = 25.$

<5.3> $\sum_{i=1}^{n} (2i - 1) = (2n - \frac{2n}{2})^2.$

<5.4> $\sum_{i=m}^{n} (2i - 1) = \sum_{i=1}^{n} (2i - 1) - \sum_{i=1}^{m-1} (2i - 1).$

<6> **The sum of geometrical progressionss of doubles and triples**

<6.1> $\sum_{i=1}^{n} 2^i = 2^n \times 2 - 2.$

<6.2> $\sum_{i=1}^{5} 2^i = 32 \times 2 - 2 = 62.$

<6.3> $2^n \times 2 - 2 = \sum_{i=1}^{n} 2^i.$

<6.4> $\sum_{i=m}^{n} 2^i = 2^n \times 2 - 2^m.$[3]

<6.5> $1 + \sum_{i=1}^{n} 2^i = 1 + (2^n \times 2 - 2).$

<6.6> $\sum_{i=1}^{n} 3^i = \frac{3^n - 1}{2} \times 3.$

<6.7> $\sum_{i=1}^{n} 3^i = \frac{243 - 1}{2} \times 3 = 363.$[4]

<7> **The sum of squares**

<7.1> $\sum_{i=1}^{n} i^2 = (\sum_{i=1}^{n} i) \cdot \frac{2}{3}n + (\sum_{i=1}^{n} i) \cdot \frac{1}{3}.$

<7.2> $\sum_{i=1}^{4} i^2 = (\sum_{i=1}^{4} i)(\frac{2}{3} \times 4) + (\sum_{i=1}^{4} i) \cdot \frac{1}{3} = 10(3 - \frac{1}{3}) + 10 \cdot \frac{1}{3} = 30.$

---

[3]This is the general formulae of the sum of the continued 'doubles' whether the first term is 2 or not.

[4]Although the text produces correct results for the sum of the continued 'doubles' and 'triples', we cannot see any generalisation of them to the other numbers.

<8> Relation between the middle number(s) and the extremes of arithmetic and geometric progressions

<8.1> If n is odd, $a_{i+1} = a_i + d$ $(i = 1, 2, ..., n-1)$, $m = \dfrac{n+1}{2}$, then

$2a_m = a_{m-j} + a_{m+j}$ $(j = 1, 2, ..., m-1)$.

<8.2> If n is odd, $a_{i+1} = q \cdot a_i$ $(i = 1, 2, ..., n-1)$, $m = \dfrac{n+1}{2}$, then

$a_m^2 = a_{m-j} \cdot a_{m+j}$ $(j = 1, 2, ..., m-1)$.

<8.3> If n is even, $a_{i+1} = q \cdot a_i$ $(i = 1, 2, ..., n-1)$, $m = \dfrac{n}{2}$, then

$a_m \cdot a_{m+1} = a_{m-j} \cdot a_{(m+1)+j}$ $(j = 1, 2, ..., m-1)$.[5]

<8.4> If n is even, $a_{i+1} = a_i + d$ $(i = 1, 2, ..., n-1)$, $m = \dfrac{n}{2}$, then

$a_m + a_{m+1} = a_{m-j} + a_{(m+1)+j}$ $(j = 1, 2, ..., m-1)$.[6]

<9> Multiplication of digits

<9.1> If $a < 10$, $a^2 = 10a - a(10 - a)$.

<9.2> $6^2 = 10 \times 6 - 6(10 - 6) = 36$.

<9.3> If $a < b < 10$, $ba = 10a - a(10 - b)$.

<9.4> $7 \times 5 = 50 - 5(10 - 7) = 35$.

<9.5> If $a < b < 10$, $ab = 10b - b(10 - a)$.

<9.6> $5 \times 7 = 70 - 7(10 - 5) = 35$.[7]

<10> The Rule of three numbers in proportion

<10.1> If $a : b = b : c$, then $ca = b^2$, $c = \dfrac{b^2}{a}$.

<10.2> If $a : b = b : c$, then $a = \dfrac{b^2}{c}$.

<10.3> If $a : b = b : c$, then $b = \sqrt{ca}$.

<11> The Rule of four numbers in proportion

<11.1> If $a : b = c : d$, then $da = cb$.

(a) If $x_1 : x_2 = x_3 : x_4$, $x_4 x_1 = x_3 x_2 = A$, then $x_i = \dfrac{A}{x_{5-i}}$.

---

[5] Rules <8.2> and <8.3> discuss the relation between the middle number(s) and the extremes of geometric progression. <8.2> discusses the case when the number of the terms is odd, and <8.3> when the number is even. The general rule is stated in <13.4>.

[6] Rules <8.1> and <8.4> discuss the relation between the middle number(s) and the extremes of arithmetical progression. <8.1> discusses the case when the number of the terms is odd, and <8.4> when the number is even. However, the general rule $2a_i = a_{i-1} + a_{i+1}$ is not discussed.

[7] Rules <9.1>, <9.3> and <9.5> are also discussed in Cashel I (pp. 18-19). Apart from the difference of the terminology, Cashel I gives different examples for the cases as well. It is worth noting that Cashel I gives a synthetic statement of the rules <9.3> and <9.5>, and the formulae can be expressed as $ab = 10a - b(10 - b)$, $a < b < 10$ which should be $ab = 10a - a(10 - b)$, $a < b < 10$. The rule <14.3> can be considered as the generalized case of the above three rules.

(b) If $x_1 : x_2 = x_3 : x_4$, then $x_i = \dfrac{x_j}{x_{5-i}} \cdot x_{5-j}$.

<11.2> If $a : b = c : d$, then $d = \dfrac{cb}{a}$.

<11.3> If $a : b = c : d$, then $a = \dfrac{cb}{d}$.

<11.4> If $a : b = c : d$, then $b = \dfrac{da}{c}$.

<11.5> If $a : b = c : d$, then $c = \dfrac{da}{b}$.

<11.6> since $10 : 30 = 2 : 6$, $6 \times 10 = 2 \times 30 = 60$.

<11.7> If $10 : 30 = 2 : x$, then $x = \dfrac{2 \times 30}{10} = 6$.

<11.8> If $x : 30 = 2 : 6$, then $x = \dfrac{2 \times 30}{6} = 10$.

<11.9> If $10 : x = 2 : 6$, then $x = \dfrac{6 \times 10}{2} = 30$.

<11.10> If $10 : 30 = x : 6$, then $x = \dfrac{6 \times 10}{30} = 2$.

<11.11> If $a : b = c : d$, then a, c and b, d represent the same things respectively.[8]

<12> **Application of the rule of four**

<12.1> If $a : b = c : d$, then $c = \dfrac{(c+d)a}{a+b}$.[9]

<12.2> If $a : b = c : d$, then $d = \dfrac{(c+d)b}{a+b}$.[10]

<12.3> If $a : b = c : d$, then $a = \dfrac{(a+b)c}{c+d}$, $b = \dfrac{(a+b)d}{c+d}$.[11]

<12.4> If $2 : 4 = c : d$, $c + d = 9$, then $c = \dfrac{9 \times 2}{2+4} = 3$.

<12.5> If $a : b = c : d$, $c < d$, then $c = \dfrac{(d-c)a}{b-a}$.

If $a : b = c : d$, $c > d$, then $c = \dfrac{(c-d)a}{a-b}$.[12]

<12.6> If $a : b = c : d$, $c < d$, then $d = \dfrac{(d-c)b}{b-a}$.

---

[8]Here the text discusses the categories of the four terms of a proportion. The category of the first term must be the same as that of the third term, as must that of the second and fourth term. In theoretical mathematics, these are unnecessary to mention. But here it is set down as a general rule, which shows the evident practical tendency of the proportional theory in the twelfth century.

[9]This rule is based on the following: if $da = cb$, then $da + ca = cb + ca$.

[10]This rule is based on the following: if $da = cb$, then $da + db = cb + db$.

[11]This rule is based on the following: if $ad = bc$, then $ad + ac = bc + ac$, $ad + bd = bc + bd$. Rules <12.1>-<12.3> could be considered as applications of the proportion by addition theorem.

[12]This rule is based on the following: if $cb = da$, then $cb - ca = da - ca$ when $c < d$, $ca - cb = ca - da$ when $c > d$.

If $a : b = c : d$, $c > d$, then $d = \dfrac{(c-d)b}{a-b}$.[13]

<12.7> If $a : b = c : d$, $a < b$, then $a = \dfrac{(b-a)c}{d-c}$, $b = \dfrac{(b-a)d}{d-c}$.

If $a : b = c : d$, $a > b$, then $a = \dfrac{(a-b)c}{c-d}$, $b = \dfrac{(a-b)d}{c-d}$.[14]

<12.8> If $2 : 4 = c : d$, $d - c = 3$, then $c = \dfrac{3 \times 2}{4 - 2} = 3$.

<12.9> The rules are the same when the categories of the terms of a proportion are considered.[15]

<12.10> If $a : b = c : d$, then $c = \sqrt{\dfrac{(cd)a}{b}}$.[16]

<12.11> If $a : b = c : d$, then $d = \sqrt{\dfrac{(cd)b}{a}}$.[17]

<12.12> If $2 : 4 = c : d$, $cd = 18$, then $c = \sqrt{\dfrac{18 \times 2}{4}} = 3$.

<12.13> If $2 : 4 = c : d$, $cd = 18$, then $d = \sqrt{\dfrac{18 \times 4}{2}} = 6$.

<12.14> If $a : b = c : d$, $b + c = p = m + n$, then m, n which satisfy $\begin{cases} m + n = p \\ mn = ad \end{cases}$ are b, c.[18]

<12.15> If $2 : b = c : 6$, $b + c = 7 = m + n$, the solutions for m, n which satisfy $\begin{cases} m + n = 7 \\ mn = 6 \times 2 = 12 \end{cases}$ are 3 and 4, and they are b, c respectively.[19]

---

[13] This rule is based on the following: if $da = cb$, then $db - da = db - cb$ when $c < d$, $da - db = cb - db$ when $c > d$.

[14] This rule is based on the following: if $ad = bc$, then $ad - ac = bc - ac$ and $bd - bc = bd - ad$ when $a < b$, $ac - ad = ac - bc$ and $bc - bd = ad - bd$ when $a > b$. Rules <12.5>-<12.7> could be considered as applications of the proportion by addition and subtraction theorem.

[15] Here again the categories of the four terms of the proportion are discussed. See the note on <11.11>.

[16] This rule is based on the following: if $ad = bc$, then $(ad)c = (bc)c$, or $(cd)a = bc^2$.

[17] This rule is based on the following: if $ad = bc$, then $(ad)d = (bc)d$, or $(cd)b = ad^2$. Rules <12.10> and <12.11> could be considered as applications of the equal ratios theorem.

[18] The two conditions for m, n are in the form of a Babylonian quadratic equation. When it is reduced into one variable equation, it turns into $m^2 + (ad) = pm$ or $n^2 + (ad) = pn$. To get the values for m, n, however, from the following example we can see that this text does not give the method. In fact, because $\begin{cases} b + c = p \\ bc = (ad) \end{cases}$, the values for b, c must be those for m, n. So this rule belongs to proportional theory rather than quadratic theory.

What is more, the following example does not consider the other case when the values for b, c are 4 and 3 respectively, i.e. when the proportion is $2 : 3 = 4 : 6$.

[19] Rules <12.14> and <12.15> are based on the rule of three and connect the proportional theory

B

<13> Relation between the middle number and its extremes
<13.1> $x^2 - 1 = (x - 1)(x + 1)$.

<13.2> $x^2 = (x - a)(x + a) + a^2$.[20]

<13.3> $5^2 = 4 \times 6 + 1^2 = 3 \times 7 + 2^2 = 2 \times 8 + 3^2 = \cdots = 1 \times 9 + 4^2$.

<13.4> If $a_{i+1} = q \cdot a_i$ $(i = 1, 2, ..., n-1)$, then $a_{j+1}^2 = a_j \cdot a_{j+2}$ $(j = 1, 2, ..., n-2)$.[21]

<14> Multiplication of a sum
<14.1> $(a + b)^2 = a^2 + b^2 + 2ab$.[22]

<14.2> $(\sum_{i=1}^{n} a_i)a = \sum_{i=1}^{n}(a_i \cdot a)$.[23]

<14.3> If $10^{n-1} < b < 10^n$, then $ab = a \times 10^n - a(10^n - b)$.[24]

<15> Calculation of fractions[25]
<15.1> If $a = b$, then $a \div b = 1$.

<15.2> If $a > b$, $a = nb$, then $a \div b = n$.

<15.3> If $a > b$, $a = nb + c$, $1 \leq c < b$, then $a \div b = n + \dfrac{c}{b}$.

<15.4> If $a < b$, then $a \div b = \dfrac{a}{b}$.

<15.5> $a \cdot \dfrac{1}{a} = a$.

<15.6> $3 \times \dfrac{1}{3} = 4 \times \dfrac{1}{4} = 1$.

<15.7> If $a < b$, then $a \cdot \dfrac{1}{b} = \dfrac{a}{b}$.

---

with the quadratic theory.

[20] Although the rules <13.1> and <13.2> here are essentially repetitions of the rules <2.3> and <2.1> respectively, they discuss the topic in a different way. First, <13.1> is the same as <2.2>, and <13.2> is the same as <2.1>. The rule <2.2> is a special case of the general rule of <2.1>, so there is some disharmony in their arrangement. Here the text takes the inverse sequence and emphasizes that <13.2> is more general than <13.1> which is more logical.

[21] Although rules <8.2> and <8.3> have discussed this topic of geometric progression, they are in fact rather special cases of <13.4>. <8.2> and <8.3> apply only to the middle term(s) of all of the terms whereas <13.4> applies to any medium terms of geometric progression. So this rule is much more general and essential than the former two ones.

[22] This is the well-known formula of the square of the sum. In al-Khwarizmi (p. 24), there is a similar rule expressed by a number and an unknown, i.e., $(10 + x)^2 = 100 + 20x + x^2$.

[23] This rule exhibits the distributive law of multiplication.

[24] Here the use of 'limit' is not the same as in other contexts. It is any number with only one numeral and one or more zeros. So here 'limit' seems to include 10, which is not the case in most other contexts.

[25] Multiplication, division and addition of the fractions are discussed in these rules. However, subtraction is not discussed.

$<15.8>$ $6 \times \dfrac{1}{12} = \dfrac{6}{12} = \dfrac{1}{2}$.

$<15.9>$ If $a = nb$, then $a \cdot \dfrac{1}{b} = n$.

$<15.10>$ $6 \times \dfrac{1}{3} = (3 \times 2) \times \dfrac{1}{3} = 2$.

$<15.11>$ If $a = nb + c$, $c < b$, then $a \cdot \dfrac{1}{b} = n + \dfrac{c}{b}$.

$<15.12>$ $8 \times \dfrac{1}{3} = 2\dfrac{2}{3}$.

$<15.13>$ $\dfrac{1}{a} \cdot \dfrac{1}{b} = \dfrac{1}{ab}$.

$<15.14>$ $\dfrac{1}{3} \times \dfrac{1}{4} = \dfrac{1}{3 \times 4} = \dfrac{1}{12}$.

$<15.15>$ $\dfrac{c}{a} \cdot \dfrac{1}{b} = \dfrac{c}{ab}$.

$<15.16>$ $\dfrac{2}{3} \times \dfrac{1}{4} = \dfrac{2}{3 \times 4} = \dfrac{1}{6}$.

$<15.17>$ If $a < b$, then $\dfrac{1}{a} : \dfrac{1}{b} = b : a$.

$<15.18>$ $\dfrac{1}{3} : \dfrac{1}{6} = 6 : 3 = 2$.

$<15.19>$ $\dfrac{b}{a} : \dfrac{c}{a} = b : c$.

$<15.20>$ $12 : 9 = \dfrac{12}{3} : \dfrac{9}{3}$.

$<15.21>$ $\dfrac{1}{a} + \dfrac{1}{b} = (a + b) \cdot \dfrac{1}{ab}$.

$<15.22>$ $\dfrac{1}{3} + \dfrac{1}{4} = (3 + 4) \cdot \dfrac{1}{3 \times 4} = \dfrac{7}{12}$.

$<16>$ Sum of arithmetic progressions

$<16.1>$ $\displaystyle\sum_{i=1}^{n} i = \dfrac{n^2 + n}{2}$.

$<16.2>$ If n is odd, $a_{i+1} = a_i + d$ $(i = 1, 2, ..., n-1)$, $m = \dfrac{n+1}{2}$, then $\displaystyle\sum_{i=1}^{n} a_i = n \cdot a_m$.

$<16.3>$ $3 + 4 + 5 = 3 \times 4$.

$<16.4>$ $3 + 5 + 7 + 9 + 11 = 5 \times 7$.

$<16.5>$ $2 + 5 + 8 = 3 \times 5$.

$<16.6>$ If n is even, $a_{i+1} = a_i + d$ $(i = 1, 2, ..., n - 1)$, $m = \dfrac{n}{2} + 1$, then $\displaystyle\sum_{i=1}^{n} a_i = (n + 1)a_m - a_{n+1}$.[26]

$<16.7>$ $2 + 4 + 6 + 8 = (2 + 4 + 6 + 8 + 10) - 10 = 5 \times 6 - 10 = 20$.

$<16.8>$ $1 + \displaystyle\sum_{i=1}^{n} (2_i - 1) = n^2$.[27]

---

[26] This rule is the deduction from rule $<16.2>$.

[27] The rule is equivalent to, but not the same as, rule $<5.1>$. In this rule, 1 is not considered as odd which is consistent with part F below, while in rule $<5.1>$, 1 is considered as an odd number.

<17> Calculation of rations

<17.1> If $a^2 + b^2 = c^2$, $a_1 : a = b_1 : b = c_1 : c = p$ then $a_1^2 + b_1^2 = c_1^2$.[28]

<17.2> If $a_i = p_i a$ $(i = 1, 2, ..., n)$, then $a = \dfrac{\sum_{i=1}^{n} a_i}{\sum_{i=1}^{n} p_i}$.

<17.3> If $b = p_b \cdot a$, $c = p_c \cdot a$, $d = p_d \cdot a$, $b + c + d = g$, then $a = \dfrac{g}{p_b + p_c + p_d}$.[29]

<17.4> Let S, P represent the money of Socrates and Plato.

$$S = 2\frac{2}{3}P = 15; \quad 2\frac{2}{3} = \frac{2 \times 3 + 2}{3} = \frac{8}{3};$$

$$15 : 8 = P : 3 = (5\,pennies + 1\tfrac{1}{4}\,obols) : 3\,pennies,$$

so $P = 5\,pennies + 1\dfrac{1}{4}\,obols$;

$$(5\,pennies + 1\tfrac{1}{4}\,obols) \times 2\tfrac{2}{3} = 2(5\,pennies + 1\tfrac{1}{4}\,obols) + \tfrac{2}{3}(5\,pennies + 1\tfrac{1}{4}\,obols)$$

$$= (11\tfrac{1}{4} + 3\tfrac{3}{4})\,pennies = 15\,pennies.$$

<17.5> $\dfrac{2}{3}(9\,obols) = 6\,obols = 3\,pennies; \quad \dfrac{9}{8} \times \dfrac{2}{3} = \dfrac{6}{8} = \dfrac{3}{4}; \quad 11\tfrac{1}{4} + 3 + \dfrac{3}{4} = 15.$

<17.6> Or if $2\frac{2}{3}P = 15$, then $P = \dfrac{15 \times 3}{2 \times 3 + 2} = 5\dfrac{5}{8}$.

<17.7> If $b > a$, then $\dfrac{cb}{a} = \dfrac{b}{a} \cdot c$.

<17.8> if $b < a$, then $\dfrac{cb}{a} = \dfrac{c}{a/b}$.

<17.9> $\dfrac{100 \times 1\,pound}{1\,shilling} = \dfrac{1\,pound}{1\,shilling} \times 100 = 20 \times 100 = 2000.$

<17.10> $\dfrac{24{,}000 \times 1\,penny}{1\,pound} = \dfrac{24{,}000}{1\,pound/1\,penny} = \dfrac{24{,}000}{240} = 100.$

<17.11> If $(\dfrac{1}{a} + \dfrac{1}{b})c = d$, then $c : d = (ab) : (a + b)$, $c = \dfrac{(ab)d}{a + b}$.

<17.12> If $(\dfrac{1}{3} + \dfrac{1}{4})c = 20$; $\dfrac{3 + 4}{3 \times 4} = \dfrac{7}{12}$; then $c : 20 = 12 : 7$, $c = \dfrac{12 \times 20}{7} = 34\dfrac{2}{7}$.[30]

<17.13> If $\dfrac{a}{b} = \dfrac{b}{c}$, then $ac = b^2$.[31]

* <17.14> If $b_1 : a_1 = b_2 : a_2 = \cdots = b_n : a_n$, then $b_i = \dfrac{(\sum_{i=1}^{n} b_i)a_i}{\sum_{i=1}^{n} a_i}$ $(i = 1, 2, ..., n)$.

* <17.15> If $b_1 : a_1 = b_2 : a_2 = \cdots = b_n : a_n$, then $b_i = \dfrac{a_i}{\sum_{i=1}^{n} a_i}(\sum_{i=1}^{n} b_i)$ $(i = $

There is a similar rule in Cashel I (p. 19). Unlike rules <5.1> and <16.8>, it uses the 'large half' to express the number of the terms, but, as in rule <5.1>, 1 is considered as an odd number.

[28] This rule is an application of the famous Pythagorean theorem to the proportional theory.

[29] Rules <17.2> and <17.3> could be considered applications of the distributive law of multiplication.

[30] Note in the margin: $c = (1 + \frac{5}{7})20 = \frac{12}{7} \times 20$.

[31] This rule is a repetition of the rule in the first sentence of <10.1>.

$1, 2, ..., n).^{32}$

<17.16> If $a_1 = 6$, $a_2 = 8$, $a_3 = 12$, $a_1 + a_2 + a_3 = 26$, $b_1 + b_2 + b_3 = 60$, then
$a_i : 26 = b_i : 60$, $b_i = \dfrac{60 a_i}{26}$ $(i = 1, 2, 3)$.

<17.17> If $a : b = c : x$, then $x = \dfrac{bc}{a}$.

<17.18> If $8 : 24 = 3 : x$, then $x = \frac{24 \times 3}{8}$; if $8 : 24 = 5 : x$, then $x = \dfrac{24 \times 5}{8}$.

C

<18> Solutions for quadratic equations

<18.1> The three compound quadratic equations are:
$\quad$ (1) $x^2 + ax = c$; (2) $x^2 + c = ax$; (3) $ax + c = x^2$.^{33}

<18.2> (1) If $x^2 + ax = c$, $x = \sqrt{\left(\dfrac{a}{2}\right)^2 + c} - \dfrac{a}{2}$.

<18.3> If $x^2 + 10x = 39$, $x = \sqrt{\left(\dfrac{10}{2}\right)^2 + 39} - \dfrac{10}{2} = 3$.^{34}

<18.4> (2) If $x^2 + c = ax$, $x = \dfrac{a}{2} - \sqrt{\left(\dfrac{a}{2}\right)^2 - c}$.^{35}

---

[32] Rules <17.14> and <17.15> are based on the proportion by addition theorem.

[33] In this text, "the thing" refers to the square of the root. Modern readers may think it easier to express "the thing" as an unknown x. But in ancient Arabic mathematics, 'the thing' can refer not only to the root of a square, such as in Khwarizmi, but also to a square itself, such as in Abu Kamil.

[34] This example exists in many Arabic algebraic texts such as in Khwarizmi (p.8) and and Abu Kamil (pp.30-32) with different terminologies for the square.

[35] The statement of the procedure of this rule is similar to rules <18.2> and <18.6>. Strictly speaking, the solution is not complete. The correct formula expressing the solutions of this type of quadratic equation should be $x = \frac{a}{2} \pm \sqrt{(\frac{a}{2})^2 - c}$.

<18.5> If $x^2 + 9 = 6x$,[36] $x = \dfrac{6}{2} - \sqrt{\left(\dfrac{6}{2}\right)^2 - 9} = 3 - \sqrt{0} = 3 - 0 = 3$.[37]

<18.6> (3) If $ax + c = x^2$, $x = \dfrac{a}{2} + \sqrt{c + \left(\dfrac{a}{2}\right)^2}$.

<18.7> If $3x + 4 = x^2$, $x = \dfrac{3}{2} + \sqrt{4 + \left(\dfrac{3}{2}\right)^2} = 4$.[38]

D

To divide a number by given ratios

* <19.1> If $\displaystyle\sum_{i=1}^{n} x_i = x$, $x_1 : a_1 = x_2 : a_2 = \cdots = x_n : a_n$, then

$$\sum_{i=1}^{n} a_i : a_i = x : x_i, \quad x_i = \left(\frac{x}{\sum_{i=1}^{n} a_i}\right) a_i.$$

For $x_i = \left(\dfrac{x}{\sum_{i=1}^{n} a_i}\right) a_i$, if the residue of xa_i is r, and the integer part of the

quotient is s, then $x_i = \left(\dfrac{x}{\sum_{i=1}^{n} a_i}\right) a_i = s + \left(\dfrac{1}{\sum_{i=1}^{n} a_i}\right) r$.

<19.2> If $x = 40$, $n = 4$, $x_2 = 4x_1$, $x_3 = 5x_2$, $x_4 = 3x_3$;

$$a_1 = 1, \; a_2 = 1 \times 4 = 4, \; a_3 = 4 \times 5 = 20, \; a_4 = 20 \times 3 = 60, \; \sum_{i=1}^{4} a_i = 85;$$

then $85 : a_i = 40 : x_i$.[39]

<19.3> $85 : 4 = 40 : x_2$, $x_2 = \dfrac{(4 \times 40)}{85} = 1\frac{75}{85}$.

---

[36] This example exists in Ibn Turk (p.101, p.166), who does not give arithmetical procedures for the solution. The equation $x^2 + c = ax$ generally has two solutions $x = \frac{a}{2} \pm \sqrt{(\frac{a}{2})^2 - c}$ when $(\frac{a}{2})^2 > c$; in the special case when $(\frac{a}{2})^2 = c$ there is one solution; and when $(\frac{a}{2})^2 < c$ there is no solution for it. These had already been discussed by Arabic mathematicians such as Khwarizmi and Abu Kamil who influenced medieval European mathematics greatly. The statement of the general rule for this equation only considers one solution $x = \frac{a}{2} - \sqrt{(\frac{a}{2})^2 - c}$ which is connected with subtraction. But the example $x^2 + 9 = 6x$ is rather special because it belongs to the case of $(\frac{a}{2})^2 = c$, and in such a case the solutions from addition and subtraction are the same, i.e. there is only one solution for the equation. So, although only subtraction is considered in the example, the result is still right.

[37] The calculation of the square root of 0 and the subtraction by 0 are novelties in ancient Arabic mathematics. From this calculation, we may see some influences of Indian mathematics on this text.

[38] This example is the same as that of Khwarizmi (pp. 12-13). The order of the three types of composite quadratic equations is the same as that of Khwarizmi and the examples of the first case and the third case are also the same as those in Khwarizmi. So there is a strong tendency to think that Part C of this text should be based on Khwarizmi or in his tradition.

[39] The diagram in the text shows that 85 is the first term of the proportional expression, 40 is the third, the empty place symbol '0' is the fourth, which represents the number corresponding to the

$$<19.4> \quad x_1 = \frac{x_2}{4} = \frac{1\frac{75}{85}\,shillings}{4} = \frac{12\,pennies + \frac{75}{85}shillings}{4}$$

$$= 3\,pennies + \left(\frac{18}{85} + \frac{1}{85} \cdot \frac{3}{4}\right) shilling$$

$$<19.5> \quad x_3 = 5x_2 = 5 \times 1\frac{75}{85} = 9\frac{35}{85}.$$

$<19.6>$ If $a = mb$, $b = \dfrac{a}{m}$.

$<19.7>$ $160 \times 5 = 85 \times 9 + 35$.

$<20>$ Multiplication of continuous ratios

$<20.1>$ If $x_i = a_i \cdot x_{i+1}$ $(i = 1, 2, ..., n-1)$ and $x_n = a_n \cdot x_0$, then

$$x_1 = x_0 \cdot a_1 \cdot a_2 \cdot ... \cdot a_{n-1} \cdot a_n, \quad x_2 = x_0 \cdot a_2 \cdot a_3 \cdot ... \cdot a_{n-1} \cdot a_n,$$

$$x_3 = x_0 \cdot a_3 \cdot a_4 \cdot ... \cdot a_{n-1} \cdot a_n, \quad x_i = x_0 \cdot a_i \cdot a_{i+1} \cdot ... \cdot a_{n-1} \cdot a_n.$$

$<20.2>$ If $a_1 = 4$, $a_2 = 3$, $a_3 = 2$, $a_4 = 1$, $x_0 = 6$, then $x_1 = 6 \cdot 4 \cdot 3 \cdot 2 \cdot 1 = 144$.

$<20.3>$ $x_2 = 6 \cdot 3 \cdot 2 = 36$.

$<20.4>$ $x_3 = 6 \cdot 2 = 12$.

$<20.5>$ $x_4 = 6 \cdot 1 = 6$.

$<21>$ To find the root

$<21.1>$ $\sqrt{a} = \dfrac{\sqrt{b \cdot a}}{\sqrt{b}}$. [40]

$<21.2>$ $\sqrt{2\frac{1}{4}} = \dfrac{\sqrt{4 \times 2\frac{1}{4}}}{\sqrt{4}} = 1\frac{1}{2}$.

$<21.3>$ $\sqrt{2} = \dfrac{\sqrt{a \cdot 2}}{\sqrt{a}}$. [41]

$<22>$ Multiplication of the columns

$<22.1>$ If $a, b < 10$, then $(a \times 10^{n-1})(b \times 10^{m-1}) = (a \cdot b) \times 10^{n+m-2}$. [42]

---

second term, and either of the ratios, i.e. 1, 4, 20 and 60 for the four men respectively, can be the second term. So the figure denotes in fact four proportional expressions. It is interesting to point out that the empty place or the symbol '0' can take four different values. Let the second term be denoted as x, then $85 : x = 40 : 0$, or $0 = \frac{40x}{85}$. It will be much clearer if we substitute y for the symbol '0' so that we have $y = \frac{40x}{85}$, a special "function" in the modern sense. Of course, it was not considered as a function at that time, although we now acknowledge it as belonging to a certain function. However, such practices are very important in the formation of the general concept of function.

[40] There is a similar rule in Khwarizmi (pp. 30-31) which is illustrated by several numerical examples.

[41] We are not sure if this explanation corresponds to the text exactly. The text here is not very clear. "The nearest square" and "the third number" may refer to the same number 2 which is in the diagram.

[42] Let a and b be two numerals belonging to two numbers respectively; their places are n and m respectively, or the number denoting a's column is n and that denoting b's column is m. The text says the numbers denoting the digit and article of the product of a and b are $n + m - 1$ and $n + m$

<23> Simplified multiplication

<23.1> If $c < 10$, $\dfrac{a}{c \times 10^n} = \dfrac{1}{d}$, then $ab = (c \times 10^n)\dfrac{b}{d}$.

<23.2> $25 \times 32 = 100(32 \times \dfrac{25}{100}) = 100 \times 8$, $8 : 32 = 25 : 100$.

<23.3> $25 \times 32 = 50(32 \times \dfrac{25}{50}) = 50 \times 16.$[43]

<23.4> If $a, b, c < 10$, then

$$(c \times 10^n + a)(c \times 10^n + b) = (c \times 10^n)(c \times 10^n) + (c \times 10^n)(a + b) + a \cdot b.$$

<23.5> $18 \times 16 = 10 \times 10 + 10(8 + 6) + 8 \times 6 = 288$.

<24> Multiplication of roots

<24.1> $\sqrt{a} \cdot \sqrt{b} = \sqrt{a \cdot b}.$[44]

<24.2> $\sqrt{40} \times \sqrt{10} = \sqrt{40 \times 10} = 20$.

<25> Calculation of ratios

<25.1> $x + x + x + \dfrac{1}{2}x + \dfrac{1}{4}x = 100$.

<25.2> $1 + 1 + 1 + \dfrac{1}{2} + \dfrac{1}{4} = \dfrac{15}{4}$, $\dfrac{100}{\frac{15}{4}} = \dfrac{\frac{400}{4}}{\frac{15}{4}} = 26\dfrac{2}{3}$.

E

<26> To determine the places of the digit and the article of the product of two numerals

<26.1> The text says there are four types of numbers, i.e. the digit, the article, the limit and the composite. The digit, article and limit in this text have two usages. On the one hand, the digit may represent any number which contains only one numeral, i.e. any of the integers from 1 to 9; the article may represent any number which contains one numeral and one zero, i.e. any of the tens from 10 to 90; and the limit may represent any number that contains one numeral and two or more zeros, i.e. $a \times 10^n$, a is a numeral and $n \geq 2$. On the other hand, they may also represent the column of the digit (the first column), the

---

respectively.

To explain this, we see that a represents the number $a \times 10^{n-1}$, and b represents $b \times 10^{m-1}$. So the product of a and b represents $(a \times 10^{n-1})(b \times 10^{m-1}) = (a \cdot b) \times 10^{n+m-2}$.

This means that the place of the digit of $a \cdot b$ is $n + m - 1$, and the place of the article of $a \cdot b$ which is 1 more than that of the digit is $n + m$.

The statement of the text is a general rule which is consistent with the modern rule of multiplication of exponents with base 10.

[43] This rule is still taught and emphasized very much in modern junior schools and is vivid in our everyday life.

[44] This rule is equivalent to <21.1>.

column of the ten (the second column) and any other column (any of the rest columns) respectively. The composite number contains at least two numerals. These four types of numbers exhaust the positive integers. So it is based on the numbers of numerals and zeros contained in a number that the text gives the four types of numbers.

<26.2> For the case of when two articles are multiplied, let a, b be the numerals. First calculate the product $a \cdot b$, then determine the places of the digit and the article of $a \cdot b$. The places of a, b are both 2, so the place of the digit of $a \cdot b$ is $2 + 2 - 1 = 3$, and that of the article of $a \cdot b$ is $2 + 2 = 4$. So the contexts are equivalent to the formula $(a \times 10)(b \times 10) = (a \cdot b) \times 100$.

<26.3> $70 \times 20 = (7 \times 2) \times 100 = 1,400$.

<26.4> For the multiplication of two composite numbers, multiply all the numerals of the multiplicand by those of the multiplier, determine the place of each numeral of the products, and then take their sum.

* <26.5-26.6> $64 \times 23 = 4 \times 3 + 6 \times 3 \times 10 + 4 \times 2 + 6 \times 2 = 12 + 180 + 80 + 1200 = 1,472$.

<26.7> By means of this rule and its examples, the rule of multiplication of the columns (<22>) is explained again in a much clearer way.

<27> Multiplication of the digit

<27.1> The text states different results of the multiplication of two digits.

<27.2> If $a, b < 10$, then $(a \times 10) \cdot b = (a \cdot b) \times 10$.

<27.3> $70 \times 7 = (7 \times 7) \times 10 = 490$.

<27.4> $(a \times 100) \cdot b = (a \cdot b) \times 100$.

<27.5> $900 \times 3 = (9 \times 3) \times 100 = 2,700$.

<28> Multiplication of the article

<28.1> If $a, b < 10$, then $(a \times 10)(b \times 10) = (a \cdot b) \times 100$.

<28.2> $70 \times 30 = (7 \times 3) \times 100 = 2,100$.

<28.3> $(a \times 100)(b \times 10) = (a \cdot b) \times 1,000$.

<28.4> $500 \times 30 = (5 \times 3) \times 1000 = 15,000$.

<29> Multiplication of the hundred

<29.1> $(a \times 100)(b \times 100) = (a \cdot b) \times 10,000$.

<29.2> $500 \times 300 = (5 \times 3) \times 10,000 = 150,000$.

<30> Multiplication involved with one repeated thousand

<30.1> $(a \times 10^n) \cdot b = (a \cdot b) \times 10^n$ (a, b are numerals and $n \geq 3$).

<30.2> $30,000 \times 6 = (3 \times 6) \times 10^4 = 180,000$.

<30.3> $(a \times 10^n)(b \times 10) = (a \cdot b) \times 10^{n+1}$ (a, b are numerals and $n \geq 3$).

<30.4> $4,000,000,000,000 \times 30 = (4 \times 3) \times 10^{12+1} = 120,000,000,000,000$.

<30.5> $(a \times 10^n)(b \times 100) = (a \cdot b) \times 10^{n+2}$ (a, b are numerals and $n \geq 3$).

<30.6> $50,000,000,000,000 \times 200 = (5 \times 2) \times 10^{13+2} = 10,000,000,000,000,000$.

<31> The column number of the numeral of a repeated thousand

<31.1> For $a \times 10^{3n}$ $(a < 10)$, the column number of a is $3n + 1$.

<31.2> For $3 \times 10^{3 \times 4}$, the column number of 3 is $3 \times 4 + 1 = 13$.

<31.3> For $(a \times 10) \times 10^{3n}$ $(a < 10)$, the column number of a is $3n + 2$.

<31.4> For $50 \times 10^{3 \times 4}$, the column number of 5 is $3 \times 4 + 2 = 14$.

<31.5> For $(a \times 100) \times 10^{3n}$ $(a < 10)$, the column number of a is $3n + 3$.

<31.6> For $(a \times 100) \times 10^{3 \times 2}$, the column number of a is $3 \times 2 + 3 = 9$.

<32> Multiplication of two repeated thousands

<32.1> $(a \times 10^p \times 10^{3n})(b \times 10^q \times 10^{3m}) = [(a \times 10^p) \times (b \times 10^q)] \times 10^{3(m+n)}$ $(p, q = 0, 1, 2)$.

<32.2> $(a \times 10^{3n})(b \times 10^{3m}) = (a \times b) \times 10^{3(m+n)}$.
$(3 \times 10^{3 \times 2})(7 \times 10^{3 \times 4}) = (3 \times 7) \times 10^{3(2+4)} = 21 \times 10^{18}$.

<32.3> $(a \times 100 \times 10^{3n})(b \times 100 \times 10^{3m}) = [(a \times 100) \times (b \times 100)] \times 10^{3(m+n)}$.
$(300 \times 10^{3 \times 4})(500 \times 10^{3 \times 2}) = (300 \times 500) \times 10^{3(4+2)} = 150,000 \times 10^{18} = 15 \times 10^{22}$.

<32.4> The text mentions briefly the case when $p = 1$ and $q = 1, 2$ in the formula of <32.1>.

<33> The name of the column of the numeral of a repeated thousand

<33.1-33.2> If $n = 3l$ (n is the column number), then $10^{n-1} = 100 \times 10^{3m}$ ($m = l - 1$).

<33.3> If $n = 12$, $l = \dfrac{12}{3} = 4$ and $m = 4 - 1 = 3$, then $10^{11} = 100 \times 10^{3 \times 3}$.

<33.4> If $n = 3l + 2$ (n is the column number), then $10^{n-1} = 100 \times 10^{3l}$.

<33.5> If $n = 11$ and $l = 3$, then $10^{10} = 10 \times 10^{3 \times 3}$.

<33.6> If $n = 3l + 1$ (n is the column number), then $10^{n-1} = 10^{3l}$.

<33.7> If $n = 10$ and $l = 3$, then $10^9 = 10^{3 \times 3}$.[45]

<34> Calculation of the concealed number

<34.1> Let n be the concealed number.

If $3n = 2m + 1$, $3(m + 1) = 2l + 1$, $l + 1 = 9p + q$ $(q < 9)$, then $n = 4p + 3$.

If $3n = 2m + 1$, $3(m + 1) = 2l + 1$, $l + 1 < 9p$, then $n = 3$.

If $3n = 2m$, $3m = 2l$, $l = 9p$, then $n = 4p$.

If $3n = 2m + 1$, $3(m + 1) = 2l$, $l = 9p + q$ $(q < 9)$, then $n = 4p + 1$.

If $3n = 2m$, $3m = 2l + 1$, $l + 1 = 9p + q$ $(q < 9)$, then $n = 4p + 2$.

If $3n = 2m$, $3m = 2l + 1$, $l + 1 = q < 9$, then $n = 2$.[46]

---

[45] Although there is no concept of exponent in this manuscript, the rules of the multiplication of two numbers show the equivalent computation to the multiplication of exponents with base 10, especially those rules of <22> and <30>-<33>.

[46] The text gives the first three rules in detail. With the fourth and fifth rule, the text does not give detailed discussions, probably to avoid repetitions, but only mentions the operations on the times of 4.

<34.2> If $n = 2$, $3 \times 2 = 6$, $3 \times \dfrac{6}{2} = 9$, $\dfrac{9+1}{2} < 9$, then $n = 2$.[47]

<34.3> If $\dfrac{ax}{1} = \dfrac{12x}{b} = \dfrac{ax + 12x}{c}$, then $c = \dfrac{12}{a} + 1$.

<34.4> If $x = 5$, $a = 1$, then $b = \dfrac{12}{1} = 12$; If $x = 5$, $a = 2$, then $b = 6$; If $x = 5$, $a = 3$, then $b = 4$.[48]

<34.5> $(x + m) - (x - m) = 2m$.[49]

<34.6> $(5 + 2) - (5 - 2) = 4$.

These rules were arrived at most probably by means of such considerations as follows.

Every number n can be expressed by either $n = 4p$ $(p > 0)$, or $n = 4p + 1$ $(p \geq 0)$, or $n = 4p + 2$ $(p \geq 0)$, or $n = 4p + 3$ $(p \geq 0)$. We consider the four forms one by one.

If $n = 4p$, $3n = 2(6p) = 2m$, $3m = 2(9p) = 2l$, then $l = 9p$. This corresponds to the third rule.

If $n = 4p + 1$, $3n = 2(6p + 1) + 1 = 2m + 1$, $3(m + 1) = 2(9p + 3) = 2l$, then $l = 9p + 3$. This corresponds to the fourth rule.

If $n = 4p + 2$, $3n = 2(6p + 3) = 2m$, $3m = 2(9p + 4) + 1 = 2l + 1$, then $l + 1 = 9p + 5$. This corresponds to the fifth and the sixth rules.

If $n = 4m + 3$, $3n = 12m + 9$, $3 \times \frac{(12m+9)+1}{2} = 18m + 15$, then $\frac{(18m+15)+1}{2} = 9m + 8$. This corresponds to the first and the second rules.

[47]The formula for expressing the statement of the text looks strange. This is because of the brevity of the expressions in this text. In fact, it is a statement by a third person of a game which is performed by two other persons. The game goes like this:

A: I am concealing a number in my mind. (The number is 2 which A knows and B does not.) Can you work out what it is?

B: Sure. You triple your number first and tell me whether the resulting number is even or odd.

A: (A will get 6 as the result.) It is even.

B: Then you divide the result into two equal parts, and take the triple of either of them. After that, you tell me again whether the result is even or odd.

A: (This time A will get 9.) It is odd.

B: Now you take the larger part of the odd number, i.e. take the number which is the nearest to the half of the odd number and is larger than this half. Having done this, tell me how many nines are contained in the larger part.

A: (A will get the larger part of 9 as 5.) Your question is absurd. In fact the larger part is less than 9.

B: So the number that you asked me to seek is 2.

[48]This problem is different from the above one. In this case, it is unnecessary to know the value of the concealed number. It is the same with the following problem.

[49]Rule <35.5> may be considered as a kind of practical application of the rule <1.1>.

F

The section from <35.1> to <44.7> (= F) is a philosophical discussion concerning the construction of numbers. The meanings of the digit, article, limit and composite are explained. In contrast to the previous five parts, the meaning of the 'limit' of a number here refers to the start of the corresponding place value (column), for example, the 'limit' of digits is 1; the limit of articles is 10 and so on, while the 'limit' of a number in the first five parts of this manuscript refers to the start of the next corresponding place value (column), and the places of the 'limit' are at least 3. For example, the limit of 25 is 100, the limit of 250 is 1000 and so on.

Sentences <39.5> and <40.2> state that 1 (the unit) is neither even nor odd. The rule <16.8> in B is consistent with this, i.e., 1 is not considered as an odd number; while in the rules from <5.1> to <5.4> in A, 1 is considered odd. Therefore, A and B are not based on the same source, whereas B and F may be.

G

<45> **Multiplication of whole units and fractions**

<45.1> Let u denote whole units, and f denote fractions, the rule states 8 species of multiplication and division with the former listed as following:

1 $u \cdot u$, 2 $f \cdot f$, 3 uf, 4 fu, 5 $u(f + u)$, 6 $f(f + u)$, 7 $(u + f)u$, 8 $(u + f)f$.

<45.2> $n(2i) = 2k$, $(2i + 1)(2k + 1) = 2j + 1$.

<46> **Multiplication and division of fractions**

<46.1> In sexagesimal numeration the result of the division of two fractions of the same kind is expressed in degrees, but the multiplication of a fraction and a degree is expressed according to the kind of fraction.

<46.2> When a fraction is divided by another fraction, the result is denominated by a higher fraction.

IX

Learning Indian Arithmetic in the Early Thirteenth Century

The most momentous development in the history of pre-modern mathematics is the shift from using roman numerals to using Indian numerals and the 'Indian way' of doing arithmetic that the use of these numerals entailed. Indian numerals were originally Sanskrit symbols that had been introduced into the Islamic world by the early ninth century, when their use was described by the mathematician and astronomer, al-Khwarizmi (ca. 825 A.D.). Al-Khwarizmi's 'On the calculation of the Indians' was, in turn, introduced to a Latin-reading public through a series of translations and adaptations produced from the early twelfth century onwards. This new kind of arithmetic became known as the algorism ('algorismus'), after the Arabic author, and the numerals were described as being either Indian or Arabic. At first there was considerable variety in the forms of numerals used, but by the early thirteenth century, they had become standardised and, with small exceptions (in particular, in the shapes of the '4' and the '5'), became the 'Arabic numerals' that are used universally today. The acceptance of the algorism within the canon of European mathematics was ensured by the magisterial *Liber abbaci* of Leonardo of Pisa (Fibonacci) in two editions (1202 and 1228), and the more popular manuals of Alexander de Villa Dei (the *Carmen de algorismo*) and of John of Sacrobosco (*Algorismus vulgaris*), both slightly later in the thirteenth century.[1]

Nevertheless, such was the novelty of Indian calculation that it took some time for it to be understood and accepted. Particularly difficult was the idea

---

[1] A recent account of the spread of the algorismus can be found in *Die älteste lateinische Schrift über das indische Rechnen nach al-Hwarizmi*, ed., trans. and comm. by M. Folkerts, with the collaboration of P. Kunitzsch, Abhandlungen der Bayerischen Akademie der Wissenschaften, phil.-hist. Klasse, n.F., 113, Munich, 1997 (English summary on pp. 163-83). The *Liber abbaci* has been edited by B. Boncompagni (Rome, 1857); a new edition by A. Allard is nearing completion. The Latin texts of Alexander de Villa Dei and John of Sacrobosco are available only in J. O. Halliwell, *Rara mathematica*, London 1841, pp. 1-26 (John of Sacrobosco) and 73-83 (Alexander de Villa Dei); modern critical editions are needed. A convenient annotated translation of the major part of Sacrobosco's text can be found in *A Source Book in Medieval Science*, ed. E. Grant, Cambridge, Mass., 1974, pp. 94-102. For arithmetical procedures in the abacus and algorism in the context of the history of numerals, see Georges Ifrah, *The Universal History of Numbers*, English version, London, 1998, pp. 556-66.

that a single symbol could be used to express an infinite range of numbers. In roman numerals, each power of ten is expressed by a different symbol (i, x, c, m etc.); in Indian calculation the symbol remains the same, but the power of ten is indicated by a new symbol for which roman numerals had no equivalent at all: the zero. An idea of the puzzlement that the zero caused can be gained from a note in a late twelfth-century manuscript now in Cambridge, in which, among the several names given to the new symbol, is the 'chimaera', the imaginary beast of mythology.[2]

A half-way stage towards assimilating Indian arithmetic is represented by the 'Gerbertian abacus', which was used for teaching arithmetic from the late tenth century onwards, and in which the Indian numerals were employed only to mark counters which were placed on an abacus board (a piece of wood or thick parchment ruled with lines). In this case, the columns of the abacus themselves indicated the decimal place and a counter for zero was not necessary. In the algorism the numerals were written directly onto the writing surface (whether this was a wax tablet, a tray sprinkled with a thin covering of sand, or a sheet of parchment—later paper), and no lines were drawn to demarcate the decimal places. Some instructions could apply equally to the abacus and the algorism, and sometimes the terminology appropriate to the abacus was carried over to the algorism.

In this article, I would like to show how the essentials of the new arithmetic were taught on the borders between England and Wales sometime in the early thirteenth century, but before the works of Fibonacci, Alexander of Villa Dei and Sacrobosco had become standard.

The evidence comes from a manuscript written, possibly, by a monk in the abbey of Tewkesbury (or a closely-related house),[3] who gathered under a single cover a collection of texts and notes which all more or less concern number. This collection is now the first 120 pages of a manuscript in Cashel (Tipperary, Ireland): G. P. A. Bolton Library (formerly Cathedral Library), Medieval MS 1. Its mathematical contents can be classified as follows:

1. Calendrical material: tables (for the years 1168-1223, and 1140-1642), short texts, some in verse, on how to calculate the various church feasts, elementary astronomy and meteorology.

2. Predictions: on the weather (p. 5); on the recovery of a sick person (the 'sphere of Pythagoras', pp. 17-18); the *Epistola Petosiris* (p. 19); on prognosticating life and death (p. 20).

---

[2]Cambridge, Trinity College, R.15.16, fol. Av, the different names of zero: 'cifra vel solfra vel nichil t. 0. cimera sipos'.

[3]This provenance is indicated by the presence of a calendar on pp. 71-6 which is derived from the calendar of Tewkesbury Abbey.

3. The algorism. The manuscript includes four works on the algorism. One occurs (pp. 111-7) in an older pamphlet (probably of the twelfth century) which has been attached to the end of the manuscript. This is the *Helcep Sarracenicum* ('Sarracen Calculation'[4]) of Ocreatus, which is distinctive for its use of the first nine roman numerals for the Indian numerals.[5] Another is a complete text on the algorism (pp. 41-58).[6] The third and the fourth are shorter introductions to the algorism (pp. 16-17 and 20-21); these are reproduced and analysed here.

1   The Table

On pp. 20-1 the function of place value is demonstrated through the use of a table (see Plates I-II). Here, each Indian numeral is copied nine times to show how it can stand for the first nine powers of ten; above each repeated Indian numeral the equivalent in roman numerals is given. The lowest digits are on the right. This may originate from the fact that in Arabic one reads from the right. But the result is that it became normal for higher digits to be written on the left, lower on the right.[7] No zeros are included in this table, which could equally show how numerals were disposed on the 'Gerbertian abacus' that was still being used at the time. One may note the cumbersome way in which the powers of ten are expressed in roman numerals; I give their equivalent in Indian numerals:[8]

i	1
x	10
c	100
$\overline{M}$	1000
x$\overline{M}$	10,000
c$\overline{M}$	100,000
$\overline{M}$ $\overline{M}$	1,000,000
decies$\overline{M}$ $\overline{M}$	10,000,000
centies$\overline{M}$ $\overline{M}$	100,000,000

---

[4] 'Helcep' is a transliteration of the Arabic word 'al-hisāb', meaning 'calculation'.

[5] This text has been analysed and edited in Burnett, '*Algorismi vel helcep decentior est* ✱ *diligentia*: the Arithmetic of Adelard of Bath and his Circle', in *Mathematische Probleme im Mittelalter: der lateinische und arabische Sprachbereich*, ed. M. Folkerts, Wiesbaden, 1996, pp. 221-331.

[6] An edition of this text by the present author is to be published in the periodical *Sciamus*.

[7] For the establishment of this norm in the West see C. Burnett, 'Why We Read Arabic Numerals Backwards', in *Ancient and Medieval Traditions in the Exact Sciences, Essays in Memory of Wilbur Knorr*, Stanford, Ca., 2000, pp. 197-202.

[8] It is customary in works on the abacus to put a tilde on the 'm', although the same sign also indicates '1000 times' the numeral: e.g. 'ī' is used for '1000', 'x̄' for '10,000' etc.

2 Some Basic Rules of Arithmetic

Indian arithmetic provides no help for the multiplication of simple digits. For these, school children nowadays learn multiplication tables (usually from $2 \times 2 = 4$, $2 \times 3 = 6$, $2 \times 4 = 8$ to $12 \times 12 = 144$). In the Middle Ages they may have done the same, but multiplication tables are also written out, the most detailed of which was composed by Victorius in Late Antiquity.[9] In texts on the algorism some convenient short cuts for multiplication are often found. This is what occurs immediately after the tables of numerals in the Cashel manuscript (p. 21). Comparison with other algorismic texts reveals that the Cashel student had some problems with understanding these short cuts.

(1) Si vis ducere digitum in se, scribe eius decuplum et aufer eum a suo decuplo per differentiam suam ad .x.

'If you wish to multiply a digit (a) by itself, write the product of a x 10 and take it away from the product of 10 times the difference between a and 10.'

What the student should have written was: 'Si vis ducere digitum in se, scribe eius decuplum et aufer *ab eo suam ductionem* per differentiam suam ad .x.': '...take away from the product (of $a \times 10$) the product of $a \times (10 - a)$'; e.g. $7 \times 7 = (7 \times 10) - (7 \times 3)$.[10] A similar mistake is made in the second rule:

(2) Si vis ducere digitum in alium, scribe decuplum minoris et aufer ipsum a decuplo suo per differentiam maioris ad .x.

'If you wish to multiply one digit by another, write the product of the lower number times 10, and take it away from the product of 10 times the difference between the higher number and 10.'

What the student should have written was: 'Si vis ducere digitum in alium, scribe decuplum minoris et aufer *ab eo ductionem maioris* per differentiam suam ad .x.': '...take away from the product (of the lesser number times 10) the product of the greater number times its difference from 10'; e.g. $7 \times 9 = (7 \times 10) - (1 \times 9)$.

Another innovation in Indian arithmetic is that of substituting numerals in the process of calculation. In the case of the 'Gerbertian abacus' one simply had to exchange one counter with another: the usual verbs in the Latin abacus treatises are 'removere' ('take off') and 'ponere' ('put down'). When calculating on a wax-tablet, a board covered with sand, or parchment, one had to erase one number and replace it by another: the usual Latin verbs are 'delere' ('destroy')

---

[9] See now Abbo of Fleury and Ramsey, *Commentary on Victorius*, ed. A. Peden, London, 2002.

[10] For the correct formulation, see Sacrobosco, *Algorismus vulgaris*, ed Halliwell, p. 12, trans. Grant, p. 98. Other versions can be found in MSS Cambridge, Trinity College, R.15.16, fol. 61r and British Library, Egerton 2261, fol. 226rb: see Burnett, 'Algorismi vel helcep', p. 305. In all these cases a single rule ('if you wish to multiply one digit by another') replaces rules (1) and (2) of the Cashel manuscript.

and 'scribere' ('write'). In this text the terminology of the abacus is still used, though this does not inevitably imply that the student was using the abacus.[11]

(3) Quotiens aliquis multiplicator multiplicat id quod supra ipsum erit, non est addendum id quod ex multiplicatione provenit ei quod supra ipsum erit sed, remoto eo quod supra ipsum erit, simpliciter ponendum est loco illius quod multiplicatione provenit.

'Whenever any multiplier (a) multiplies that (digit) which is above it (b), the product (c) should not be added to b, but rather b should be taken away and c should simply replace b.'

Next comes a rule about a number series:

(4) Si fiat ascensio per impares numeros de proximo ad proximum, si ab unitate incipiatur, ut .i. iii. v. 7., maioris maior medietas in se multiplicata omnium propositorum summam reddit.

'If there is a series of consecutive odd numbers beginning from one—e.g. 1, 3, 5, 7—the product of the larger half of the highest number multiplied by itself is the same as the sum of all the numbers in the series.' The term 'greater middle' refers to the larger of the two unequal parts that an odd number is divided into: in this case 4, since $7 = 3 + 4$; $4 \times 4 = 16 = 1 + 3 + 5 + 7$. This, too, has equivalents in Sacrobosco and the Egerton and Trinity manuscripts.[12] Note the mixture of roman and Indian numerals in this rule.

The last two rules are parallel to the first two, but this time deal with the multiplication of 'articles', i.e. numbers followed by one or more zeros:[13]

(5) Si vis ducere articulum in se, vide a quo digito de(s)cendat et quot unitates proveniunt ex multiplicatione illius digiti in se, tot cente (sic) provenient ex multiplicatione illius articuli in se.

'If you wish to multiply an article by itself, see how many decimal places it is from a unit, and the number that arises from the multiplication of the simple unit in itself will be the number of hundreds that will arise from the multiplication of the article in itself.' E.g. in the case of 700×700, '7' is two decimal places away from the units; one multiplies the digits ($7 \times 7 = 49$), and one makes them into hundreds. The word 'cente' is not attested elsewhere. 'Cente(ni)' ('hundreds') is not strictly accurate here, since the 'article' could be any power of ten higher than the unit. Thus it is worth considering whether 'cente' is not

---

[11] The verb 'ponere' is used, as well as 'scribere', in the *Helcep Sarracenicum* in the Cashel manuscript: see Burnett, 'Algorismi vel Helcep', p. 241.

[12] See Sacrobosco, *Algorismus vulgaris*, ed Halliwell, p. 19, trans. Grant, p. 100 ('progressio'), and Burnett, 'Algorismi vel helcep', p. 309.

[13] The terminology 'digit' (literally 'finger') and 'article' (literally 'knuckle') derives from the representation of numbers in finger-calculation. Note that, in my translations, 'digit' means any single number used in a calculation, whereas 'number' means the whole number (which could consist of several digits).

a truncation of 'centeni', but rather the student's deliberate attempt to find a way of referring to 'powers of ten'.

(6) Si vis ducere articulum in alium, vide a quo digito uterque descendat, et quot unitates provenient a multiplicatione unius digiti in alium, tot .c. etc.

'If you wish to multiply one article by another, see how many places each one of them is from a digit, and the number of units that arises from the multiplication of the simple digits in themselves, will be the number of hundreds etc.'

These rules are relevant both to the 'Gerbertian abacus' and to the algorism. They have parallels to the 'six rules of multiplication' of Sacrobosco[14] which in turn are similar to some rules in the Egerton and Trinity manuscripts. The Cashel rules differ from all three of these other sources by separating 'multiplication of a number by itself' (i.e. squaring) from 'multiplication of a number by a different number'; such a separation is not arithmetically necessary. This feature, however, is also found in the *Helcep Sarracenicum* in the same Cashel manuscript.[15]

3 The Algorism

The text on pp. 16-17 deals with the algorism itself. It appears to be independent of that on pp. 20-21. For it uses the zero and is more consistent in employing the terminology of the algorism ('delere' and 'scribere') rather than that of the abacus. It gives succinct rules for adding, subtraction, multiplication and division, which are common to all algorisms, but adds a paragraph on how to calculate the highest common factor and the lowest common multiple of several numbers, which is not found in other early algorisms.

(1) Cum numero numerum addere volueris, cui alium addere volueris, prescribas, addendum autem ei supponas, et sic ut numerus prime differentie sub numero prime, numerus secunde sub numero secunde, numerus tertie sub numero tertie sit differentie, et sic secundum ordinem, si plures fuerint differentie, ut semper differentie sibi respondeant. Ex coniunctione ergo suppositi et suprapositi minor numerus quam 10 vel maior vel tantum .x. proveniet. Si minor quam .x. excrescat, idem in loco suprascripti numeri ponatur, deleto suprascripto. Si tantum .x., cifra in loco suprascripti posita, unitas pro .x. in sinistriori proxima transferatur. Si autem maior quam .x., numero superexcreto denario in loco suprascripti posito, unitas similiter in sinistriori loco proximo pro denario transferatur; pro singulis etiam denariis singule unitates in sinistram transferantur partem.

---

[14]See Sacrobosco, *Algorismus vulgaris*, ed Halliwell, p. 12-13, trans. Grant, p. 98.

[15]This peculiarity is noted in Burnett, 'Algorismi vel Helcep', pp. 241 and 242.

'When you wish to add a number to a number, you put the 'adder' on top, and the 'addend' underneath, in such a way that the digit (of the 'addend') in the first decimal place is under the digit (of the 'adder') in the first decimal place, the digit in the second under the digit in the second, the digit in the third under the digit in the third, and so on, if there are more decimal places; the decimal places (of 'adder' and 'addend') will always match each other. The sum of the higher and lower digits will be less than ten, exactly 10, or more than 10. If the sum is less than ten, that digit should be put in the place of the higher digit, which is erased. If exactly 10, a zero should be put in the place of the higher digit, and a one standing for 10 should be moved to the next decimal place to the left. If more than then, the number in excess of 10 should be put in the place of the higher digit, and again a one standing for 10 should be moved to the next decimal place to the left. (As a general rule) every one that is moved to the left stands for a 10.'

(2) Cum numerum a numero subtrahere volueris, ordine predicto numeros scribas, et deinceps numerum differentie a numero differentie sibi paris tollas, si fieri potest. Quod si fieri nequit, solam unitatem a sinistra differentia demas, que unitas erit tibi pro denario, et sic ab alio denario que oportuerit tollas. Residuum autem in loco suprascripti ponas, et sic habebis propositum. Cave tamen, si a centenario vel a millenario vel deinceps unitas demenda fuerit vel consimilis, ab ultima differentia unitas dematur, et pro singulis cifris novenarius ponatur. Ad ultimum autem pervento, ab eo numerus propositus quasi a denario subtrahatur, residuum autem in loco suprascripti ponatur.

'When you wish to subtract one number from another, write the numbers in the way described before (for addition), and then take the digit in a particular decimal place from the digit in the same decimal place, if that can be done. If that cannot be done, take one from the decimal place on the left and treat it as a 10, and thus take what you need from another[16] 10. Put the remainder in the place of the higher digit, and thus you will have the answer. Make sure, however, in the cases when ones are to be taken from 100s and 1000s and so on, that the one is taken from the last decimal place (to the left), and for each zero a nine is substituted. But when one comes to the last decimal place (to the right), the relevant digit is subtracted from it as if from 10, and the remainder is put in the place of the higher digit.' E.g. if 7 is subtracted from 1000, 9s are substituted for two of the zeros and the 7 is taken from the remaining 10, to produce 993. The text is made less clear by the use of 'ultimus' to describe both the last decimal place on the left, and the last one on the right.

(3) Cum autem numerus per numerum multiplicandus fuerit, multiplicandus prescribas, multiplicatorem autem subscribas, non tamen ut in modis predictis, sed sic ut primus multiplicatoris numerus sit sub ultimo multiplicandi. Deinde

---

[16]I.e. 10 plus the unit in the same decimal place.

vero ab altiori numero tam multiplicatoris quam multiplicandi incipiatur secundum ordinem suum; numerus ex multiplicatione proveniens ponatur. Multiplicato ergo ultimo multiplicandi per quamlibet multiplicatoris, retrahantur
figure in dextram partem, prima in dextram differentiam proximam et quelibet
succedat loco alterius, ne tamen differentiam proximam aliqua transiliat.

'When one number is to be multiplied by another, write the multiplicand
on top, the multiplier underneath, but not in the same way as before (for addition and subtraction), but in such a way that the first digit (i.e. lowest) of the
multiplier is under the last digit (i.e. highest) of the multiplicand. Then one
should start from the highest digit, both of the multiplier and the multiplicand,
in order; the product should be substituted. Then, when the last digit of the
multiplicand has been multiplied by each[17] of the digits of the multiplier, the
digits (of the multiplier) are moved to the right, the first digit into the next decimal place to the right and each subsequent one into the place of its neighbour,
making sure that none jumps over a decimal place.'[18]

(4) Cum autem numerus per numerum dividendus fuerit, subscribatur dividendus et subscribatur divisor, nullo modorum predictorum, sed sic ut ultimus
divisoris sit sub ultimo dividendi, si fieri potest. Quod si fieri non poterit, in
dexteram protrahantur partem. Et postea fiat divisio a sinistra incipiendo differentia. Notandum tamen numerum ultimum totiens a sibi superscripto esse
demendum quotiens quilibet sequencium a residuo suprascripto sibi demi possit.
Ex directo ergo ultimi divisoris denominatio ponenda est iuxta quam divisio fiat,
et deinceps protrahantur differentie dextrorsum et alia tunc ponatur denominatio super ultimum numerum divisoris tracta (?) et sic deinceps ut compleatur
propositum.

'When one number is to be divided by another, one should write the dividend
on top and the divider underneath, but not in any of the previous ways, but
rather in such a way that the last (i.e. highest) digit of the divider is under the
last (i.e. highest) digit of the dividend, if possible (if that is not possible, the
digits are dragged to the right). Then the division should made starting from
the decimal place on the left. Note that the last digit should be taken away from
the digit above it as many times as each of the digits following it can be taken
away from the remainder (?) written above them. The quotient that results
from the division should be placed directly above the last digit of the divider;
then the digits should be dragged into the decimal places on their right, and
another quotient should then be placed and drawn (?) above the last[19] digit of
the divider, and so on until the answer is completed.'

---

[17]'Quamlibet' should mean 'whichever you like' or 'any'; 'quamque' would be expected.
The feminine form implies 'figura' which first appears in the next phrase.

[18]This injunction is necessary, presumably, because there might be a temptation to miss
out decimal places with zeros in them.

[19]'Proximum' ('next') would be expected.

(5) 1 2 3 4 5 6 7 8 9
10. 100. 1000. 10000. 100000.
20. 200. 2000. 20000. 200000. 2000000.
30. 300. 3000. 30000. 300000.
40. 400. 4000. 40000. 400000. et sic deinceps secundum quamlibet
figurarum quemlibet numerum scribe; attendens idoneus eris.

These are examples of writing Indian numerals, which one would have expected to have come first in this text. The student is told: 'Write each number using these numerals (figure). If you pay attention, you will be competent!'

(6) Si propositi fuerint tres[20] numeri et scire volueris maximum numerantem, illos vide si primi duo sint primi ad se invicem aut compositi. Si primi sunt ad se invicem, multiplica unum in alium et qui inde producetur erit minimus numeratus ab illis. Postea vide si ille scilicet qui educitur ex ductu primi in secundum et tertius sint primi ad se invicem aut compositi. Si compositi, quere maximum numerantem illos, divide scilicet unum per alium et minimus divisor erit maximus numerans utrumque. Quere postea minimos in proportione, divide scilicet primo unum postea alium proximum numerantem illos, et illi cum denominatione exeunte erunt .iiii. numeri proportionales. Multiplica ergo primum in quartum, et qui inde producetur erit minimus numeratus a tribus pro(p)ositis numeris. Vide ergo si ille et quartus si primi ad se invicem an compositi et tunc operandum ut supradiximus.

'If you have three numbers and want to know the highest common factor, look at them and see if the first two numbers are prime in respect to each other or composite. If they are prime to each other, multiply one by the other and the produce of this is the lowest common multiple. Then see if the product of the multiplication of the first by the second is prime in respect to the third, or composite. If composite, find the highest common factor, i.e. divide one by the other and the lowest divider will be the highest common factor. Then find the numbers smallest in proportion, i.e. divide first one, then the next one which is a factor of them, and those, added to the denomination (?) that results, will be four numbers in proportion. Then multiply the first and the fourth, and the result will be the lowest common multiple of the three numbers. See, then, if that number and the fourth one are prime or composite to each other, and then proceed as before.'

Here we have something unusual for early algorisms. Unfortunately the abbreviated nature of the text, and possibly some errors on the part of the Cashel student, make it difficult to work out exactly what arithmetical procedures are being described. The subject-matter is the highest common factor ('maximus numerans') and the lowest common multiple ('minimus numeratus'), both of

---

[20]The abbreviation in the manuscript would normally be read as 'tibi'; however, 'tres' is necessary for the context.

which have to be known when adding fractions with different denominators. If the denominators have common factors (i.e. are 'composite' in respect to one another), then it is necessary to find the highest common factor. If they do not have common factors (i.e. if they are 'prime' in respect to each other), then one must find the lowest common multiple by multiplying them together. What is unclear is why it is necessary to arrange the numbers concerned in a 'proportion' of four terms (which should mean $a : b = c : d$).

4   Conclusions

These texts give summaries of procedures, as an *aide-mémoire* written by a student of arithmetic, rather than fully-explained instructions. No examples are included. It is possible that texts **1** and **2** come from a different source from text **3**, since the former can apply equally to the abacus and the algorism, whereas the latter is specific to the algorism. The procedures described in **2** and **3** are brought together in John of Sacrobosco's *Algorismus vulgaris*, which may therefore represent a later stage in the teaching of the algorism. Both the Cashel student and Sacrobosco use 'cifra' (as a noun in the first declension) for 'zero' rather than the indeclinable 'cif(f)re' (a transliteration of the Arabic $sifr$) distinctive of earlier English writers,[21] and much of the terminology is shared by both authors: e.g. ordo, differentia, excrescere, transferre, delere, figura, etc. Nevertheless, the phraseology is sufficiently different between the Cashel and Sacrobosco text to suggest that the one is not dependent on the other. Rather, they both represent a common English tradition of the twelfth and early thirteenth century, whose richness and diversity has only recently begun to be appreciated.

---

[21] For the use of 'cif(f)re' by Adelard of Bath, 'Ocreatus' and in the *Liber ysagogarum Alchorismi*, see Burnett, 'Algorismi vel helcep', pp. 236-7.

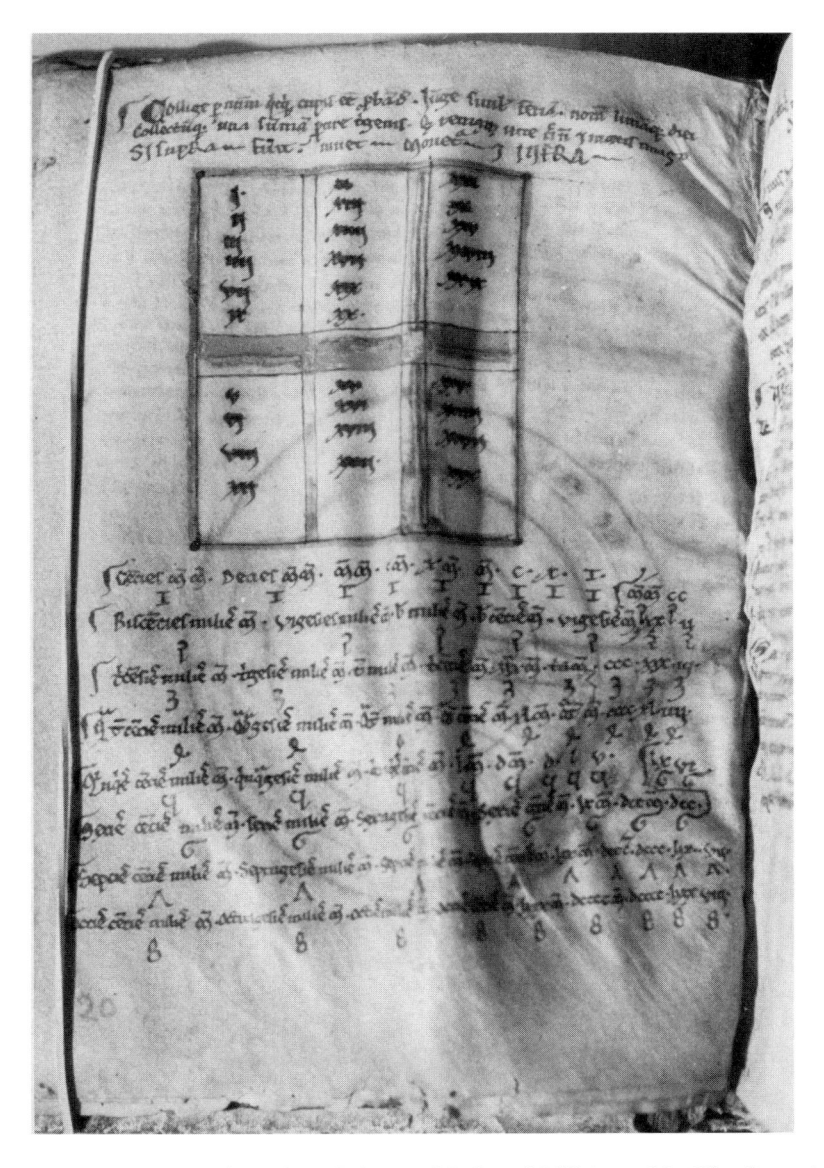

Plate I. Cashel, G. P. A. Bolton Library, Medieval MS 1, p. 20. The first eight lines of the table illustrating Indian numerals and the principle of place-value. Above this is a table for prognosticating whether a sick man will live or die, from the numerical values of the letters of his name, and those of the 'Moon' and the day.

Plate II: Cashel, G. P. A. Bolton Library, Medieval MS 1, p. 21. The last line of the numerical table, the basic rules for arithmetic, and the key to the numerical equivalents of Latin letters in onomancy.

X

Latin Alphanumerical Notation, and Annotation in Italian, in the Twelfth Century: MS London, British Library, Harley 5402

The use of the Latin letters in their alphabetical order as numerals, on the model of the notation for numerals which is normal in Greek, Arabic and Hebrew, has been observed in a group of closely related works written by a certain 'Stephen' and an "Abd al-Masīḥ of Winchester'.[1] Two of these works are dated respectively to 1121 and 1127, and were copied in Antioch. The fullest information is given in respect to the second of these: a translation of the medical compendium, *al-kitāb al-malakī* (*Liber regalis dispositionis*) of 'Alī ibn al-'Abbās al-Majūsī, of which the respective books were translated by Stephen 'the disciple of philosophy' and copied in Antioch on various days during the year 1127. This Stephen is known, from references to him in other medical works, to have originated in Pisa, and it is apparent that he expects his audience to be in Italy. While the numeral system in all these texts is virtually the same, no key is provided. They were evidently written in a context in which Latin alphanumerical notation was expected to be understood. In MS British Library, Harley 5402, however, there is a planetary table written in Latin alphanumerical notation which is accompanied by a key to this notation. The evidence for the date and provenance of this manuscript lies in the annotations in a mixture of Latin and Italian that refer to the date 1160 and to the use of the tables of Lucca. This may confirm that the audience of Stephen's translations and the associated works was indeed in Italy, and that they needed to be told how the alphanumerical notation worked.

* * *

I am very grateful for the help of David Brancaleone, Tzvi Langermann, David King, Giulio and Laura Lepschy, Nicholas Mann, Raymond Mercier, David Pingree, Faith Wallis, and Francesca Ziino. This article is also the fruit of innumerable inspiring communications with Paul Kunitzsch about the transmission of numeral forms and other aspects of the transfer of Arabic learning into Latin.

1 See C. Burnett, "Abd al-Masīḥ of Winchester', in *Between Demonstration and Imagination: Essays on the History of Science and Philosophy Presented to John D. North*, eds L. Nauta and A. Vanderjagt, Leiden, 1999, pp. 159–69, id., 'Antioch as a Link between Arabic and Latin Culture in the Twelfth and Thirteenth Centuries', in *L'Occident et le Proche-Orient au temps des croisades: traductions et contacts scientifiques entre 1000 et 1300*, ed. A. Tihon, I. Draelants, and B. van den Abeele, Turnhout, 2000, pp. 1–78, and id., 'The Transmission of Arabic Astronomy via Antioch and Pisa in the Second Quarter of the Twelfth Century', in *Perspectives on Science in Medieval Islam*, ed. J. Hogendijk and A. Sabra, Cambridge, Ma. (in press).

Several stages can be discerned in the production of MS Harley 5402.[2] The most primitive stage is represented by fols 1–69, on which a single copyist has written the following works:

   a. Fols 1r–15r.

      Pseudo-Ptolemy, *Iudicia*. This is the earliest form of a Latin text on judicial astrology which presumably derives from an Arabic text (this Arabic text has not yet been identified). Its language betrays some influence of a Gallo-Romance vernacular.[3] In this manuscript alone the work is attributed to 'Alkanderinus'.[4]

   b. Fols 15v–16r.

      Two unidentified tables, the subject of this note.

   c. Fols 17r–69r.

      The astrological collection of the ninth-century Arabic astrologer, Sahl ibn Bishr, consisting of *Introductorium, 50 Precepts, Iudicia, Elections*, and *Liber temporum*. In this manuscript the author is called 'Chelbebrith'.

These texts have been copied neatly by a professional scribe in a dull brown ink; red ink has been used for rubrication and the drawing of diagrams (on the first folio only, yellow and blue initials alternate). Items a and c are two of the earliest Latin astrological texts to be translated from Arabic. They are the major sources for Raymond of Marseilles' *Liber iudiciorum*, which was written ca. 1140, and can be found in other manuscripts of the twelfth century: the first work was apparently included in MS Chartres, Bibliothèque municipale, 213, a manuscript destroyed in the Second World War,[5] the second is in MS Munich, Bayerische Staatsbibliothek,

---

  2 P. Lehmann, 'Mitteilungen aus Handschriften', *Sitzungsberichte der Bayerischen Akademie der Wissenschaften, Philos.-hist. Abteilung*, 1930.2, Munich 1930, p. 27, states that the MS was written in the thirteenth and fourteenth centuries, and this dating has been followed by A. G. Watson, in *Catalogue of Dated and Datable Manuscripts c. 700 – 1600 in the Department of Manuscripts, The British Museum*, London, 1979, p. 168 (Watson does not mention the date '1160', but only the '1421' added by Nicholas of Cues). The later dating is plausible for fols 70–109, but I would question the thirteenth-century date for the first two stages described here. Manuscripts of scientific works used by practitioners are often dated too late because they do not show the more traditional features preserved in carefully copied manuscripts of the Schools and the Church. It is true that *e-caudata* is not used by either scribe of fols 1–70, but the first scribe uses the upright form of the 'd' alongside the curved-backed 'd'; the lines are ruled in plummet rather than dry point, but the writing is above the top line. A. Krchňák, 'Die Herkunft der astronomischen Handschriften und Instrumente des Nikolaus von Kues', *Mitteilungen und Forschungsbeiträge der Cusanus-Gesellschaft*, 3, 1963, pp. 109–80 (176–7) accepts that the earliest texts in the manuscript 'einige Jahrzehnte früher [als 1160] enstanden sind', while the cataloguers (R. Danzer, A. Krchňák, R. Haubst) in the same volume of the *Mitteilungen*, pp. 81–4 date fols 1–69 to the twelfth century.

  3 Among the words related to the vernacular are 'ialnus' ('yellow'), 'brunus' ('brown'), 'iardinum' ('garden'), and 'ingaudiare' ('to rejoice'); note also the phrase 'si autem nosse desideras quando iter fieri non debeat et ut vulgo loquamur distornetur ...'

  4 The name 'Alkanderinus' has been almost completely obliterated in the rubric to the text, but it must have been legible when the running heads were added (see p. 82 below) because the running heads are taken in all other cases from the original rubrics; 'Alkanderinus' appears on the top of every recto of the text.

  5 See Burnett, 'The Contents and Affiliation of the Scientific Manuscripts Written at, or Brought to, Chartres in the Time of John of Salisbury', in *The World of John of Salisbury*, edited M. Wilks, Studies in Church History, Oxford, 1984, pp. 127–60 (at pp. 135–6).

clm 13021 (written between 1163 and 1168); both are found in Paris, BNF, lat. 16208 (late twelfth century). In both texts roman numerals are used throughout.

The date and place of the copying of these texts is not indicated. However, soon after their copying, folios 1–70 fell into the hands of another man whose annotations, in a very informal and uneven hand in black ink, fill up the folio left incomplete after the explicit of Sahl ibn Bishr's work (fol. 69r). They start with a date (1160 A.D.), and consist of the following items, which are transcribed in full:[6]

    1) Anni domini .Mclx. correbas ciclum· lune .xix. annos· et corre· da .i. in .xix.

        Luna inueniendam· Multiplica Ciclum· lune per quinque· si'est
        infra .Ƴ0. (= 60^7) manda illa memorie, si'uero fuit Ƴ0. proice totum,
        si uero super Ƴ0. deme Ƴ0. et relicuum· manda memoria,
        querite peracto (for 'epactam'?), multiplica item ciclum· per .Ƴ., et numerum qui
                                   inde
        deciderit iunge numerum dierum· ad kalendas· januarii· usque in die
        quam querit· sit totum per .Ƴt (= 60) diuide· et de'omnia Ƴt. unitatem
        faciat· iungat illat (sic) unitas a'numero quem memoria· man-
        dasti· et'oc numerum sic acouatum (for 'aceruatum'?) si'quit numerum
                                 infra .Ƴt (= 60)
        remansit· adat illi et tunc proferat tot esse dies luna et tot·
        esse puntos.

        Os dies et punto· multiplica super .xii et numerus
        quibus inde prouenerit hec'sunt gradis (sic) separationis solis· a luna.
        Mutatur hec'pacta (for 'epacta'?) semper· ad'kalendas· januarii.

    2) Ego deus qui'aparui abraam· yzaac· et jacop jn'deo
        patre· omnipotente· et nomen meus adonai· non abscondi· eius.

 *   3) La'rasione· de'le littere Julii· Inperator· dicoti come uuoli·
        fare· per'cognoscere· ille· tutta ora guarda in'eli'uersi·
        e'la prima ⊕ ce'troui dalo incomincamento del'capo'si'segna·
        el'prima ⊕ dalo incomincamento disopra· o'dinanti ⊕ o'darieto·
        quale e', e'dunque piu'presso delā cumincata. (?) disopra si'segna· e'dunque
        giunge insieme· quelle· dinanti ⊕ con'quelle· dinanti ⊕ et'com-
        puta insieme' et sic fas de'ceteris· et'in'istum primum uersum· uide·
        qualis littere super abundat· et secundum hoc judicat· set si'fuisset pari· et pari,
        prende· quali sunt· alato· la coppia.

    4) /fol. 69v/ Sciatis quod'tabule iste facte sunt super ciuitas luce· Et da angle
                                        da'occidente
        usque ad'illam abet (i. e. habet) G<radus> .xxxiv. ed'ora'directa· es G .xv., si'uos
                                     autem fuiset in'alium

---

  6 In the realizations of abbreviations letters not in the manuscript are printed in italics. Capi-
    tals are as in the manuscript. A *punctus medius* represents the sole punctuation used in the
    manuscript; numerals and abbreviations (e.g. '.xpi.' for 'Christi', '.G.' for 'gradus' etc.) usually
    have a *punctus* before and after them. Other punctuation has been added to show the division
    of the phrases. For clarity's sake words which have been combined into one have been broken
    up into their constituent units with '; for the original see Plates I–II.
  7 The calculation (see Appendix) shows that the scribe was using 'figure indice' (the oriental
    forms of Arabic numerals) in which Ƴ = 6, rather than the western forms of Arabic numerals
    (which became the norm) in which Ƴ = 5.

locum da′oriente· aut a′occidente· uidetis longitudo terre ubi′sit· si′plus autem
fuiset longitudo de .xxxiv. G., uos′estis orientālem· Et′scias medium cursum solis
quod abet infra′terra ubi′sit et luce· et minues eum de′medium cursum solis quod
traere-
tis de′tabule· Et si′minus est longitudo′terrarum uos′estis occidentalem· crescite
medium curssum· sicut nos′dixit[8] quod minuetis· Ec′facite ad omnium planetarum.

Si′locum solis uoluis sciat anni .xpi. et in′calende marsum semper′facias caput annum·
et′introite semper′in′tabule cum anni .xpi. trapassati· et′cum′mensi· et′cum′iorni
trapasati.

et′cum′oras· et′si′tantum non′abet, de′anni concogliti· prendite de′anni spansi
et′copite nu-
merum· et′scriuite ali′anni concogliti· li .S.<igna> e′li .G.<radus> e′li .M.<inuta>
e′li .S.<ecunda> e′sic
ali spansi faciat· et sic a′mensit· (i. e., menses) et′a′iornis· et′a′oras· et′a′partes oras·
et postea
gungit totum insimul· et fac signit (i. e., signa) et .G. et .M. et .S. Et ohc est medius cursum· solis· et sic faciat ad omnium planetarum. Minuet de′medium
cursum solis· au′cum (apicem?) solis· et quo′remanes est· par solis· intra cum′parte
solis· in
regula directa· et prende cos′est incontra· \depreparatione solis/[9] de .G. et .M. et .S.
ohc est adob′solis.
Respice· si′part solis est minus de .vi. S., minues adob solis· de′medium
cursum· e′si′part solis plus es de .vi. S., eres ladob solis supra medium cursum
et′quod adueneris tibi decresere aut minuere ohc est locum solis.

Si′locum lune queris· trae medium cursum· e′trae partem lune· sicut dixit·
de′sol· minues medium cursum de′sol· de′medium cursum lune· e′si′minus
est medium cursum lune· mitte ei .xii. signos· et′minues medium cursum
sol de′medium cursum lune· et quod remanes addoppia· et ohc est lon-
gitudo· doppio· intra quon′istum (i. e. cum isto) doplum in′regula directa· et accipe
quod inueneris· in′punto .G. et .M. et quod inueneris in pres .M.
et scribe· persimul· ambi· Et postea relice[10] sil′doblament fuis .vi. S., me-
nima laddobament del′punt de′parte lune· e′si′minus fuis lo′doplum· de .vi. S.,
gungite lo′adobbament del′punt sobra la′part de′luna· e′dunc aueretis partem adobata· intra con′illam in′regulam directam et′prendite quod inueneritis in′re-
gula de′circulo minor· et scribe desobtum· pres· ce′auetis scrittum persimul·
et postea prendite quod est desubtum· adobament de′parte· et respice
pres· qualis′partem· est de .lx., e′prendite talis partem de′circulo minor· e′gun-
gite· ad adobament de′parte· et istum gungime<n>t· abet nomen adobament·
di′ueritatem·
respice partem lune· si′est minus de .vi. .S., tollite adobbament di′ueritatem
di′medius cursum lune· et si′plus est parte de .vi. S., gungite adobamentum di′ueri-
tatem supra medium cursum lune· et ohc est locum lune.

While the precise interpretation of these items remains problematic, they can be
recognized as consisting of: a handy way of finding the position of the Moon (no. 1;

---

8 The 'nos' looks more like 'uos', but the sense requires 'nos'.
9 'depreparatione solis' added above (= 'de equatione Solis'?).
10 possibly 're<s>pice'.

see Appendix), a religious formula (no. 2), a variant of the common form of divination known as 'the Victorious and Vanquished' (no. 3),[11] and some instructions for using astronomical tables (no. 4). The tables have been drawn up for Lucca, and it is reasonable to suppose (but by no means certain) that the annotator was writing in Lucca.

What is interesting about these notes is their vernacular character. The annotator pays no attention to Latin grammar, and in his word divisions he regularly joins prepositions, and the conjunctions 'et' and 'si', to the words that follow. He also combines the personal pronoun with its verb ('uosestis', 'nosdixit'), and occasionally joins relative pronouns, adverbs and the negative 'non' to the following word. His spelling is inconsistent, and he writes (presumably) as he speaks. While item n. 4 has traits of Gallo-Romance (e. g., in its omission of final vowels in *doblament, punt, part*, etc.) the majority of the vernacular forms in these notes (especially in item n. 3) are close to the Tuscan dialect that one would expect to have been spoken in the region of Lucca.[12]

There are very few texts written in Italian in this period, and this text does not seem to have been noticed in scholarship on Italian philology.[13] We may have here a rare example of the kind of language in which Jewish scholars or Arabic speakers 'interpreted' texts for the Latin-educated translators, who then put them into grammatical Latin.[14] Moreover, the instructions for the use of the tables of Lucca (item no. 4) are very similar to those for the tables for Pisa in British Library, Arundel 377.[15] These are headed 'Tractatus magistri Habrahe de tabulis planetarum'. 'Habrahe' is the Jewish scholar Abraham ibn Ezra who was also the author of another text in which he advocated the use of the tables of Pisa: the *Book of the Foundations of Astronomical Tables*. The tables of Pisa, themselves, must have been drawn up before 1150, perhaps with the aid of Abraham.[16] The use of the 'figure indice' (see n. 7 above) is also a characteristic in the tables of Pisa and the Latin works of Abraham ibn Ezra.

---

11 See Burnett, 'The Eadwine Psalter and the Western Tradition of the Onomancy in Pseudo-Aristotle's *Secret of Secrets*', *Archives d'histoire doctrinale et littéraire du moyen âge*, 55, 1988, pp. 7–21, reprinted in id., *Magic and Divination in the Middle Ages*, Aldershot, 1996, article XI.

12 I owe this judgement to Giulio and Laura Lepschy.

13 B. Migliorini, in *Storia della lingua italiana*, Florence, 1985, can only point to a few fragments of Italian in legal documents and confessionals between ca. 960 and the late twelfth century, when the first examples of Italian verse begin to appear. The evidence of scientific manuscripts, such as Harley 5402, does not seem to have been exploited by Italian philologists. The mixed Latin and Italian of our annotator recalls the kind of language that merchants used in their letters, according to Boncompagno writing in 1215 (*apud* Migliorini, *Storia*, p. 91): 'Mercatores in suis epistolis verborum ornatum non requirunt, quia fere omnes et singuli per idiomata propria seu vulgaria vel corruptum latinum ad invicem sibi scribunt et rescribunt'.

14 It is interesting to note that one of the most substantial documents written in Italian in the twelfth century is a religious poem written by a Jew in Hebrew script; see Migliorini, *Storia*, p. 111.

15 See Arundel 377, fols. 36v–37r. The main difference is that Lucca is given a longitude of 34° in place of Pisa's 33°.

16 J.-M. Millás-Vallicrosa, 'El magisterio astronómica de Abraham ibn 'Ezra en la Europa latina', in id., *Estudios sobre historia de la ciencia española*, Barcelona, 1949, pp. 289–347 and R. Mercier, 'The Lost Zīj of al-Ṣūfī in the Twelfth Century Tables for London and Pisa', in *Lectures from the Conference on al-Ṣūfī and Ibn al-Nafīs*, Beyrut and Damascus, 1991, pp. 38–72.

In Hebrew documents, however, Abraham is said to have been not in Pisa, but in Lucca in the 1140s.[17] Abraham wrote in Hebrew, and the Latin works to which his name is attached have evidently been written down in Latin from his dictation (presumably in the vernacular or in ungrammatical Latin), by other, anonymous scholars.[18] The coincidences of time, place, and language between what we know about Abraham, and the annotator of the Harley manuscript are striking, but not conclusive. It can at least be said that our annotator was using tables deriving from those of Pisa and instructions based on those written by early users of the tables of Pisa. The date '1160' at the beginning of his annotations may be the radix of the version of the tables that he was using. But, considering the nature of the note, it would seem more likely that it is the current date or soon after.

This same annotator has rubricated the texts of Sahl ibn Bishr,[19] as can be seen both from the informal nature of the hand in which these rubrications have been written, and from their language: e. g., fol. 50r, 'De bestie qualis uinces cursu' (cf. the opening of Sahl's chapter: 'Et si interrogatus fueris de bestiis qualis bestia uincet in cursu'); fol. 51v, 'Si de'rex ueniet uel principe' (cf. Sahl: 'Et si interrogatus fueris de epistola utrum egrediatur a principe uel a rege'); fol. 56v, 'de'ora accepiendi comendationis' (cf. Sahl: 'Cum uolueris eligere horam in acceptione accommodationis pecunie ...'); fol. 57r, 'de'ora edificendi domus'; fol. 60r, 'de bonam oram belli', etc.

This 'Luccan annotator' probably acquired the first 70 folios of Harley 5402, and added notes in Lucca in, or soon after, 1160. It is, further, likely that at the time of its acquisition the manuscript also contained the verses accompanying the instructions for the Victorious and the Vanquished (item no. 3 above), and a set of astronomical tables drawn up for the meridian of Lucca, related to those drawn up for Pisa.

This annotated manuscript then came into the possession of another person interested in astrology, who copied out several texts on the subject, beginning by filling the last folio of the quire (fol. 70) and continuing by adding further quires (fols 71–104). He also filled in a space on fol. 16r with notes on the compass directions associated with the planets:

> Saturnus habet ex partibus mundi desteram septentrionis
> Iupiter habet occidentem
> Mars habet meridiem
> Solis fortitudo in plagis circulli in oriente
> Venus in destera orientis
> Mercurius fortitudinem habet in septentrione
> Luna vero fortitudinem habet in destra occidentis.

The orthography of the Latin of these notes may indicate that this scribe too is Italian, but his Latin is much more correct than the earlier annotator, whose hand does not appear on fols 70–104. The later (thirteenth or fourteenth-century) date of

---

17 See J. L. Fleischer, 'R. Abraham ibn 'Ezra u-melactō ha-siprutīt be-ereṣ Anglia', in Oṣar ha-Ḥayyīm, 7, 1931, pp. 189ff.

18 See C. Burnett, The Introduction of Arabic Learning into England, London, 1997, pp. 47–8 and 56–8.

19 The rubrication of Pseudo-Ptolemy's Iudicia has been done by the scribe of the text; he apparently did not get as far as filling in the rubrics for Sahl's text.

this hand is indicated not only by its appearance, but also by the fact that it uses the usual Western forms of the Hindu-Arabic numerals, which became common during the course of the thirteenth century.

Another scribe drew up a list of contents on a bifolium of music used as end-papers (fols 1*–2*), and wrote the running heads for the whole manuscript. The titles as running heads, and in the list of contents, correspond; they have evidently been taken from the rubrics of the works themselves. Eventually the manuscript was acquired by Nicholas of Cues, from whose collection it arrived (together with several other Cusan manuscripts) in the Harley collection of the British Library.[20]

* * *

Having established the likely period of the copying of the earlier portions of this manuscript, we may now turn to the two tables written by the earliest scribe on fols 15v–16r. The first table gives the lords of the houses, exaltations, triplicities, terms, and decans; the scribe has (piously or facetiously?) filled a space in the table by adding 'dominus vobiscum' along with the other 'domini'. The numerals in the table are entirely roman.

The table on fol. 16r, on the other hand, is written in Latin alphanumerical notation, except for the first column in which roman numerals are used (see Plate III). The table has been ruled in brown ink, but all the letters and numerals (except for two corrections) have been written in red. This table occurs in one other manuscript: on fol. 84r of Pommersfelden 66, written in the early fourteenth century (see Plates IV and V).[21] In the following transcription (Table 1), the alphabetical numerals in the Harley copy are represented as small capitals or lower-case letters as in the manuscript, and the dotless 'ı' represents the single stroke, sometimes extended downwards,[22] for the 'i' in the manuscript. The Hindu-Arabic numeral equivalents have been added alongside the alphabetical numerals. The correct numerical values are given in Hindu-Arabic numerals in brackets. The variants in the Pommersfelden manuscript are given in alphanumerical values in brackets.

One should note that, in the Harley manuscript, small capitals (i. e. capitals of the same size as lower-case letters) have occasionally been used by for 'g', 'r', 'q' and 'l', and always for 'k', 'm' and 'n'. In the case of 'l' this is possibly to avoid confusion with 'ı', two instances of such confusion occur in this table (see notes 26 and 30). It is noticeable that lower-case 'g', 'r', and 'q' are almost identical with the 'figure indice' for '5', '3' and '9'; the use of capitals for these letters would be a sensible expedient in a context in which both alphanumerical notation and Hindu-Arabic numerals ('figure indice') were used.[23] In the case of the remaining confusable letter – p = '2' in the 'figure indice' – replacement by the capital form is of no help. The letters 'm' and 'n'

---

20 See *Mitteilungen und Forschungen* ... 3 (n. 2 above), pp. 81–4.

21 I owe this reference to Fritz Saaby-Pedersen who kindly sent me copies of the relevant pages of the Pommersfelden manuscript. The date '1306' has been written in the same hand as the text, in the bottom margin of fol. 84v and the planetary positions for March, 1306, have been added on fol. 85r.

22 The latter form is used regularly by this scribe when an 'i' comes at the end of a word.

23 The two numeral systems are found together in the *Liber Mamonis* of Stephen the Philosopher, in MS Cambrai, Bibliothèque municipale, 930 (see Burnett, 'The Transmission of Arabic As-

may have been written as small capitals to ensure that they were not confused with the roman numerals 'iii' and 'ii'. Some copying errors suggest that the table was copied from another table which already used alphanumerical notation (see notes 26, 30, 32 and 33). In the Pommersfelden manuscript small capitals are used less frequently and without any apparent rule, while the scribe has added the Hindu-Arabic numerical equivalents in the first four and last three lines.

Table 1

	o159	RC 83	q1(Q1) 79	h 8	nf 46	mG (mg) 37
	Saturnus	Iupiter	Mars	Venus	Mercurius	Caput
.c.	Na 41	Ke (KG) 15 (17)	La 21	d[24] (o)4	h 8	Lf 26
.cc.	lc 23	md 34	Nb 42	t (h) 100 (0)[25]	f (kf) 6 (16)	Ke (ke) 15
.ccc.	e 5	o1(oa) 59 (51)	pc 63	d 4	1d[26] (ld) 24	d[27] 4
.cccc.	Nf 46	ph 68	e 5	t (h) 100 (0)	Mb 32	M 30
.d.	le[28] (lh) 25 (28)	b 2	lf 26	d 4	N 40	K1 19
.dc.	K 10	K1 19	nf (mg) 46 (47)	t (h) 100 (0)	b 2	h (h1) 8
.dcc.	oa 51	mh 38 (36)	ph 68	d 4	K 10	md 34
.dccc.	Mc 33	tM 130 (53)[29]	K 10	t (h) 100 (0)	kh 18	lc 23
.dcccc.	Ke 15	Q 70	Ma 31	d (h) 4	1f[30] (lf) 26	Kv (kb) 12[31]
.m.	of 56	d 4	ob 52	t (h) 100 (0)	md 34	a 1
.mc.	Mh 38	la 21	Qc (qc) 73	d 4	Nb[32]	lG[33] (lg) 27
.mcc.	l 20	Mh 38	Re (Ke) 85 (15)	t (h) 100 (0)	d 4	Kf 16
.mccc.	b 2	oe 55	Mf 36	d 4	Kb 12	e 5
.mcccc.	vc (nc) 43	Qb (qb) 72	og 57	t (h) 100 (0)	L 20	Ma 31
.mccccc.	le 25	f (bf) 6	qh 78	d 4	Lh 28	l 20

.i.	.ii.	iii.	iiii.	v.	vi.	vii.	viii.	viiii.
.a.	b.	c.	d.	e.	f.	G.	h.	1.
x.	xx.	xxx.	xl.	L.	lx.	lxx.	lxxx.	lxxxx.
K.	L.	M.	N.	o.	p.	q.	r.	s.
c.	cc.	ccc.	cccc.					
t.	v.	x.	y.					

tronomy', n. 1 above), where no capitals are used; the Harley example may imply or prefigure a later development in the use of the two systems together.

24 A letter has been erased after 'd'.

25 The scribes of the Harley and Pommersfelden manuscripts have each apparently mistaken a special for zero in their sources as a letter. The key to the alphanumerical notation does not include the symbol for zero, but presumably it was a shape like '6', which was commonly used in Latin astronomical works, and could have been misinterpreted as either a 't' or a 'h'.

26 In the Harley manuscript a lower-case 'l' (20) has been misinterpreted as 'i' (9).

27 'd' has been written over an erasure.

28 'e' has been changed in red ink from or to 'i' (9).

29 The corruption here is difficult to explain, but in the case of 'm', the decimal position is wrong (30 for 3).

30 Again, in the Harley manuscript a lower-case 'l' (20) has been misinterpreted as 'i'.

31 The Harley scribe writes a 'b' as a 'v'.

32 'b' corrected in black ink from 'h' (8).

33 'G' corrected in black ink from 'b' (2).

The origin of the two tables in the Harley manuscript is difficult to ascertain, since neither of them has a heading. They do not belong to either of the two texts that surround them. The first one summarises in tabular form commonly occurring astrological information concerning the various lordships of the planets and has nothing surprising or unusual about it (except for the 'dominus vobiscum'). The purpose of the second one, however, is, in the words of David King,[34] 'to plug into a set of perpetual tables of the kind we know from Ammonios, al-Zarqallu, and various other Islamic sources, then Zacuto', the first row of numbers being 'the standard periods (returns to the same longitude) of the planets', and the roman numerals being hundreds of years. Thus, the current year of each planet's period is equivalent to the remainder, when the roman numerals in the vertical column are divided by the planet's period in the first horizontal line. These remainders are the values given in the table, and their correctness can easily be checked. This table, as far as I know, is not found in the best-known sets of Latin astronomical tables of the twelfth century, such as the tables of al-Khwārizmī and al-Battānī, and the Toledan, Pisan or London tables, nor in the Hebrew tables of Abraham bar Ḥiyya or Abraham ibn Ezra.

In the Pommersfelden manuscript this table precedes an unidentified almanac written in the Western forms of the Hindu-Arabic numerals. It was not copied from the Harley manuscript since it does not reproduce all the mistakes in that manuscript; it is, on the whole, more correct. Moreover, the scribe gives some clues on how to use the table, entitling it 'Tabula inveniendi annos planetarum expansos', and adding in the margin of the following page (fol. 84v):

> In Saturno divide mille centum per .l.viiii. et remanent .xxxviii.[35] Deinde iunge residuum de annis Christi cum illis .xxxviii. et tunc (?) divide per .l.viiii. Et cum numero qui remanserit infra .59. intra in tabulas et procede computando unus duo tres etc. quousque pervenias ad residuum quod tibi remansit infra .59. Et si .59. tantum tibi remanserint, cum illis intra in tabulam Saturni.
>
> In Iove, divide .1100. per .83. et remanent .21. Deinde fac ut supra in Saturno. In Marte, divide per .79. In Venere, per .8. In Mercurio, per. 46. In Capite, per 35, et in omnibus fac ut supra fecisti in Saturno.

These notes show that the scribe knew the purpose of the table. It is not clear, however, whether he copied them from his exemplar.[36] The provenance of the Pommersfelden manuscript and its sources are still to be explored.

* * *

Harley 5402, then, aside from giving a rare example of the twelfth-century Italian vernacular, provides a valuable testimony to the explanation and use of Latin alphanumerical notation. While the place and time of the writing of the relevant text cannot be ascertained with certainty, it is significant that it should occur in a manuscript annotated in 1160, or soon after, in a Latin-Italian macaronic language, in which there

---

34 I owe this information to a letter that he kindly sent me.

35 This is the value found in the square with coordinates 'Saturn' and '.mc.' in the table.

36 The same scribe, however, in the lines in which he has inserted the Hindu-Arabic numerical equivalents, has added together the numbers in each line: e. g. in the top line the sum of the numbers is 312, which the scribe describes as 'anni omnium planetarum insimul pro radicibus'. Adding the numbers has no astronomical significance.

X

is a reference to the presence of tables for Lucca. Stephen the 'disciple of philosophy', who regularly uses the alphanumerical notation, also refers to astronomical tables for which he has written instructions, and Henry Bate, writing in the thirteenth century, refers to the 'Pisan and Winchester' tables, whose titles nicely bring together the places associated with the two scholars, Stephen of Antioch and Pisa and 'Abd al-Masīḥ of Winchester.[37] The extant tables of Pisa are thought to be based on the lost tables of Baghdad drawn up by 'Abd-al-Raḥmān ibn 'Umar al-Ṣūfī (d. 986), which are not known among Arabic writers in the West.[38] It is conceivable that scholars in Antioch were responsible for introducing al-Ṣūfī's tables into Italy, along with alphanumerical notation, and that these tables were, in turn, worked on by several scholars in Pisa and Lucca. It is their shoulders that we feel we are looking over when we read MS Harley 5402.

* * *

Appendix

I owe to David Pingree the solution of the calculation of the position of the Moon in the first item in the Italianate text above (p. 78). He writes as follows:

"The lunar computation is an interesting adaptation of an Indian method found, e. g., in the Brāhmapakṣa [see Pingree, 'History of Mathematical Astronomy in India', in *Dictionary of Scientific Biography*, ed. C. C. Gillispie, vol. 25, New York, 1978, pp. 533–633 (555–80)]. The epact each year is 11 *tithi*s (a *tithi* is a 30th part of a synodic month, measured by an increase of 12° in the elongation between the Sun and the Moon). The epact means that in a solar year the Moon makes 13 rotations and travels a bit further; the 13 rotations take a little more than 354 days. Also, in 19 years there are 235 synodic months, so that the longitude of the Moon returns to 0° if it began at 0°.

The epact of 11 *tithi*s is divided into two parts; one of 5 *tithi*s, the other 6. The computation is sexagesimal, so that when $5 \times y$ (year-number) $= 60$, the 60 is dropped (intercalary months are added to bring the longitude of the Moon back to near 0°). Add to the product of $5 \times y$ modulo 60 the sum of $6 \times y$ + the days ($\approx$ *tithi*s) lapsed of the current year; the result is the accumulated epact plus the lapsed *tithi*s of the current year. To convert it to a sexagesimal number, divide by 60 and put down 1 for each 60. Since 1 *tithi* = 12°, the total number of *tithi*s multiplied by 12 is the degrees of elongation between the Sun and the Moon. These degrees of elongation added to the longitude of the Sun are the approximate longitude of the Moon.

One must begin from the known longitude of the Sun and the Moon on the kalends of January – ideally from their conjunction on a kalends of January."

---

37 This point is further explored in my "Abd al-Masīḥ of Winchester' (n. 1 above).
38 See Mercier, 'The Lost Zīj of al-Ṣūfī' (n. 16 above).

frig erit in illo mse. Si uenus 7 mars: fili erit pluuia multa. 7 tonitrua.
7 fulmina. hac in hieme. In estate erit ros multus 7 calor; Explicit.

Anni dni. Mely. correlas ciclu. lune. xix. annof. 7 corre da 1. vi x ix.
I una inueniuda. multiplica ciclu. lune per qnq; fiest
infra. 40 manda ill' memorie fuero fuit 40. piet totu
fiuero fup 40. deme 40. 7 relicuu. manda memoria
quente pacto multiplica re ciclu. p. 4. 7 numeru qd
deciderit iunge numeru dieru. addt. Januariu u' q' die
qua querit. fit totu p. 40 diuid 7 de dia 40. unitate
faciat iungat illa unita anumero q memoria man
dasti. 7 oc numeru fic acouatu fiet numeru infra. 40
remasit. adat illi re pferat tot ee die luna 7 tot.
ee punto'
of die 7 punto. 4 multiplica super xii 7 numeruf
qb; ide puenerit hec fe gradi feparationi' folif aluna
mutatur hec pacta fep. addt. Januarii.

Ego deus qui a patru abraam. yzaac 7 iacop iudeo
patre. oiporente 7 nom meu adonai. no abfco di eius.

+ Larafione. dele littere Iulii. Inparor. dico ti come uuoli.
fare. pegnofcere. ille. turta ora guarda ineliuerfi.
el prima [+] cetrou dalo meonu cuneto delapofifagna.
el prima [+] dalo meonuca rito difopra. odinati. [+] odarico.
quale edugi pnupreffo delt cuii cata dofopra fifegna. edugi
guige infieme. guelle. dinati [+] coquelle dinati. [+] ecco
puza infione. 7 fic fas decctori 7 miftii primu uorfu. uide.
quali litte fup ahudat. 7 fecudu hoc Iudicat. fet fi tuiffy pari 7 pari
predo. quali fut. alaro. la coppia.

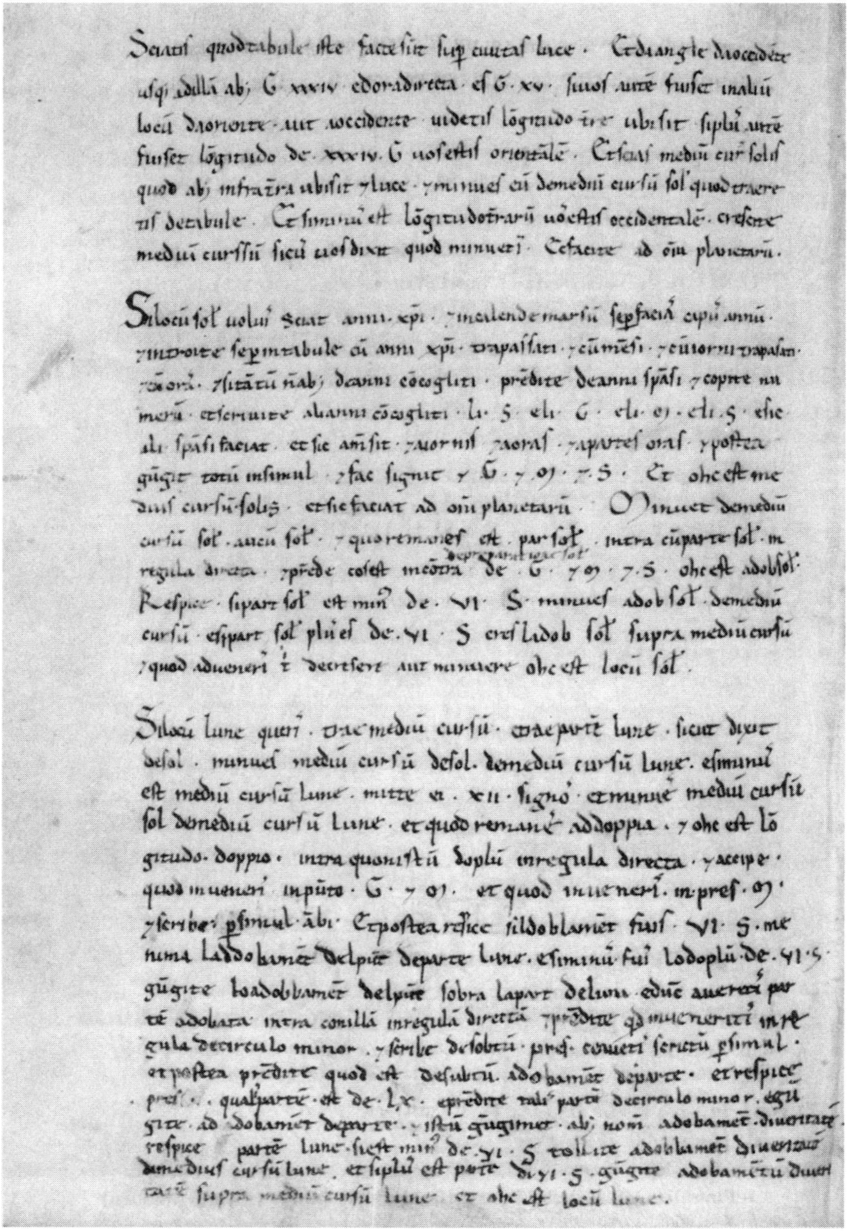

Plate II: MS British Library, Harley 5402, fol. 69v (with permission)

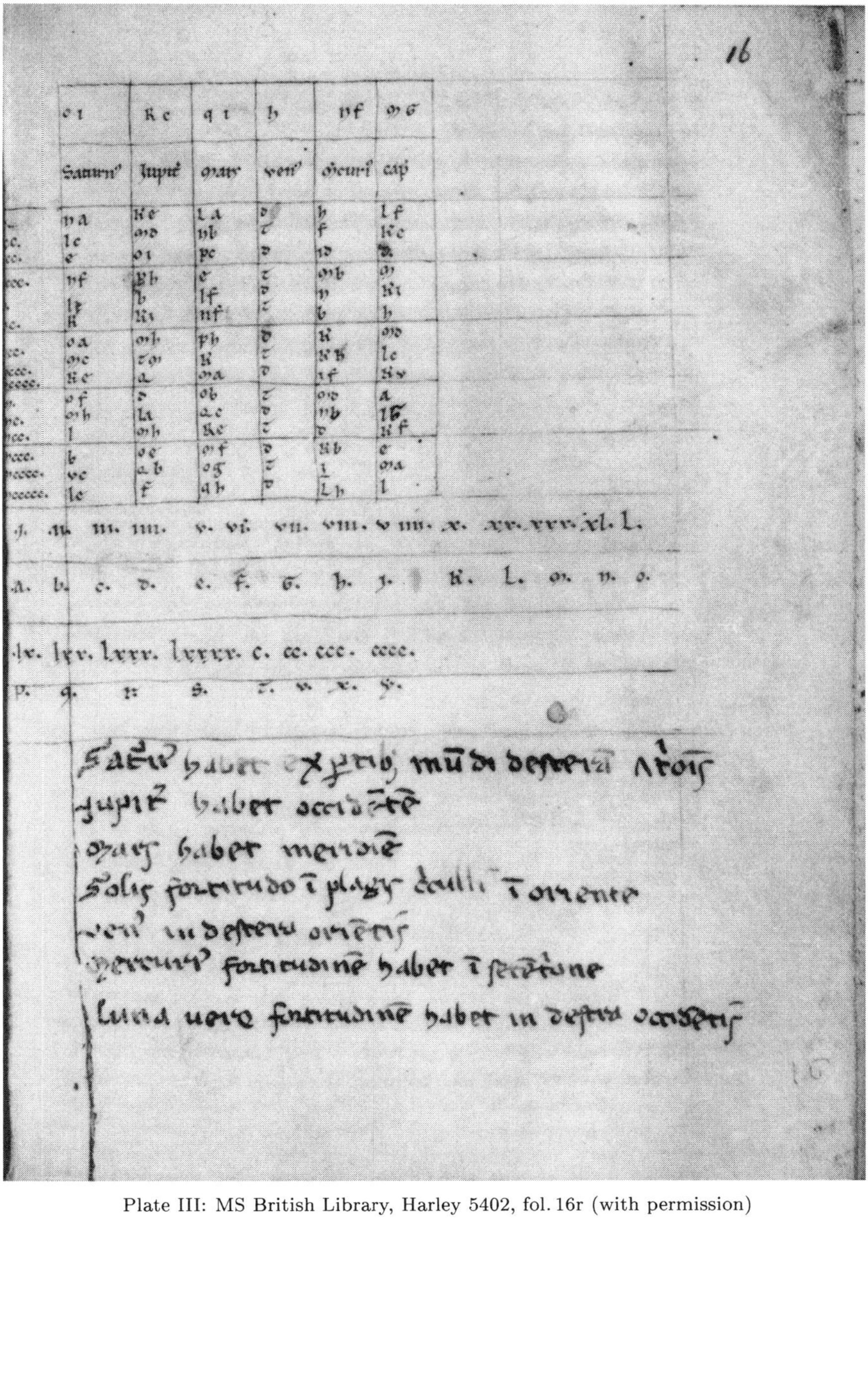

Plate III: MS British Library, Harley 5402, fol. 16r (with permission)

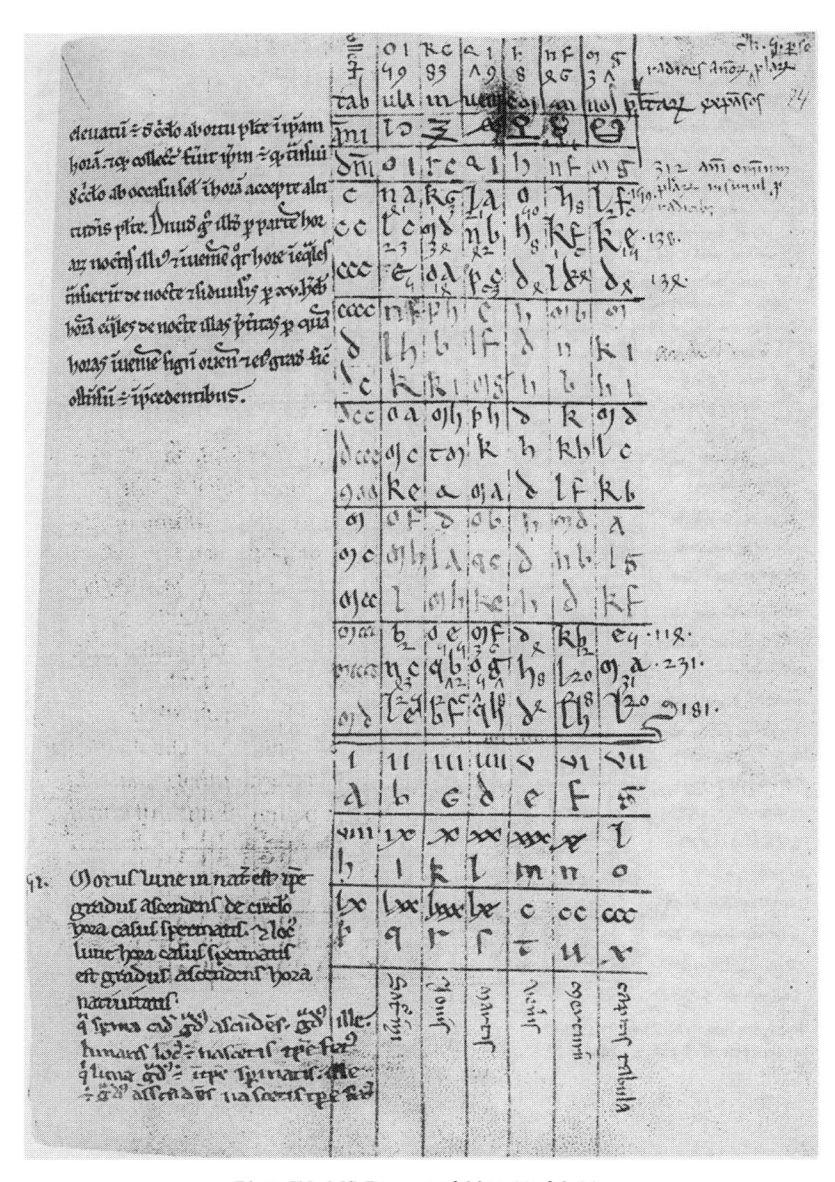

Plate IV: MS Pommersfelden 66, fol. 84r

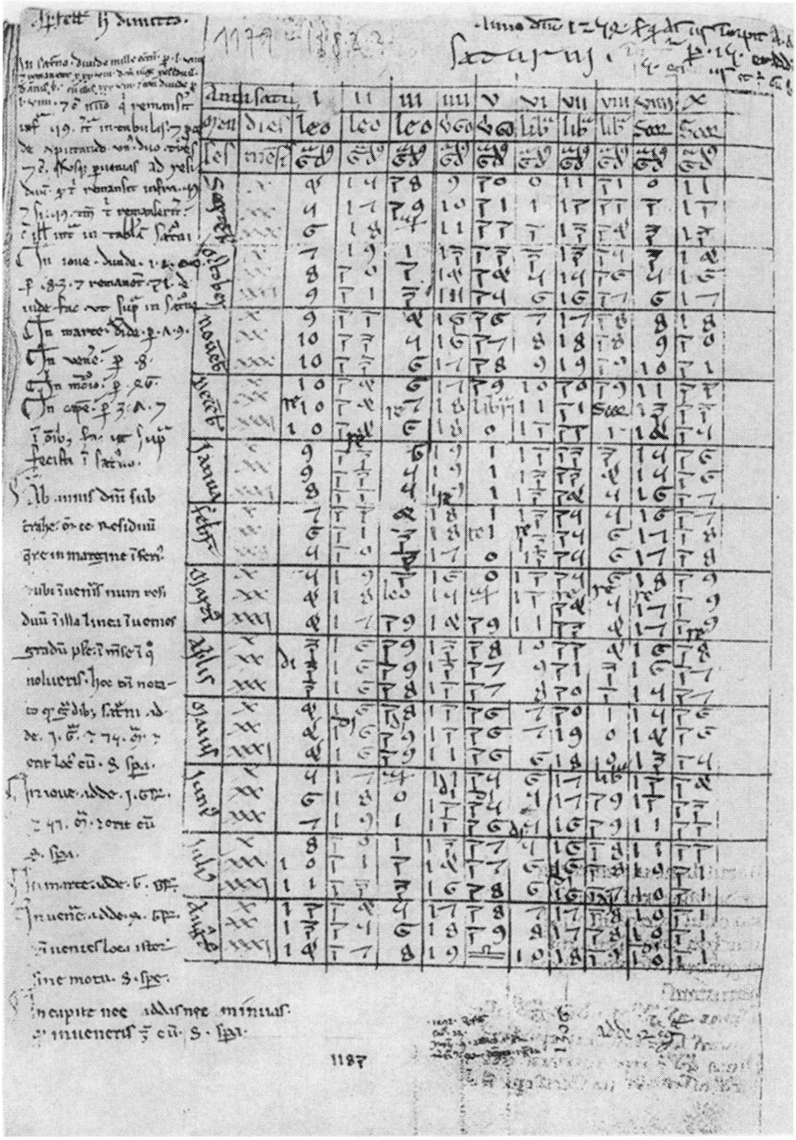

Plate V: MS Pommersfelden 66, fol. 84v

XI

FIBONACCI'S 'METHOD OF THE INDIANS'

I n the preface to his *Liber abbaci* (1202),[1] Fibonacci describes his discovery of the 'art of the Indians' in the following words:

1. When my father, appointed by his homeland, held the post of public *scriba* (notary or representative) in the custom-house of Bejaia for the Pisan merchants frequenting it, he arranged for me to come to him when I was a boy and, because he thought it would be useful and appropriate for me, wanted me to spend a few days there in the *abbaco* school, and to be taught there. **2.** Here I was introduced to that art (the *abbaco*) by a wonderful kind of teaching that used the nine figures of the Indians. **3.** Getting to know the *abbaco* pleased me far beyond all else and I set my mind to it,[2] to such an extant that I learnt, through much study and the cut and thrust of disputation, whatever study was devoted to it in Egypt, Syria, Greece, Sicily and Provence, together with their different methods, in the course of my subsequent journeys to these places for the sake of trade. **4.** But I reckoned all this, as well as the algorism and the arcs of Pythagoras, as a kind of error in comparison to the method of the Indians (*modus Indorum*). **5.** Therefore, concentrating more closely on this very method of the Indians, and studying it more attentively, adding a few things from my own mind, and also putting in some subtleties of Euclid's art of geometry, I made an effort to compose, in as intelligible a fashion as I could, this comprehensive book, divided into 15 chapters, demonstrating almost everything that I have included by a firm proof, **6.** so that those seeking knowledge of this can be instructed by such a perfect method (in comparison with the others), and so that in future the Latin race may not be found lacking this (knowledge) as they have done up to now.[3]

---

1. André Allard has demonstrated that this preface is likely to have belonged to the original version of the *Liber abbaci*, and that the paragraph that precedes this in the printed edition, which includes the dedication to Michael Scot, was added in 1228: see his article in this volume.   *

2. 'Intellexi ad illam', which I tentatively translate as 'I set my mind to it', is the only obscure phrase in this paragraph. There may be a corruption here.

3. '**1.** Cum genitor meus a patria publicus scriba in duana Bugee pro Pisanis mercatoribus ad eam confluentibus constitus preesset, me in pueritia mea ad se venire faciens, inspecta utilitate et commoditate futura, ibi me studio abbaci per aliquot dies stare voluit et doceri. **2.** Ubi ex mirabili magisterio in arte per novem figuras Indorum introductus, **3.** scientia artis in tantum mihi

There is a certain bravado here. Fibonacci wants it to appear that he is doing something quite innovatory in the Latin world. But the words he uses are nevertheless perplexing. What he experienced in Bejaia was a method of learning to calculate that used 'the nine figures of the Indians', which he later calls simply 'the method of the Indians'. And he contrasts this with all the forms of calculation that he came across in other places throughout the Mediterranean (both the Islamic and the Christian world), which he considers 'quasi error'-a deviation, a false track. We are not told what these other forms of calculation consisted of.[4] Fibonacci does, however, make a further claim: that 'the algorism and the arcs of Pythagoras' known to his Latin audience are just as erroneous. The meaning of these terms is clear. The algorism is the calculation with Indian numerals based ultimately on the text on the subject by Muhammad ibn Musa al-Khwarizmi, which became known to the Latin world in the early twelfth century.[5] The 'arcs of Pythagoras' are the Gerbertian abacus, or the abacus 'with

pre ceteris placuit et intellexi ad illam quod quicquid studebatur ex ea apud Egyptum, Syriam, Greciam, Siciliam et Provinciam cum suis variis modis, ad que loca negotiationis *causa* (tam Boncompagni) postea peragravi, per multum studium et disputationis didici conflictum. **4.** Sed hoc totum etiam et algorismum atque arcus Pictagore quasi errorem computavi respectu modi Indorum. **5.** Quare, amplectens strictius ipsum modum Indorum et attentius studens in eo, ex proprio sensu quedam addens et quedam etiam ex subtilitatibus Euclidis geometrice artis apponens, summam huius libri quam intelligibilius potui in .xv. capitulis distinctam componere laboravi, fere omnia que inserui certa probatione ostendens, **6.** ut *ex tam* (extra Boncompagni) perfecto pre ceteris modo hanc scientiam appetentes instruantur, et gens Latina de cetero, sicut hactenus, absque illa minime inveniatur': LEONARDUS PISANUS (Fibonacci), ed. B. Boncompagni, *Il Liber Abbaci di Leonardo Pisano*, Rome, 1857, p. 1 (I have silently changed the punctuation and added section numbers; I mark corrections to Boncompagni's text in italics). The annotated English translation of R. E. Grimm (*The Autobiography of Leonardo Pisano*, «Fibonacci Quarterly», 11, 1976, pp. 99-104), though occasionally helpful, is marred by misconstructions of the Latin. Unfortunately, the most recent English translation - L. E. SIGLER's *Fibonacci's Liber Abaci: A Translation into Modern English of Leonardo Pisano's Book of* Calculation, New York etc., 2002 - is unintelligible for this paragraph. I understand 'studium' in the sense of Italian 'studio' as a place of study and teaching, and 'abbacus' as the Italian 'abbaco', i.e. the art of calculation in general. It is to this art, rather than specifically to calculation according to the Indian method, that Fibonacci is referring in sentences 2 and 3 of the English translation.

4. The most common forms of numerals used by merchants in the Mediterranean in the Middle Ages were derived from Greek alphanumerical notation: namely those called by the Arabs 'rumi' ('[Byzantine] Greek'), 'al-qalam al-fasi' (the 'script of Fez') and 'al-rasm al-zimami' (from the Arabic word 'zimam' meaning 'register'): see ROSA COMES, *Arabic, «Rumi», Coptic, or merely Greek Alphanumerical Notation?, The Case of a Mozarabic 10th Century Andalusi Manuscript'*, «Suhayl», 3, 2002-3, pp. 157-185. The Latin cannot be made to imply that Fibonacci found the 'Indian method' anywhere except in the school in Bejaia.

5. For more details, and the most up-to-date bibliography see P. KUNITZSCH, *The Transmission of Hindu-Arabic Numerals Reconsidered*, «The Scientific Enterprise in Islam», eds J. P. Hogendijk and A. I. Sabra, Cambridge, Mass., 2003, pp. 3-21 and C. BURNETT, *Indian Numerals in the Mediterranean Basin in the Twelfth Century, with Special Reference to the "Eastern Forms"*, in *From China to Paris: 2000 Years Transmission of Mathematical Ideas'*, eds Y. Dold-Samplonius et al., Stuttgart, 2002, pp. 237-288. The standard work of KARL MENNINGER (*Zahlwort und Ziffer. Eine Kulturgeschichte der Zahl*, 2nd revised ed., 2 vols, Göttingen, 1958) remains valuable, while GEORGES IFRAH, *The Uni-*

apices', whose invention was attributed to Gerbert d'Aurillac (d. 1003). The 'arcus' refer to the columns of the abacus, or to the semicircles which join every three columns, and the abacus is frequently stated to be the invention of Pythagoras.[6] Fibonacci avoids calling it the 'abacus' presumably because he uses the word in the more general sense of 'calculation'; this was the sense that 'abbaco' was to have in Italy from his time onwards.[7]

Fibonacci's claim is surprising. For do not both the algorism and the Gerbertian abacus use, precisely, 'the nine figures of the Indians'? In fact, this claim is similar to two earlier statements that referred, respectively, to the algorism and the abacus. In the mid-twelfth century, 'magister Iohannes', a mathematician in Toledo, had promised to compose a work on the algorism. He wrote of

[...] especially that kind of numeral which is written with the letters of the Indians and used above all by the Saracens: the good al-Khwarizmi, with remarkable intelligence, discovered this kind and we, with the help of God, have decided to deal with these numerals a little, according to the measure of our inexperience, either at the end of this work or in another place, if God wills and we live long enough[8].

As for the abacus, we have, in 976, contemporary with the first appearance of the 'Gerbertian' abacus, the statement by Vigila, in the monastery of Albelda in the Rioja (Asturias), that

We must know that the Indians have a most subtle talent and all other races yield to them in arithmetic and geometry and the other liberal arts. And this is clear in the 9 figures with which they are able to designate each and every degree of each order (of numbers). And these are the forms.[9]

versal History of Numbers, English version, London, 1998, pp. 577-591, provides a readily accessible introduction.

6. '...ab antiquis mensa pytagorica; a modernis autem vel abax vel abacus nuncupatur' ('[The abacus] is called by the Ancients, the Pythagorean table, by the moderns, 'abax' or 'abacus'): TURCHILLUS, Reguncule super abacum, ed. E. Narducci, «Bullettino di bibliografia e storia delle scienze matematiche e fisiche», 15, 1882, pp. 111-54, at p. 135.

7. See W VON EGMOND, Practical Mathematics in the Italian Renaissance: a Catalogue of Italian Abbacus Manuscripts and Printed Books to 1600, Florence, 1980, and E. ULIVI, Scuole e maestri d'abaco in Italia tra medioevo e rinascimento, in Un ponte sul Mediterraneo: Leonardo Pisano, la scienza araba e la rinascita della matematica in Occidente, ed. E. Giusti, Florence, 2002, pp. 121-59.

8. '...maxime illud genus numeri quod fit per litteras Indorum quo Sarraceni maxime utuntur quod probus Alchoarismi mirabili ratione invenit, de quibus, nos auxiliante Deo, vel in fine huius operis vel in alio loco, si Deus voluerit et vita comes fuerit, aliquantulum pro modulo imperitie nostrae tractare decrevimus': MAGISTER IOHANNES, De differentiis tabularum, ed. J. M. Millás Vallicrosa, in Una obra astronómica desconocida de Johannes Avendaut Hispanus, «Osiris», 1, 1936, pp. 451-475, and reprinted in idem, Estudios sobre historia de la ciencia española, I, Barcelona, 1949, pp. 263-288 (see p. 274). For the authorship of this text see C. BURNETT, John of Seville and John of Spain: a mise au point, «Bulletin de philosophie médiévale», 44, 2003, pp. 59-78 (at p. 63-67).

9. 'Scire debemus [in] Indos subtilissimum ingenium habere et ceteras gentes eis in arithmetica et geometrica et ceteris liberalibus disciplinis concedere et hoc manifestum est in novem figuris quibus designant unumquemque gradum cuiuslibet gradus quarum haec sunt formae': Escorial d.I.2, fol. 9v, see IFRAH, Universal History (n. 5 above), p. 579.

The Indian origin of the numeral forms that were used in the early abacus texts was subsequently ignored. The earliest examples of abacus boards, from the late tenth century, include the information that 'Gerbert <gave> to the Latin world the numbers of the abacus and their shapes' ("numeri" and "figurae") without indicating where he got the shapes from.[10] By the early twelfth century an English author, Turchillus could write that the 'figures came from Pythagoreans, while their names are Arabic'.[11] In the texts on the algorism, however, the figures are clearly stated to have come from the Indians. The earliest Latin version states that:

[…] the numeration of the Indians with 9 letters by which they have set out all their numbers, for the sake of ease and brevity.[12]

In an unpublished early thirteenth-century algorism in a manuscript in a manuscript now in Cashel, Ireland, we read:

There are nine shapes by which all numbers are represented, with the addition of the circle. They are Indian letters, and are of this kind: 123456789. 'Cifra' or 'the circle' is 0.[13]

Outside the context of the algorism, too, the numerals are decribed as 'Indian': for example in the translation of Abu Maʿshar's *On the Great Conjunctions* (probably made by John of Seville in Toledo in the second quarter of the twelfth century), a high number is written out in words, and then, 'per figuram Indicam' or 'per figuras Indicas' (the Arabic is *bi l-sura al-hindiya*),[14] and in a Munich manuscript of the late twelfth century, two forms of numeral are given, of which the first is described as 'Toledan' and the second 'Indian' ('Toletane f(igure)' and 'Indice f(igure)').[15]

10. 'Gerbertus Latio numeros abacique figuras': see M. FOLKERTS, *Frühe Darstellungen des Gerbertischen Abakus*, «Itinera mathematica», *Studi in onore di Gino Arrighi per il suo 90° compleanno*, eds. R. Franci, P. Pagli and L. Toti Rigatelli, Siena, 1996, pp. 24-41 (at p. 30).

11. Has autem figuras, ut domnus Guillelmus R. testatur, a Pytagoricis habemus, nomina vero ab Arabibus: TURCHILLUS, *Reguncule super abacum, op. cit.*, p. 136.

12. '[…] numero Indorum per IX litteras, quibus disposuerunt universum numerum suum causa levitatis atque adbreviationis': *Dixit Algorizmi* in *Die älteste lateinische Schrift über das indische Rechnen nach al-Ḫwārizmi*, ed., trans. and comm. by M. Folkerts, with the collaboration of P. Kunitzsch, *Abhandlungen der Bayerischen Akademie der Wissenschaften*, phil.-hist. Klasse, n. F., 113, Munich, 1997, p. 28.

13. 'Sunt autem figure .ix. quibus omnes numeri representantur, cum circuli addicione. Et sunt littere Indorum et huiusmodi: 123456789. Cifra vel circulus 0': MS Cashel, GPA Bolton Library, Medieval MS 1, p. 41.

14. ABU MAʿSHAR, *On the Great Conjunctions*, I, lines 164-6, discussed in C. BURNETT, *The Strategy of Revision in the Arabic-Latin Translations from Toledo: The Case of Abū Maʿshar's* On the Great Conjunctions, in *Les Traducteurs au travail: leurs manuscrits et leurs méthodes*, ed. J. Hamesse, Turnhout, 2001, pp. 51-114 and 529-40 (at pp. 66-7). The variants are those of the two versions of the Latin text.

15. MS Munich, Bayerische Staatsbibliothek, Clm 18927, fol. 1r: Plate 3 in Burnett, *Indian Numerals in the Mediterranean Basin* (n. 5 above).

Fibonacci *might* have wished to distance his art-as the genuine Indian art-from that of the Arabs ('Algorism' was clearly an Arabic name, and the algorism was also called 'helcep sarracenicum'-the 'Saracen calculation'[16]), but if so, it is surprising that he points out that he learnt it in the Arabic context of the town of Bejaia, and did not pretend to have picked up further afield.

When Fibonacci was writing this preface, then, the Indian numerals had been used by Latin scholars for well over a century. There had been considerable variation in the *forms* of these numerals. The shapes that appear on the abacus tables, and were meant to be written on the abacus counters, are distinct from those used in the algorism, and there is considerable variation between the forms used in the algorism and other contexts in the twelfth century, especially between 'Western forms' and 'Eastern forms'.[17] Fibonacci, as the leading mathematician in Pisa, would have been aware of both these forms of numerals. For the Eastern forms are found especially in manuscripts associated with Pisa (e.g. in copies of the instructions for the astronomical 'Tables of Pisa'). Moreover, they are used in the earliest manuscript of the Greek-Latin translation of Euclid's *Elements* (made in ca. 1160): MS Paris, BNF, lat. 7373-a manuscript written in Tuscany. As Busard has shown, this version of the *Elements* was known to Fibonacci, who may have added an appendix to the translation found in this very manuscript.[18] On the other hand, the earliest manuscripts of the mathematical translations made in Toledo in the middle to late twelfth century were written probably in the vicinity of Padua at the turn of the thirteenth century, and use the Western forms in a cursive form also used by a notary in Perugia from 1184 until 1206.[19] The ease in which these Western forms of the numerals are written in these manuscripts suggests that they were of common currency among schol-

---

16. For the text 'Helcep Sarracenicum', written in the second quarter of the twelfth century, and the reference to 'algorismi vel helcep' in MS Cambridge, Trinity College, R.15.16, fol. 3r, see BURNETT, *Algorismi vel helcep decentior est diligentia: the Arithmetic of Adelard of Bath and his Circle*, in *Mathematische Probleme im Mittelalter: Der lateinische und arabische Sprachbereich*, ed. M. Folkerts, Wiesbaden, 1996, pp. 221-331.

17. For tables of these forms see G. F. HILL, *The Development of Arabic Numerals in Europe, Exhibited in Sixty-Four Tables*, Oxford, 1915, A. ALLARD, *Le Calcul indien (Algorismus)*, Paris and Namur, 1992, p. 252, and BURNETT, *Indian Numerals in the Mediterranean Basin* (n. 5 above), pp. 265-267.

18. See H. L. L. BUSARD, *The Mediaeval Latin Translation of Euclid's* Elements *Made Directly from the Greek*, Stuttgart, 1987, pp. 18-20 and the article by Menso Folkerts in this volume.

19. See A. BARTOLI LANGELI, *I notai e i numeri (con un caso perugino, 1184-1206)*, in *Scienze matematiche e insegnamento in epoca medioevale: Atti del convegno internazionale di studio, Chieti, 2-4 maggio 1996*, eds P. Freguglia, L. Pellegrini and R. Paciocco, Naples, 2000, pp. 225-254, and discussion in BURNETT, *Indian Numerals in the Mediterranean Basin* (n. 5 above), pp. 254-255.

ars in Northern Italy in the early thirteenth century, and should have been known to Fibonacci.

It may, nevertheless, be argued that Fibonacci (for whatever reason) re-introduced Indian numerals from the Islamic world. For the forms of the numerals in the early manuscripts of Fibonacci's works, while clearly being Western, differ a little from those used in the North Italian Toledan texts and by the Perugian notary, most obviously in that the '2' and '3' look as if they have been written upside down, and the '4' is written with both a hook and a loop (rather than just a loop).[20] The closest analogues to these distinctive forms in Latin manuscripts are in, respectively, the earliest Latin manuscript to reproduce Arabic numerals (the Albelda manuscript of 976 A.D.), and the earliest Latin version of al-Khwarizmi's *Indian Arithmetic* (see Table).[21] And rather than to suggest that Fibonacci revived these archaic Latin forms, it is possible to argue that he followed the models of these forms, i.e. the numerals as they were found in Western Arabic manuscripts, which continued to be written in these forms.[22] He might even have thought that he was restoring genuine Indian forms that had become corrupted in the Latin tradition. It would be rash, however, to draw too sharp a conclusion at this stage of research from these manuscripts of Fibonacci's works, the earliest of which was written considerably after the composition of the *Liber abbaci*.

There is nothing, then, innovative about Fibonacci's use of Indian numerals as such. But is his implication that the 'Indian method' used by him is somehow superior to the methods employed in the Gerbertian abacus and the algorism?

The 'Indian method' relies on numerals having different values according to their position-i.e. place value. This is a very difficult concept to grasp for someone used to Roman numerals, or even to Greek, Arabic or Hebrew alphabetical numerals, and had to be explained in the clearest

20. For examples of these numerals see *Un ponte sul Mediterraneo* (n. 7 above), pp. 64, 66, 71, 72, 75, 81, and 114, R. FRANCI, *Il* Liber Abaci *de Leonardo Fibonacci*, «Bollettino della Unione Matematica Italiana», 8, 2002, pp. 293-328 (see figures 1, 4, 6 and 7) and p. 113 below. Similar numerals can be also be found on the first 17 folios of Florence, Laurenziana, S. Marco 194, a fourteenth-century codex containing astronomical tables for Pisa and Novara.
21. The second parallel has already been pointed out by Paul Kunitzsch in his *The Transmission of Hindu-Arabic Numerals* (n. 5 above), p. 16, but Kunitzsch's conclusion is that 'the numerals in the Leonardo manuscripts follow the forms current in the known Latin arithmetical texts...they do not show the intrusion of new Arabic influence resulting from Leonardo's oriental travels and his personal contacts with trade centers in the Arab world.'
22. Examples are the numerals in MS Tunis, University Library, 2043 (1611 A.D.), illustrated in Ifrah *Universal History* (n. 5 above), p. 535. The upside-down 2 (but not the three) also appears on the 'Spanish' astrolabe sold by Maitre Sylvie Teitgen on 28 June 1998 (descriptions by Anthony Turner and David King).

possible language. But we find the same kind of explanations in the abacus, algorism and Fibonacci:

1) Turchillus, having given a picture of the abacus with its ten 'arcs' ('arcus'), and the forms of the numerals, writes: 'Sciendum est tamen quod omnes he supradicte figure secundum diversas positiones, id est locationes, diversa significant. Si enim igin in primo formule arcu locetur, ibidem significat tantummodo unum, si in secundo, ibidem .x., si in tertio ibidem centum, si in quarto ibidem mille, si in quinto ibidem .x. milia...si in decimo ibidem milies mille milia. Similiter, si andras in primo arcu statuatur, ibidem significat tantummodo duo, si in secundo...('One must know that all these abovementioned forms signify different (amounts) according to different positions, i.e. locations. For if the counter for 'one' is placed in the first 'arc' it only indicates 'one', if in the second, '10', if in the third '100', if in the fourth '1000'...if in the tenth 1,000,000,000. If the counter for 'two is placed in the first 'arc' it signifies only itself, if in the second... etc.)[23]

In the algorism this is demonstrated most commonly by a table, in which the numeral is repeated in each column.[24] In Fibonacci, we read, immediately after the row of Indian numerals is given:

With these nine figures, then (with the addition of this sign '0' which in Arabic is called 'zephirum') any number you like can be written, as shall be shown below. For a number is a poured-out collection or aggregation of units, which ascend through their degrees to infinity. The first degree (*gradus*) consists of units, which are from one to ten, the second consists of tens, which are from ten to 100... Thus a figure / numeral found in the first degree represents itself, i.e. if there is a figure 1 in the first degree, it represents 1; if 2, 2; if three, 3, and so on through the ranks so that the figure of nine represents 9. But the nine figures which are in the second degree represent as many 10s as they represent unities in the first: i.e. if the figure 1 occupies the second degree, it denotes 10, if 2, 20, if three, 30, if nine, 90. A figure in the third degree denotes as many 100s as it denoted 10s in the second and units in the first degree etc.[25]

23. TURCHILLUS, *Reguncule super abacum, op. cit.*, p. 136.

24. Three such tables can be found in a mathematical manuscript, Paris, BNF, lat. 15461, one of the manuscripts of works of Toledan origin copied in Northern Italy at the turn of the thirteenth century: see R. LEMAY, *The Hispanic Origin of Our Present Numeral Forms*, «Viator», 8 (1977), pp. 435-462, Figure 9.

25. 'Cum his itaque novem figuris, et cum hoc signo 0, quod Arabice zephirum appellatur, scribitur quilibet numerus, ut inferius demonstratur. Nam numerus est unitatum perfusa collectio sive congregatio unitatum, que per suos in infinitum ascendit gradus. Ex quibus primus ex unitatibus, que sunt ab uno usque in decem, constat, secundus ex decenis, que sunt a decem usque in centum, fit. ... Figura itaque que in primo reperitur gradu se ipsam representat, hoc est, si in primo gradu fuerit figura unitatis, unum representat, si binarii, duo, si ternarii, tria, et ita per ordinem que secuntur usque si novenarii, novem. Figure quidem que in secundo gradu fuerint, tot decenas representant quot in primo unitates; hoc est, si figura unitatis secundum occupat gradum, denotat decem, si binarii, viginti, si ternarii, triginta, si novenarii, nonaginta. Figura namque que in tertio fuerit gradu, tot centenas denotat quot in secundo decenas, vel in primo unitates...': FIBONACCI, *Liber abbaci*, ed. Boncompagni, *op. cit.*, p. 2.

The only difference here is in the terminology. Whereas the abacus texts speak of 'arcus' or columns, and the algorism of 'differentiae' or 'limites',[26] Fibonacci speaks of 'gradus' or 'degrees'.

Then there is the question of the 'zero', which is said to be most distinctive of the Indian system of arithmetic. This symbol was not listed among the nine Indian figures mentioned by Vigila of Albelda, nor is it included at the heads of columns of the earliest depictions of abacus tables. But this may be because it was not considered to be a number, but rather the sign of the absence of a number. Therefore, it is generally described separately from the other numbers. Fibonacci, as in the early algorisms, also gives the nine 'Indian figures' in the Arabic order (with 9 on the right), and only mentions the zero in the following sentence (as we have seen above). On the next page he explains how to use the zero:

Si autem septuaginta tantum scribere voluerit, ponat in primo gradu 0, et post ipsum ponat figuram septenarii, sic: 70... Si quinquaginta tantum scribere volueris, in primo et in secundo gradu ponas zephyra, et in tertio figuram quinarii, hoc modo: 500. Et sic cum duobus zephyris quemlibet centenariorum numerum scribere poteris (If he wished to write seventy only, he places 0 in the first degree, and after it he places the figure of seven, thus: 70...If you wish to write five hundred only, you place zeros in the first and second degrees, and the figure of five in the third, in this way: 500. Thus with two zeros you can write any number in the hundreds.)

One would expect the zero not to be used on the Gerbertian abacus, because the columns alone 'held' the place of the number. However, the zero is, in fact, referred to, and depicted, from an early date, probably already in ca. 1000. It is described in the last verse of a poem on the abacus numerals: 'Hinc sequitur sipos; est qui rota nempe vocatus' ('Sipos' follows, which, for obvious reasons, is called "the wheel"), and a rimmed and spoked wheel is depicted adjacent to the verse.[27] A century later, the use of this 'rotula' on the abacus table is described in detail by Ralph of Laon,[28] and Adelard of Bath. The later states clearly that to 'show the number 90,707 one should 'add the character *siposcelentis* to the empty (columns of the) thousands and the tens.'[29]

---

26. For the range of terms used in early algorisms see ALLARD, *Le Calcul indien (Algorismus)* (n. 17 above), pp. 253-265.

27. The poem is edited from several manuscripts by M. FOLKERTS in *Frühe westliche Benennungen der indisch-arabischen Ziffern und ihr Vorkommen*, in *Sic itur ad astra: Studien zur Geschichte der Mathematik und Naturwissenschaften. Festschrift für den Arabisten Paul Kunitzsch*, eds M. Folkerts and R. Lorch, Wiesbaden, 2000, pp. 216-233. For the rimmed wheel see the version in MS Trier, Stadtbibliothek, 1093/1694, fol. 198r, illustrated in C. BURNETT, *The Abacus at Echternach in ca. 1000 A.D.*, «Sciamus», 3, 2002, pp. 91-108 (at p. 106).

28. RADULFUS LAUDUNENSIS, *De abaco*, ed. A. Nagl, *Abhandlungen zur Geschichte der Mathematik*, 5, Leipzig, 1890, pp. 86-133 (at p. 107).

29. 'Adde...deceno et (a MS) milleno vacantibus siposcelentis caracterem': B. BONCOMPAGNI, *Regulae Abaci di Adelardo di Bath*, «Bullettino di bibliografia e storia delle scienze matematiche e fisiche», 14, 1881, pp. 91-134, at p. 99.

My contention is that essentially there is not much difference between the abacus and the algorism. On the abacus calculation with Indian numerals was performed by placing counters with those numerals marked on them (and, where relevant, the zeros) in the relevant columns; in the algorism the same calculations were performed using a surface that could be written upon (whether a table of sand or a piece of parchment). It is quite possible (though we do not yet have the proof of this) that the 'Gerbertian' abacus was devised as a teaching tool, intended to mimic the calculation with Indian numerals on a sand-board that Christians had come across in Spain.

There are, of course, differences in procedures between the abacus, the algorism and the *Liber abbaci* of Fibonacci. These have been very clearly set out in an undervalued book by Suzan Benedict.[30] But Benedict shows that there are also striking continuities in procedures from the abacus (in the treatise of Ralph of Laon that she uses) to Fibonacci and beyond. A writer in the mid-twelfth century might say that the algorism is 'more fitting' (*decentior*) than the abacus and rhythmomachy,[31] but this need not refer to the better *arithmetic* of the algorism. It may rather refer to social advancements, in which more parchment, and even paper, was being used, and teaching involved writing out texts and taking notes, rather than using the material objects, as it had done in Gerbert's day (the abacus, the 'chess board' used for rhythmomachy; the globes and hemispheres used for astronomy).[32] Just as no serious scholar was using the sand-board any more, so no student of arithmetic should be using the old-fashioned abacus any more: he now had an ample supply of paper, or scraps of parchment, or (most likely) his wax tablets.

The continuity from the abacus to Fibonacci can also be seen from Fibonacci's articulation of a large number by means of arches over every three numbers (*virgula in modum arcus*).[33] This is clearly an imitation of the grouping of numbers on an abacus board, and Fibonacci uses the same word-*arcus*-as he had used in pouring scorn on the procedures of the abacus.

---

30. S. R. BENEDICT, *A Comparative Study of the Early Treatises Introducing into Europe the Hindu Art of Reckoning*, University of Michigan, 1914.

31. 'Est autem et circa hius artis practicam qui dicitur liber operationis, et tractatus super abacum et Rimachiam. Algorismi vero vel helcep-id est numerandi-ipsis decentior est diligentia' ('Concerning the practical side of arithmetic there are what is called "a book on how arithmetic is practised", and texts on the abacus and rhythmomachy. But the study of the algorism or "helcep"–i.e. numeration–is more fitting than these'): Trinity College, R.15.16, fol. 3r.

32. For Gerbert's predilection for using models and instruments for teaching see the account in RICHERUS, *Historiarum libri quattuor*, III, c. 49-54, ed. G. Waitz, *Monumenta Germaniae Historica, Scriptores rerum germanicarum in usum scholarum*, Hannover, 1877, pp. 102-104.

33. *Liber abbaci*, ed. Boncompagni, p. 4.

But even though the procedures changed, the basic operation 'per novem figuras Indorum' remained the same: Indian numerals were used, calculation was done on the basis of symbols having place value, and the zero was essential to the system. It is hardly likely that any of the new sources of arithmetic or algebra that Fibonacci may have introduced could have been regarded by Fibonacci or any of his readers as 'Indian' rather than 'Arabic'. Why, then, should Fibonacci claim to discard the abacus and the algorism in favour of the 'modus Indorum'?

One reason may simply be to advertise the quality of his book. 'Indian' had the connotations of ancient wisdom. Hermann of Carinthia in the mid-twelfth century had introduced 'Abidemon the Indian' as the teacher of Hermes, and the 'most ancient writer on astrology'.[34] In this Hermann had followed the ninth-century Arabic astrologer Abu Ma'shar, who had traced a tradition of wisdom from India, via Persia, to Baghdad. It was true that Indian astronomical tables as well as the Indian numerals did reach Baghdad from India in the late eighth century, and one can document several elements of Indian doctrine in Arabic, and hence in Latin, works of astronomy, astrology and magic.[35] Bartholomew of Parma, the follower of Michael Scot, when paraphrasing a passage from a translation of Abu Ma'shar's work, added to the Arabic astrologer's reference to the Indians as authorities, that they are 'our predecessors and wiser than any of our contemporary scholars'.[36] Addressing a readership who had this image of India, one can perhaps see how effective an appeal to the 'method of the Indians' could be. But I am bound to say that Fibonacci is rather unfair to his Latin predecessors, who had already prepared the way for his *Liber abbaci*.

34. '...a tempore Abidemon, Indorum regis, cuius auditores Hermes et Astalius': Hermann of Carinthia, *De essentiis*, ed. C. Burnett, Leiden, 1982, p. 82.

35. Cfr. D. Pingree, *The Indian and Pseudo-Indian Passages in Greek and Latin Astronomical and Astrological Texts*, «Viator», 7, 1976, pp. 141-195.

36. 'Indi priores nobis et ante nos modernos sapienciores': Bartholomeus Parmensis, *Tractatus spere*, pars tertia, 4.2, ed. C. Burnett, in *Seventh Centenary of the Teaching of Astronomy in Bologna 1297-1997*, eds. P. Battistini et al., Bologna, 2001, p. 176. Bartholomew goes on to refer (on p. 177) to the 'dicta antiquorum librorum sapientum Indorum que reperta sunt ab Albumasar' ('the sayings of the ancient books of the wise men of India, which were discovered by Abu Ma'shar').

1.Fibonacci MSS	כ	8	7	6	4	۳	ع	ح	1
2.Western Arabic	כ	8	1	6	4	۳	ع	ح	1
3.Albelda	כ	8	7	ხ	Ꮞ	۳	ع	ح	1
4.Latin al-Khwarizmi	९	8	٩	σ	Ꮞ	۳	ع	7	]
5.Toledan texts	ه	ხ	ᄉ	6	Ꮞ	ى	3	z	1
6. Raniero	ك	8	ヽ	σ	ყ	٨	3	z	ι

Fibonacci's Indian Numerals.

This table compares the forms of the Indian numerals typical of the early manuscripts of Fibonacci's works with those in other Arabic and Latin sources. The rows give, respectively, the numerals in:

1) MS Siena, Biblioteca publica comunale, L.IV.20 (2nd half of the thirteenth century), reproduced in R. FRANCI, Il Liber Abaci de Leonardo Fibonacci, «Bollettino della Unione Matematica Italiana», 8, 2002, pp. 293-328 (see figures 1, 4, 6 and 7).

2) Arabic MS, Tunis, University Library, 2043 (1611 A.D.), from Ifrah, Universal History (n. 5 above) p. 535.

3) MS Escorial, d.I.2 (976 A.D.), reproduced in Ifrah, Universal History (n. 5 above), p. 579.

4) MS New York, Hispanic Society of America, HC 397/726 (Spain, 13th century), reproduced in Die älteste lateinische Schrift über das indische Rechnen nach al-Ḥwārizmi, ed., trans. and comm. by M. Folkerts, with the collaboration of P. Kunitzsch, Abhandlungen der Bayerischen Akademie der Wissenschaften, phil.-hist. Klasse, n.F., 113, Munich, 1997, table 1.

5) MS Paris, Bibliothèque nationale de France, lat. 15461 (Padua region, early 13th century).

6) Notarial documents written by Raniero of Perugia between 1184 and 1206, reproduced in A. Bartoli Langeli, 'I notai e i numeri (con un caso perugino, 1184-1206)', in Scienze matematiche e insegnamento in epoca medioevale: Atti del convegno internazionale di studio, Chieti, 2-4 maggio 1996, eds P. Freguglia, L. Pellegrini and R. Paciocco, Naples, 2000, pp. 225-54.

ADDENDA AND CORRIGENDA

I. The abacus at Echternach in ca. 1000 A.D.

For more on Bibliothèque nationale de Luxembourg, Ms. 770, see Luc Deitz, 'Ein Boethius-Fund und seine Bedeutung, BnL, Ms. 770, und die Echternacher Klosterschule um das Jahr 1000', in Nova de veteribus. *Mittel- und neulateinische Studien für Paul Gerhard Schmidt*, eds A. Bihrer and E. Stein, Munich and Leipzig, 2004, pp. 247–91, and Thomas Falmagne, *Die Echternacher Handschriften bis zum Jahr 1628 in den Beständen der Bibliothèque nationale de Luxembourg*, 2 vols, Wiesbaden, 2009, II, pp. 376–7. For a clear explanation of the use of this kind of abacus and an edition and translation of the text of Hermannus Contractus, see Martin Hellmann, 'Der Rechenlehrer Herimannus, mit Edition der *Regulae, qualiter multiplicationes fiant in abaco*', in *Hermann der Lahme, Gelehrter und Dichter (1013–1054)*, Reichenauer Texte und Bilder 11, Heidelberg, 2004 (2nd ed. 2005), pp. 33–71.

II. Abbon de Fleury, *abaci doctor*

p. 130, n. 3. A colour reproduction of the whole of Oxford, St John's College 17, together with a commentary by Faith Wallis, is now freely available on-line under the title The Calendar and the Cloister: Oxford – St John's College MS 17 (http://digital.library.mcgill.ca/ms-17/help.htm)

III. *Algorismi vel helcep decentior est diligentia*

p. 249. 'totus' in the sense of 'that number of' also appears in the glosses to Boethius's *De institutione musica* associated with Adelard of Bath in Oxford, Trinity College, 47, fol. 79v: 'id est toto loco quot...'

V. Indian numerals in the Mediterranean Basin in the twelfth century

p. 249. The Arabic and Latin texts described in Lévy 2001 have now been
edited by Tony Lévy and Charles Burnett in 'Sefer ha-Middot: a Mid-
Twelfth-Century Text on Arithmetic and Geometry Attributed to Abraham
Ibn Ezra', *Aleph* 6, 2006, pp. 57–238.

p. 249, n. 31. See Charles Burnett, 'Weather Forecasting, Lunar Mansions and
a Disputed Attribution: the *Tractatus pluviarum et aeris mutationis* and
Epitome totius astrologiae of "Iohannes Hispalensis"', in *Islamic Thought
in the Middle Ages: Studies in Text, Transmission and Translation, in
Honour of Hans Daiber*, eds Anna Akasoy and Wim Raven, Leiden and
Boston, 2008, pp. 219–265.

p. 260, item 24. It would be more accurate to say: 'The distinctive "Palermitan"
forms are used in the remaining tables (fols. 53v–54, 57v–60v, 66v, 69v–
70v, 81v, 86v, 89r–90r), with the exception that, on fol. 63r, the Western
forms are used (a later addition: note that some tables are left unfilled: fols.
71r–v, 76r).

VII. Why we read Arabic numerals backwards

p. 202. The following bibliographical reference has been omitted: Burnett
(1999) = C. Burnett, 'Antioch as a Link between Arabic and Latin Culture
in the Twelfth and Thirteenth Centuries', in *Occident et Proche-Orient:
contacts scientifiques au temps des croisades*, ed. A. Tihon, I. Draelants,
and B. van den Abeele, Louvain-la-Neuve, 2000, pp. 1–78.

Addendum: The same 'back-to-front' order of numerals as in Arundel 206
occurs in the horoscope for 1178 associated with the *Microcosmographia* of
a certain 'William', dedicated to William of the White Hands, archbishop of
Reims. The date of the horoscope is written as 'm.c.87.' and Mars's degree
is given as '.91. Libra'. See J.R. Williams, 'The *Microcosmographia* of
Trier MS. 1041', *Isis* 22 (1934–5), pp. 106–35 (at p. 107).

VIII. The Toledan *regule* (*Liber Alchorismi*, part II)

The following modifications to the mathematical translation have been provided by Ji-Wei Zhao:

<17.14> If $b_1 : a_1 = b_2 : a_2 = \cdots = b_n : a_n$, then $b_p = \dfrac{(\sum_{i=1}^{n} b_i) a_p}{\sum_{i=1}^{n} a_i}$ $(p = 1, 2, \cdots, n)$.

<17.15> If $b_1 : a_1 = b_2 : a_2 = \cdots = b_n : a_n$, then $b_p = \dfrac{a_p}{\sum_{i=1}^{n} a_i} (\sum_{i=1}^{n} b_i) (p = 1, 2, \cdots, n)$.

<19.1> If $\sum_{i=1}^{n} x_i = x$, $x_1 : a_1 = x_2 : a_2 = \cdots = x_n : a_n$, then

$$\sum_{i=1}^{n} a_i : a_p = x : x_p, \; x_p = (\frac{x}{\sum_{i=1}^{n} a_i}) a_p \; (p = 1, 2, \cdots, n).$$

For $x_p = (\dfrac{x}{\sum_{i=1}^{n} a_i}) a_p$, if the residue of $x a_p$ is r, and the integer part of the quotient is s, then

$$x_p = (\frac{x}{\sum_{i=1}^{n} a_i}) a_p = s + (\frac{1}{\sum_{i=1}^{n} a_i}) r.$$

<26.5-26.6> $64 \times 23 = 4 \times 3 + 6 \times 3 \times 10 + 4 \times 2 \times 10 + 6 \times 2 \times 100 = 12 + 180 + 80 + 1200 = 1,472$.

IX. Learning Indian arithmetic in the early thirteenth century

p. 17, n. 5. This edition, though prepared, has not yet been published.

X. Latin alphanumerical notation, and annotation in Italian, in the twelfth century

p. 78. Dr Pär Larson of the CNR – Opera del Vocabolario Italiano, is preparing a new edition of passage 3 for the third (posthumous) edition of Arrigo Castellani's *I più antichi testi italiani* (Bologna: Pàtron, 1st ed. 1973, 2nd ed. 1976).

XI. Fibonacci's 'method of the Indians'

p. 87, n. 1. The article of Allard never appeared in the volume.

INDEX NOMINUM

This word index includes all proper names of premodern persons, of places, of peoples, dynasties and sects. In cases where different spellings of the proper name are found, the lemma is generally under the English form (i.e., 'John of' rather than 'Iohannes de' or 'Jean de'). Brackets are added when the index refers to a manuscript form of a name that is not readily identifiable. In Arabic names 'al-' is disregarded for alphabetical purposes. This index also includes titles or incipits of anonymous texts. An 'n' following a page number indicates that the reference can be found in a note on that page. Additions (*add.*) to the text (in the section *Addenda and corrigenda* at the end of the volume) are indexed as relevant after the page number of the article concerned.

INDEX MANUSCRIPTORUM

All the manuscripts mentioned in this volume are listed in alphabetical order of the name of the place in which they are now located. As guides to the names of the Western and Islamic libraries I have used respectively P.O. Kristeller, *Latin Manuscript Books Before 1600*, 4th ed., revised and enlarged by Sigrid Krämer, Munich, 1993, and F. Sezgin, *Geschichte des arabischen Schrifttums*, VI, Leiden, 1978. Additions (*add.*) to the text (in the section *Addenda and corrigenda* at the end of the volume) are indexed as relevant after the page number of the article concerned.

INDEX TERMINORUM MATHEMATICORUM

This index gives a selection of numbers, and of terms and topics related to arithmetic. Additions (*add.*) to the text (in the section *Addenda and corrigenda* at the end of the volume) are indexed as relevant after the page number of the article concerned.